INTENTIONALITY, MIND, AND LANGUAGE

edited and with an introduction by Ausonio Marras

The concept of intentionality has a long history in philosophical inquiry. Philosophers now see it as a valuable conceptual tool for reformulating and analyzing traditional as well as contemporary issues in many areas of philosophy, from the philosophy of science to the philosophy of language. Marras brings together for the first time an important sample of the writings of prominent (and largely contemporary) analytic philosophers which deal with intentionality.

Divided into four sections, the essays treat Franz Brentano's statements on intentionality to differentiate between the language used to describe psychological phenomena and the language of the physical sciences; the unity of science; the interrelatedness of intentionality in thought and language; and the nature of meaning, language, and conceptualization.

Supplementing the text is the previously unpublished "Rosenthal-Sellars Correspondence on Intentionality."

The contributors are: Bruce Aune, Gustav Bergmann, Rudolf Carnap, Roderick M. Chisholm, James W. Cornman, Gottlob Frege, Carl G. Hempel, Jaakko Hintikka, Leonard Linsky, David R. Luce, W. Gregory Lycan, Ausonio Marras, Thomas Nagel, Hilary Putnam, W. V. O. Quine, David M. Rosenthal, Bertrand Russell, Gilbert Ryle, Wilfrid Sellars, and Robert C. Sleigh.

AUSONIO MARRAS is associate professor of philosophy at the University of Western Ontario. He has also edited *Agent, Action, and Reason* (with R. W. Binkley and R. N. Bronaugh, 1971).

Intentionality, Mind, and Language

Intentionality, Mind, and Language

EDITED BY

Ausonio Marras

UNIVERSITY OF ILLINOIS PRESS

Urbana, Chicago, London

Preface

Since it was revived by Franz Brentano nearly a century ago, the medieval concept of intentionality has played a very central role in such philosophical movements as phenomenology, existentialism, and neo-Scholasticism. It was not until recent years, however, that the concept of intentionality proposed itself to the attention of philosophers in the analytic tradition as a powerful conceptual tool for reformulating and analyzing, in an interesting and fruitful way, many of the traditional as well as contemporary problems, issues, and theses in nearly all areas of philosophy, ranging from metaphysics to logical theory, from the philosophy of science to the philosophy of language.

The purpose of this anthology is to bring together a significant sample of the writings of various, for the most part contemporary, analytic philosophers who have dealt with some of these issues either by making explicit use of the concept of intentionality or by expressing views that are in some interesting way directly related to the claims of the "intentionalist viewpoint." Few anthologies exist in print which attempt to present a cluster of problems in different areas of philosophy from the perspective of a single given concept; none, to my knowledge, attempts to do so from the perspective of the concept of intentionality. Hence the felt need for an anthology of this sort.

The readings included in this volume are not intended to exhaust the range of topics or points of view that could reasonably be included in an anthology of this title. In contrast with a somewhat more customary practice in the editorial field, I thought it preferable to sacrifice variety and comprehensiveness to thematic unity and detailed treatment of selected topics. This policy, which is especially evident in Parts I and III, may justify, I hope, my neglect of certain important topics (particularly some of ontological and epistemological nature), as well as my omission of certain important points of view within the topics selected (e.g., P. Geach's analogical theory of judgment in *Mental Acts*—which, however, even if judiciously edited, would have preempted much of Part III). A more detailed acknowledgment of the important omissions will be found in the bibliographical essay at the end of this volume.

I am grateful to Charles E. Caton for his encouragement and advice in the initial planning of this anthology, and also to Thomas M. Lennon, who brought to my attention the correspondence on intentionality between David M. Rosenthal and Wilfrid Sellars (published here for the first time—see the Supplement). Less tangible but no less significant is my debt to my teacher Romane L. Clark, who first exposed me to most of the issues treated in this collection, and to such philosophers as Roderick Chisholm, Wilfrid Sellars, James W. Cornman, and Leonard Linsky, who through their writings provided guidance and inspiration in planning this anthology as well as writing the introduction.

I am also grateful to the University of Western Ontario for a generous grant which covered the clerical costs; to Elizabeth I. B. Moysey for her careful proofreading and preparation of the Index; and to the staff of the University of Illinois Press for their valuable editorial assistance.

A. M.

Contents

Intentionality, Mind, and Language

Introduction

Like many philosophical terms, the term 'intentionality' is used in more than one sense. In a primary and more traditional sense, the term 'intentionality' (or the cognate adjective 'intentional') is used to designate a distinguishing characteristic of mental as distinct from physical phenomena. In a chapter of *Psychologie vom empirischen Standpunkt* (1874) Franz Brentano wrote:

> Every mental phenomenon is characterized by what the scholastics of the Middle Ages called the intentional (and also mental) inexistence (*Inexistenz*) of an object (*Gegenstand*), and what we would call, although in not entirely unambiguous terms, the reference to a content, a direction upon an object (by which we are not to understand a reality in this case), or an immanent objectivity. Each one includes something as object within itself, although not always in the same way. In presentation something is presented, in judgment something is affirmed or denied, in love [something is] loved, in hate [something] hated, in desire [something] desired, etc.

This intentional inexistence is exclusively characteristic of mental phenomena. No physical phenomenon manifests anything similar. Consequently, we can define phenomena by saying that they are such phenomena as include an object intentionally within themselves.[1]

The concept of *intentional inexistence*, to which Brentano makes reference, is related to certain complex ontological issues of medieval ancestry, pertaining to the question of what sort of existence or "mode of being" is being ascribed to the objects of

---

[1] Franz Brentano, *Psychologie vom empirischen Standpunkt* (Vienna, 1874; reprint, Hamburg: Felix Meiner, 1956–59), vol. I, book II, ch. i. An English translation of this chapter may be found in R. Chisholm, ed., *Realism and the Background of Phenomenology* (Glencoe, Ill.: Free Press, 1960).

mental phenomena, especially when these "objects," like unicorns and round squares, are actually nonexistent. These issues,
which have a central position in the discussions of phenome-
nologists such as Meinong and Husserl, have been for the most
part neglected by contemporary analytic philosophers,[2] and will
not therefore be discussed in the present anthology.

Whatever the importance of these ontological issues, Brentano's point can be appreciated independently of a discussion
of these issues. For whatever the ontological status of the
"objects" of mental phenomena may turn out to be, the fact
remains, if Brentano is right, that all and only mental phenomena have the characteristic of "being directed" upon an
object which may or may not actually exist. Brentano's point
can perhaps be stated in this way. The mind can be regarded as
a faculty of awareness, that is a faculty, or *capacity*, which can
be exercised in a variety of ways, for example in judging, doubting, perceiving, hoping, loving, desiring, despairing, and so on.
All these "acts" or modes of awareness, and therefore awareness itself, are always *about* something or other. Awareness always bears, or purports to bear, a reference to an object. One
cannot be said to think without thinking about something, nor
to desire without desiring something. This referential characteristic, this "aboutness," is, for Brentano, an essential mark
of awareness. To be aware is always to be aware of an object,
whatever the nature or "ontological status" of the object may
turn out to be. Awareness, in a word, is the appearing, or being
present, of an object to a subject. Perhaps from '*cogito*' we are
not entitled to infer '*ergo sum*'; but we are surely entitled to
infer '*ergo aliquid cogito.*' That is, 'I think' (in the broad sense
of 'being aware') means 'I am being presented with an object,'
or 'something appears to me.' This apparently unique feature
of awareness, this "aboutness" or "object directedness," is what
Brentano meant by the term 'intentional,' and what he claimed
to be the distinguishing characteristic of mental as opposed to

[2] A notable exception in this respect is Gustav Bergmann. See, e.g., "Acts" and
other essays in his *Logic and Reality* (Madison, Wis.: Univ. of Wisconsin Press,
1964).

physical phenomena. Mental phenomena are thus said to be essentially *intentional*.

It is customary for many contemporary analytic philosophers to restate philosophical theses about the nature of a given kind of thing by reference to the logical features of the language we use in talking about things of that kind. In conformity with this approach to philosophical problems by way of language, Brentano's thesis has been reformulated in recent years by R.M. Chisholm and others in such a way as to show that the *sentences* we must use in describing mental or psychological phenomena form a logically distinct class of sentences, irriducible to any class of sentences we must use to describe non-mental or nonpsychological phenomena. The use of the term 'intentional' has accordingly been extended to apply not only to certain kinds of phenomena (namely the mental ones) but also to those sentences belonging to the class of sentences that seem to be required to describe mental phenomena. Membership in the class of intentional sentences is determined by a sentence's success in meeting certain so-called "criteria of intentionality," that is, by a sentence's exhibiting certain logical properties (so-called "intentional properties") not exhibited by the sentences we must use to describe nonmental phenomena.

The attempt to specify criteria of intentionality has proved to be no easy task. In "Sentences about Believing" Chisholm formulates three criteria of intentionality for (noncompound) declarative sentences (see pp. 32 ff.). These criteria are intended to provide singly sufficient and jointly necessary conditions of intentionality: that is, any sentence which meets at least one of the criteria will have to be counted as intentional, and any sentence which meets none of the criteria will have to be counted as nonintentional. It seems, however, that Chisholm's criteria fail in their intended purpose in at least two ways: (1) sentences we must use to express logical modalities (e.g., 'It is necessary that . . .'), which surely describe nothing mental or psychological, turn out to be intentional by Chisholm's third criterion; (2) cognitive (epistemic) sentences such as 'John knows the president of France,' which, according to

Chisholm, are about mental or psychological states, turn out to be nonintentional by failing to meet any of Chisholm's criteria.

In view of these difficulties, J.W. Cornman has proposed in his article "Intentionality and Intensionality" that (a) we dispense with Chisholm's third criterion, thus overcoming problem (1), and that (b) we cease to regard cognitive sentences as expressing any kind of "mental activity" (and therefore as needing to be reckoned as intentional), thereby overcoming problem (2). In my own paper "Intentionality and Cognitive Sentences," while I endorse Cornman's first proposal, I reject the second proposal by arguing that cognitive sentences ought to be regarded as intentional and can be so regarded if Chisholm's first two criteria be supplemented by the following condition: "Any sentence which entails a sentence that is intentional by either of these criteria is itself intentional."

In "Notes on the Logic of Believing" Chisholm makes a fresh start in his attempt to provide criteria of intentionality by seeking to show that psychological prefixes (such as 'Jones believes that'), when combined in certain ways with the logical quantifiers ('For every x,' 'There exists an x'), exhibit certain patterns of entailment not exhibited by other nonpsychological prefixes (such as 'It is true that,' 'It is morally indifferent whether'). On Chisholm's analysis, one of the distinguishing entailments for psychological prefixes is from 'There exists an x such that s believes that x is F' to 'S believes that there exists an x such that x is F.' In the ensuing discussion (selection 6), R.C. Sleigh argues that this entailment does not hold (in which case the psychological prefixes would remain undifferentiated from the prefix 'It is morally indifferent whether'), but fails to win Chisholm's assent. In his reply to Sleigh, Chisholm defends his original claim and, in closing, proposes an alternative "mark of intentionality" for sentence prefixes: A prefix M is intentional if, for every sentence p, the result of prefixing 'M' to 'p' is a logically contingent sentence (p. 95).

In the concluding paper of Part I of this anthology ("On 'Intentionality' and the Psychological"), W.G. Lycan critically examines various attempts to formulate criteria of intentionality, including Chisholm's last-mentioned criterion, and concludes his essay by proposing a criterion of his own, which incorporates suggestions from other authors.

II

This brief survey of various, often belabored, attempts to formulate criteria of intentionality prompts the following question: What is the *point* of these attempts; why exactly do we need distinguishing criteria for intentional sentences? One thing should be clear: the point of seeking distinguishing criteria for intentional sentences cannot be to have at our disposal a rule by means of which we may *test* a sentence in order to *tell* whether or not it is intentional, that is, about the mental. On the contrary, in the various attempts to formulate criteria of intentionality the presupposition seemed to be that we can tell by *independent means* (either by preanalytical intuition or as a result of philosophical analysis) which sentences are about the mental and which are not. Indeed, the criteria were in each case shaped in the light of such an independent understanding of the distinction between the mental and the nonmental: and when a sentence which was regarded as (not) about the mental turned out to be nonintentional (intentional) by a proposed criterion, that was taken to be a sufficient reason for revising the criterion. Thus Cornman, for example, had rejected Chisholm's third criterion in view of his belief that neither modal sentences nor cognitive sentences are about the mental.

What, then, is the point of looking for such criteria? The point is at least the following: if we can show that there *are* such criteria, then we shall have shown that intentional sentences form a logically distinct and irreducible class of sentences, so that any attempt to *analyze* or *translate* intentional sentences into nonintentional sentences, e.g., sentences about behavior or neurophysiological activities, must be bound to fail. Thus Chisholm's linguistic thesis, if correct, would constitute a refutation of any form of linguistic monism as, e.g., such linguistic

theses of physicalism as logical behaviorism or logical materialism. For example, Chisholm's thesis of linguistic dualism entails the denial of the physicalistic thesis defended by C.G.
Hempel in "Logical Analysis of Psychology"—the thesis,
namely, that "all psychological statements which are meaningful . . . are translatable into propositions which do not involve
psychological concepts, but only the concepts of physics" (pp.
123). Conversely, Hempel's thesis entails the denial of Chisholm's linguistic thesis. Moreover, it entails the denial of
Brentano's metaphysical thesis of psychophysical dualism: for
if two statements have the same meaning, they must describe
the same fact. It is understandable, therefore, that a defender
of Chisholm's thesis should, as Chisholm himself does in the
latter part of "Sentences about Believing," attempt to refute
physicalistic theses similar to Hempel's. If such theses can be
shown to fail, Chisholm can justly take comfort in this fact, for
a necessary (though, as we shall see, not sufficient) condition
for the correctness of psychophysical dualism would then be
satisfied.

Whereas Chisholm's linguistic thesis of intentionality is obviously incompatible with translatability forms of linguistic
physicalism such as Hempel's logical behaviorism, it is by no
means clear that it is also incompatible with other forms of
physicalism such as Carnap's in "Logical Foundations of the
Unity of Science" (p. 132 ff.). Hempel's thesis of translatability
was a corollary of the early logical empiricist principle that a
statement is meaningful only if it is conclusively and intersubjectively verifiable. On the (questionable) assumption that
only physicalistic statements about publicly observable things
and events are verifiable in this sense, it follows that, if a psychological statement has any meaning at all, it must have the
same meaning as some intersubjectively verifiable physicalistic
statement. But the principle of conclusive intersubjective *veri-
fiability* is a dogma that has long been rejected by the logical
empiricists themselves, including Hempel, in favor of a more
liberal criterion of *confirmability* in terms of physicalistic or
observable "thing-predicates." The new criterion does not require that we have, even in principle, a set of logically necessary

and sufficient conditions in terms of behavior or other physical states for the correct application of psychological terms, and therefore does not require that psychological statements be *translatable* into physicalistic statements. It only requires that the criteria of application for psychological terms (which are construed by Carnap as *dispositional* terms like 'soluble' or 'magnetized') be specifiable on the basis of empirical regularities among observables, and that the satisfaction of such criteria provide only a certain degree of confirmation for psychological statements. One way in which these empirical criteria can be specified is by means of what Carnap has called "reduction statements" (pp. 140), that is, statements having the following general form: "When such and such observable test conditions are realized, an organism is in such and such psychological state if and only if such and such observable behavioral response or physical state is present." Since a reduction statement does not in general have the form of an equivalence '. . . ≡ –', it does not purport to give a *complete* definition of a psychological term, but only a *partial* criterion of application for the term in question, relative to a given experimental situation. Thus, although by Carnap's method psychological statements are *reducible* to statements in the physicalistic or observational language, they are not *translatable* into such statements. Carnap's reducibility thesis, therefore, does not seem to be inconsistent with Chisholm's thesis of linguistic dualism.

The question to be raised, now, is: What is the philosophical significance of Chisholm's thesis of linguistic dualism, vis-à-vis the traditional psychophysical problem? Chisholm himself raises a similar question in the last section of "Sentences about Believing": "Let us suppose for a moment that we *cannot* rewrite belief sentences in a way which is contrary to our linguistic version of Brentano's thesis. What would be the significance of this fact? I feel that this question is itself philosophically significant, but I am not prepared to answer it. . . ." (p. 50). Chisholm, however, does give us an indication of what his answer would be. For he goes on to say: "I think that, if our linguistic thesis about intentionality is true, then the followers of Brentano would have a right to take some comfort in this fact. But if someone were to say that this linguistic fact indicates that there is a ghost in the machine I would feel sure that his answer to our question is mistaken" (*ibid.*). Thus Chisholm seems to believe that although his linguistic thesis does not *entail* the metaphysical thesis of psychophysical dualism, it does nevertheless provide *some* evidence for it. It is difficult, however, to see in what way, exactly, Chisholm's linguistic thesis provides any evidence for psychophysical dualism. If we were to assume that logically distinct (nonsynonymous) expressions must refer to factually distinct kinds of things, then it would follow that psychological (intentional) statements and nonpsychological (nonintentional) statements, being logically distinct or nonsynonymous, would refer to distinct sorts of phenomena, for example, psychological and nonpsychological phenomena respectively. But then on this assumption Chisholm would have to claim that his linguistic thesis provides not merely *some* evidence for, but a *conclusive* proof of metaphysical dualism. At any rate, the assumption in question does not seem to be tenable. For although identity of meaning does entail identity of reference it is not true that distinctness of meaning entails distinctness of reference: witness the fact that the statements 'The spaceship landed on the moon' and 'The spaceship landed on the earth's only natural satellite,' though different in meaning, refer to one and the same event. It is possible, therefore, that Chisholm's (as well as Carnap's) linguistic thesis is consistent with metaphysical *monism*—be this materialism, idealism, or a Spinozistic neutralism. Chisholm's thesis, for example, appears to be compatible with the form of materialism or physicalism defended by Thomas Nagel in "Physicalism." Nagel's theory, which is a somewhat more sophisticated version of the so-called "identity theory" of materialism associated with the names of U.T. Place and J.J.C. Smart,[3] identifies a person's having a certain

---

[3] U.T. Place, "Is Consciousness a Brain Process?," and J.J.C. Smart, "Sensations and Brain Processes"; both in V. C. Chappell, ed., *The Philosophy of Mind* (Englewood Cliffs, N.J.: Prentice Hall, 1962).

sensation or thought with the person's body being in a certain physical state, without thereby identifying the meanings of the linguistic expressions which describe these events.

A common assumption of the philosophers we have been considering has been that psychological (intentional) statements have a *categorical* form, and are used to *state* or *report* facts of a certain kind—whether these facts be mental (e.g., Chisholm) or physical (e.g., Nagel). One philosopher who disagrees with this assumption is G. Ryle. For Ryle, psychological statements are logically distinct from physical statements not because these statements may be used to report different kinds of facts but because psychological statements do not report any facts at all. Psychological terms like 'know', 'believe', 'admire', etc., Ryle claims, signify dispositions, and to say that something has a disposition for Ryle is not to report the occurrence of an episode, physical or nonphysical, but to make a lawlike statement whose form is, despite appearances, like that of a *hypothetical*, not categorical, statement, and whose function is not to describe but to explain and "licence inferences." Thus Ryle believes that psychological statements and statements describing physical occurrences are logically distinct because they belong to different logical *categories*. To ask such questions as "Are mental episodes identical with or different from physical episodes?" is to ask something logically improper, in the sense that it embodies a category mistake. Therefore, on Ryle's account, the traditional psychophysical problem cannot even be properly formulated; the metaphysical theses of psychophysical dualism and of monism are, Ryle claims, pseudo-problems.

III

That the concepts of speech and thought are in some important sense *analogically related* is a well-known thesis whose prephilosophical origin is perhaps to be found in the quaint Biblical reference to thinking as "saying things in one's heart."

The core of this analogy, as described, for example, by Bruce Aune in "Thinking," consists in the supposition that there exists some sort of *isomorphism* or formal similarity between speech and thought (or, in general, between overt "verbal episodes" and "mental episodes"). This formal similarity may be exhibited in a variety of ways. For example, both verbal episodes and mental episodes may be said to *refer* or *be about* something or other. Both verbal and mental episodes are reportable by means of indirect speech constructions: we may have 'John *said* that p' and 'John *thought* that p'. Sentences reporting verbal episodes, as well as sentences reporting mental episodes, satisfy Chisholm's criteria of intentionality: the statements 'John *believes* the cat is on the mat' and 'John *says* the cat is on the mat' are alike in that neither they nor their contradictories imply that the cat *is*, or is not, on the mat, or even that there is any cat at all. And what I judge, as well as what I say, may be said to be *true* or *false*, interesting or uninteresting, justified or unjustified; and whatever principles of reasoning seem to govern verbal activities also seem to govern mental activity.

The formal assimilation of mental and verbal episodes can be pursued on a finer level, as is done by P. Geach in *Mental Acts* and, along similar lines, by A. Kenny in *Action, Emotion, and Will*.[4] According to Geach and Kenny, a mental episode may be described as being "made up" of conceptual elements or "mental utterances," appropriately related to one another, in a sense *analogous* to that in which a verbal episode is made up of words or physical utterances, grammatically and syntactically related. The form of the "predicative" or "descriptive" content of a mental episode (the "thought," as it were, as distinct from the "modal"—assertive or affective, etc.—component which differentiates the various species of mental episodes), may then be described as follows: X's thought that a_1, a_2, . . . a_n stand in a relation R_n to one another consists of X's mental utterances a_1, a_2, . . . a_n and ϕ_n standing to one another in a relation ψ to

[4] *Mental Acts* (London: Routledge and Kegan Paul, 1957); *Action, Emotion, and Will* (London: Routledge and Kegan Paul, 1963).

be interpreted by reference to whatever verbal episode in X's language plays the same formal role as that played by a (non-assertive) utterance of 'R_n (a_1, a_2, . . . a_n)' in English.

The description of the form of a mental episode can then be completed, as is shown by Kenny, by prefixing a new (predicate) operator, 'J', 'V', etc., which specifies the modal component of the mental episode. Thus the symbol 'J' is interpreted by reference to *assertions*, the symbol 'V' by reference to expressions of intention or desire, etc. For example, the schema 'J ψ(a_1, a_2, . . . a_n, ϕ_n)' says that someone's mental utterances a_1, a_2, . . . a_n, and ϕ_n are so related to one another that a formally analogous physical occurrence of the judgment would consist of the assertion 'R_n (a_1, a_2, . . . a_n)'.

To describe in more detail the analogy between speech and thought would be not as difficult as to *explain* the apparently remarkable fact that such an anology should exist at all. This latter task, in effect, is what Sellars undertakes in the selection from "Empiricism and the Philosophy of Mind." Essentially Sellars attempts to explain the analogy by giving us a theory in which the notion of a verbal episode is taken as *primitive* and the notion of a mental episode is characterized in a derivative way, as an anological *extension* of the notion of a verbal episode. The question might be raised whether the primitiveness which Sellars assigns to the notion of a verbal episode is merely a methodological priority, a matter of practical convenience, or whether it implies a conceptual primacy of the notion of a verbal episode. Couldn't one, for example, have used the notion of *thought* as primitive, and characterized speech as an analogical extension of it? After all, is thought *internalized speech*, or is speech *externalized thought*? Does it matter what we say? Judging from the controversy between Chisholm and Sellars on this question (Selection 12), it would seem that it does matter what we say. The bone of contention between the two philosophers seems to be whether the concept of thought is more basic than the concept of language. Chisholm seems to answer this question affirmatively, Sellars negatively. It is not easy to characterize exactly the grounds of

disagreement between the two philosophers; the following remarks, however, might provide an approximation to a correct
understanding of their differences.

One way to begin is to ask what makes a given utterance a
meaningful string of words or sounds. Both Sellars and
Chisholm would agree that an utterance is meaningful if it is
not only syntactically well formed but also has certain *semantical* properties which are analogous to the intentional
properties of reference or aboutness typical of mental episodes.
Now Chisholm's contention is that semantical talk by which
semantical properties are ascribed to verbal episodes is *covertly
psychological*, in the sense that it involves an implicit reference
to *mental* episodes. For to say that a person's utterance was
meaningful, or meaningfully uttered, Chisholm believes, involves us in saying that the person *knew* what the words he used
mean, and that he used them with the *intention* of conveying
such and such, etc. (cf. "Sentences about Believing," pp. 43 ff.).
The semantical locution ' ". . ." means p' is to be analyzed,
Chisholm claims as ' ". . ." expresses t and t is about p', where
t is a thought. Thus for Chisholm it is only because of the
intentional properties of mental episodes that meaningful verbal
episodes can be said to have the semantical properties they
have.

Now a *first* characterization of Sellars' thesis would be to say
that it is exactly the converse of Chisholm's: for Sellars, it is
only because of the semantical properties of verbal episodes
that mental episodes can be said to have the intentional properties they have. To see how this is so Sellars invites us to imagine
a stage in prehistory in which humans are limited to a primitive
Rylean language of public objects, devoid of theoretical discourse, psychological discourse, and semantical discourse. Given
such a language, Sellars' claim is that two resources would have
to be added to this primitive language in order that its users
might come to acquire the concept of a mental episode as we
now have it. These two resources are semantical discourse and
theoretical discourse. Semantical discourse would enable our
mythical speakers to characterize verbal episodes in semantical

terms, i.e., in terms which are analogous to those in which we *now* talk about the intentionality, reference, or aboutness of thoughts. Theoretical discourse would enable our fictional speakers to envisage a new kind of episodes, namely mental episodes, which are *like* verbal episodes in having certain formal properties analogous to the semantical property of aboutness of verbal episodes, but which are *not* verbal episodes. The theoretical motivation for postulating such entities is the remarkable fact that people seem to be capable of (nonhabital) intelligent behavior even when no detectable verbal productions accompany such behavior. For example, the fact that a person is often able to assert the conclusion of a long and complex deductive argument without going through each step outloud or on paper suggests that he must have done so in some covert manner and that, in particular, he must have made a silent use of the same principles of reasoning which are known to govern overt verbal behavior. To account for this remarkable fact, Sellars' fictional speakers (now equipped with the necessary semantical and theoretical resources) develop a theory to the effect that "overt verbal behavior is the culmination of a process which begins with 'inner speech'" (p. 209). Thus for Sellars mental episodes are postulated as *theoretical* entities, on the model of overt verbal episodes. This should clarify Sellars' claim that the intentionality of thought is to be *explained* by reference to the categories of semantical discourse about language, and not the other way around, as Chisholm would want. Sellars can make his claim, of course, because the *model* used in constructing the notion of thought is semantical discourse itself.

A virtue of Sellars' theory is that it enables us to give a qualified endorsement to Chisholm's claim about the primacy of thought—the claim that it is only "because" of the intentional properties of mental episodes that verbal episodes can be said to have the semantical properties they have. For, according to Sellars' theory, verbal episodes are the *culmination*, or causal manifestation, of thoughts, and without thoughts, verbal episodes (as distinct from mere marks and noises) would not *exist*. But although thought is *causally* or *existentially* prior to speech,

speech is *conceptually* (i.e., in the order of conceptual reconstruction) prior to thought, for, in Sellars' theory, the concept of thought is modeled on (i.e., is an analogical extension of) the concept of speech. To summarize Sellars' position we can say, somewhat aphoristically, that in the *order of being* it is correct to say that *speech is externalized thought,* while in the *order of conceiving* it is correct to say that thought is *internalized speech.*

While Chisholm would surely agree with the former claim, he would probably disagree with the latter. For he claims that if the people of Sellars' myth were to give "just a little bit of thought to the semantical statements they make, (they would) then see that these semantical statements *entail* statements about the thoughts of the people whose language is being discussed" (p. 248). Thus Chisholm seems to be claiming that the priority of thought is assertable independently of any *theoretical* considerations such as Sellars,' but is discoverable simply as a result of a little conceptual reflection. Thus Chisholm seems to be claiming that thought is not just causally, but also *conceptually,* prior to language. This claim, of course, is what Sellars is *committed* to denying, for if he wishes to introduce the theoretical framework of thoughts on the basis of a semantical conception of linguistic behavior, then, as he has said in another context: "It must be possible to have a conception pertaining to linguistic behavior which . . . is genuinely independent of concepts pertaining to mental acts. . . . Otherwise the supposed 'introduction' of the framework would be a sham."[5]

The crucial question, then, is this: is it possible to have an adequate conception of the semantical characteristics of language prior to having a conception of thoughts and other mental episodes?

One answer to this question may be found in the latter part of Aune's selection "Thinking." Aune claims that although the concept of overt speech as we *now* have it cannot be elucidated without appeal to the concept of silent speech, nevertheless if we start from a reduced "protoversion" of overt speech

[5] *Science and Metaphysics* (New York: Humanities Press, 1968), p. 71.

(whose elucidation does not require reference to mental episodes), we can, by a series of stages, build up the richer concept of overt speech we now have, and together with it, the concept of thoughts and silent episodes.

The question might be raised, however, if in order for *us now* adequately to characterize (in retrospect as it were) the protolanguage Aune envisages, we need or need not make use of semantical concepts. If we do not, then one may just wonder whether Aune's protolanguage is a language at all: no matter how primitive, a language which like Aune's protolanguage is to be conceived as possessing sufficient resources to generate *within it* (by however gradual a development) the semantical discourse by which we *now* characterize overt speech, must surely be rich enough as to possess at least some basic semantical properties, whether its *users* were *aware* of this fact or not. On the other hand, if we do need to appeal to semantical concepts in order to characterize Aune's protolanguage, then we are back with the original question: can those semantical concepts be elucidated without invoking the concept of a mental episode?

Sellars himself attempts to answer this question in the affirmative in his essay "Notes on Intentionality," to which Bergmann's paper on the same subject ("Intentionality") ought to be compared. To say (a) ' "Es regnet" (in German) means that it is raining' is to say, for Sellars, that (b) ' "Es regnet" expresses the proposition that it is raining.' Whereas for Bergmann a proposition is a "type" which only mental episodes (awareness) can exemplify or token, and the word 'express' signifies a "unique relation" between the linguistic item and the proposition it expresses, for Sellars propositions can be tokened both by mental and verbal episodes, and moreover the word 'express' does not signify any relation at all. Sellars construes (b) as (c) ' "Es regnet" is a token of ·it is raining·,' where the phrase in dot-quotes is "a common noun which applies to items in any language that play the role played in our language by the sign-design that occurs between the dot-quotes" (p. 325). Note that on this analysis of meaning no reference is made to mental episodes. Indeed, Sellars points out, as soon as we realize that "it is

just as appropriate to speak of what *mental* sentences mean"
(my emphasis), and if we use '«Es regnet»' to stand for "the
kind of mental act that occurs in the minds of German-speaking
people and finds its overt expression in candid utterances of 'Es
regnet'," then for Sellars it is just as sensible to say (d) '«Es
regnet» (in the minds of German speakers) means it is raining'
as it is to say (a). Accordingly, (d) would be analyzed in a way
analogous to (a), namely as (c), except that '«Es regnet»' would
replace 'Es regnet' in (c). Whether Sellars' account of meaning
is successful is still, however, a controversial question. For further discussion on this and other issues emerging from the
readings in Part III, the more ambitious reader is invited to
read the collection of letters between D.M. Rosenthal and Sellars in the Supplement.

IV

Suppose that a given object, for example, the planet Venus, has
a certain property, for example, the property of being larger
than the moon. Suppose further that one and the same object
may be referred to by more than one name, and that the object
we have called 'Venus' is also called 'the Morning Star'. Then if
it is true that V*enus* has the property of being larger than the
moon, it must also be true that the *Morning Star* has that very
same property—for Venus and the Morning Star are one and
the same object.

This reasoning seems entirely uncontroversial, and merely illustrates our understanding of the concept of identity: to say
that x and y are the same object (or that x is identical with y)
is to say that whatever property x possesses y also possesses.
This was referred to by Leibniz as the *principle of indiscernibility of identicals*, and amounts to the claim that numerical
identity entails qualitative identity. What makes this principle
so uncontroversial is this kind of consideration: to say of an
individual that it has a certain property is to say that it has that
property by virtue of being the individual it is, and not by virtue

of being referred to by one name or another. That is, the central point about identity is this: any property which is truly ascribed to a given individual under one designation is truly ascribed to the same individual under a different designation.

Assume now that all statements that can be meaningfully formulated in our language ascribe properties to individuals; then it follows from what has already been said that a true statement remains true *no matter how* (by what designation) the individual spoken of in that statement is referred to, so long of course as the designating expression we use does indeed refer to that individual. It is in the light of this reasoning that the principle of indiscernibility of identicals has been reformulated, in a way relevant to language, as follows: "Given a true statement of identity, one of its two terms may be substituted for the other in any true statement and the result will be true."[6] This principle, known as the *principle of substitutivity of identicals,* is so basic to our understanding of identity and singular reference that, some have claimed, it cannot be properly challenged.

However, as G. Frege and Russell have shown in their essays "On Sense and Nominatum" and "On Denoting" respectively, this principle gives rise to paradoxes of the following sort: Given the true statement (1) 'Scott = the author of *Waverley*,' the principle of substitutivity allows us to substitute 'Scott' for 'the author of *Waverley*' in (2) 'George IV wished to know whether Scott was the author of *Waverley*' and thereby infer (3) 'George IV wished to know whether Scott was Scott.' And yet, to quote Russell's witticism, "an interest in the law of identity can hardly be attributed to the first gentlemen of Europe" (p. 369). The same paradox arises in any context introduced by such intentional prefixes as 'X believes that,' 'X hopes that,' etc., as well as in contexts introduced by modal prefixes such as 'It is necessary (possible) that.' Contexts in which an application of the principle of substitutivity gives rise to paradoxes of this sort have been called "nonextensional contexts";[7] an important

[6] W.V.O. Quine, *From a Logical Point of View* (Cambridge, Mass.: Harvard Univ. Press, 1961), p. 139.
[7] For a definition of 'extensionality', see J.W. Cornman's essay "Intentionality and Intensionality," this volume, p. 56.

problem in contemporary semantics and logical theory has been how to "make sense" of nonextensional contexts and, in particular, of contexts introduced by intentional prefixes. It is with this problem that the readings in Part IV of this anthology are mainly concerned. The problem, essentially, is how to work out a semantics for our language, in terms of such concepts as the meaning and/or reference of expressions, adequate to explain why the principle of substitutivity fails to apply in intentional contexts (if indeed it does fail), and how, consequently, such contexts are properly to be interpreted.

Frege attempted to resolve the paradox by drawing a distinction between the *sense* (connotation, conceptual meaning) and the *nominatum* (denotation, object named) of designating expressions, and by stipulating that in certain special (so called "oblique") contexts, such as the intentional contexts we have been considering above, an expression acquires a new sense and a new nominatum; in particular, the nominatum of an expression in such contexts is identical with the sense the same expression has in ordinary contexts. Given this semantical account of designating expressions, it is not surprising that the argument in the preceding paragraph should fail: it fails, according to Frege, not because the principle of substitutivity is unsound, or not universally applicable, but because what counts as a true identity statement depends on a particular context. Thus the argument fails because of a fallacy of equivocation, that is, because the terms of the identity statement in premise (1) do not have the same nominata as the terms in premise (2).

Unfortunately, Frege's "way out" of the paradox has certain undesirable consequences. For Frege's stipulation that the nominatum of an expression (the *entity* named by the expression) in an oblique context be *identical* with the sense of the same expression in ordinary contexts requires us to regard senses as *entities* which are *meant* in ordinary contexts and which are *named* in oblique contexts. But as soon as senses are regarded as nameable entities, it must be possible to provide a name for the sense of an expression in an *oblique* context, which in turn, like all names, must have a sense which can also be named by a further expression in a linguistically higher context; and so on indefinitely. Frege's account, therefore, seems to require the undesirable postulation of an infinite hierarchy of names and nameable entities or senses. This result is undesirable especially because, as we shall soon point out, it is by no means clear what sort of "reality" these entities or senses are conceived to possess.

Russell's resolution of the paradox under discussion is in terms of his famous "theory of descriptions." According to this theory, descriptive phrases of the form 'the so-and-so,' unlike proper names like 'Scott,' have no meaning when taken in isolation, but can only be defined contextually in such a way that the descriptive phrase disappears altogether. For example, the subordinate sentence in premise (2) of our puzzle, 'Scott was the author of *Waverley*,' is, according to Russell's theory, an abbreviation for 'One and only one individual wrote *Waverley*, and that individual was Scott.' The descriptive phrase has disappeared, and hence there is no term to be replaced by 'Scott' in (2). Thus the puzzle disappears.

In effect Russell's solution, like Frege's, leaves the principle of substitutivity unchallenged but makes explicit the fact that the terms that may be interchanged must be interpreted as "proper names" or "purely designating" devices, i.e., must have the same logical form and function as the individual constants in a calculus, which replace the free variables in open sentences (predicates or "propositional functions") to form statements. Thus the principle of substitutivity is in general applicable only in contexts of the logical form 'C(x)'. Since contexts containing descriptions do not, according to Russell's theory, have this form, the principle of substitutivity is in general not applicable in these contexts. It is, however, applicable in those special cases where the descriptive phrase has a "primary" as distinct from a "secondary" occurrence in a sentence (pp. 374): in these cases, presumably, the descriptive phrase is construed as functioning like a proper name or an individual constant. A detailed discussion of Russell's views on these matters is found in L. Linsky's "Substitutivity and Descriptions" (pp. 415 ff.).

Carnap's treatment of the paradox, and consequently, his interpretation of intentional contexts, rests on a formal development of Frege's suggestion that designating expressions have both a *nominatum* and a sense. These Carnap calls 'extension' and 'intension' respectively, and are extended to apply not only to singular designating expressions, but also to other "designators" such as declarative sentences and predicates ("predicators").[8] Designators of the same type, for Carnap, may stand in two distinct relations to one another: they may have the same extension but not the same intension or they may have the same intension and the same extension. Unlike Frege, Carnap explains these relations without explicit appeal to extralinguistic entities. Two designators have the *same extension* if they are *equivalent*, that is, if the equivalence sentence ('. . .≡- - -') connecting them is *true*; two designators have the *same intension* if they are *L-equivalent*, that is, if the equivalence sentence connecting them is *L-true* (analytically true). The concept of truth and L-truth are taken as formally undefined primitives. Carnap then distinguishes three kinds of contexts: extensional contexts, intensional contexts, and contexts which are neither extensional nor intensional. A context is extensional if each designator occurring in it may be replaced by an equivalent designator *salva veritate*; a context is intensional if it is nonextensional and each designator occurring in it may be replaced by an L-equivalent designator *salva veritate*.

It can now be seen how Carnap would resolve the paradox that perplexed Frege and Russell. Contexts introduced by intentional prefixes ('John believes that . . .' or 'George IV wished to know whether . . .') are neither extensional nor intensional. This can be seen, for example, as soon as we try to replace the expression '$(12 > 11)$' in 'John believes that $(12 > 11)$' with either an equivalent expression, e.g. '(the number of planets > 11),' or an L-equivalent expression, e.g. '$(3^2 + 3 > 11)$'; in either case the original sentence may change its truth value and thus the context is neither extensional nor in-

[8] See *Meaning and Necessity*, 2nd ed. (Chicago: Univ. of Chicago Press, 1956), ch. 1.

tensional. Therefore, substitution of 'Scott' for the equivalent term 'the author of *Waverley*' in 'George IV wished to know whether Scott was the author of *Waverley*' fails because the context is neither extensional nor intensional.

What relation, according to Carnap, must then obtain between two expressions in order for them to be interchangeable in intentional contexts *salva veritate*? This relation, which must obviously be stronger than L-equivalence, is what Carnap calls '*intensional isomorphism*' (see p. 388 for a precise definition). Thus '$(3^2 + 3 > 11)$' could not replace '$(12 > 11)$' in the above context because these expressions, though L-equivalent, are not intensionally isomorphic. On the other hand, '$(XII > XI)$' *can* replace '$(12 > 11)$' in that context because these expressions *are* intensionally isomorphic. We shall not discuss here the question whether even such a strong condition as intensional isomorphism is adequate to warrant interchange of expressions in intentional contexts. Some have argued that it is not. Hilary Putnam, in his paper "Synonymity and the Analysis of Belief Sentences," discusses certain objections raised by A. Church and B. Mates, and modifies Carnap's definition of intensional isomorphism in the attempt to meet Mates' objection.

We must, however, say a word about the *semantical* implications of Carnap's theory. Carnap, as we recall, requires that designators in general, and individual terms (e.g., names) in particular, have both an extension and an intension. Since intensions are required in Carnap's analysis of belief-sentences (intensional identity being a necessary condition for intensional isomorphism), it is natural to inquire what these intensions might be. It is clear enough what the extension of the individual expression "the author of *Waverley*' might be: it is the man who wrote *Waverley*, namely Scott. But what sort of entity is the *intension* of that expression? If intensions be entities, what are the identity-conditions for intensions? True, Carnap does not *formally* require that we understand intensions by reference to (nonlinguistic) entities, though *informally* he tells us that, e.g., the intensions of individual expressions are individual *con-*

cepts. Indeed, as we recall, the phrase 'the same intension' was explicated in terms of L-equivalence and the latter in terms of L-truth, which was taken as a primitive, formally undefined concept.[9] But if the concept of L-truth is to be given a semantically adequate explication, then a precise explication must be given of the semantical relation between L-true expressions and the world. After all, what is the L-true (intensional) sentence 'It is necessary that the author of *Waverley* is identical with the author of *Waverley?*' true *of?* Surely not of the flesh and blood individual who wrote *Waverley*, namely, Scott, for it is not necessary that Scott is identical with the author of *Waverley*. The sentence, then, must be true of an intensional entity, namely (for Carnap), of the individual concept The-Author-of-*Waverley*. Lacking *semantically* clear identity conditions for intensions, it is not surprising that some philosophers, like W.V.O. Quine, should regard intensions as "creatures of darkness."

It is Quine, eminently, who has raised the most sustained objections against admitting nonextensional contexts, that is, contexts in which the principle of substitutivity fails to apply. Failure of substitutivity in a context is, for Quine, evidence that the context is "referentially opaque," in the sense that individual terms occurring within that context fail to refer to anything at all. For suppose that the term 'the author of *Waverley*' in 'George IV wished to know whether Scott was the author of *Waverley*' referred to something. What can this something be? Can it be the author of *Waverley*, i.e., Scott? Surely not, Quine argues, because George IV did not wish to know whether Scott was *Scott*. In the absence of any other plausible entity as the reference of 'the author of Waverley' in the above sentence, this expression must fail to refer to anything in that context. Thus, for Quine, failure of substitutivity of a term in a given context is tantamount to failure of reference of that term in that context. Furthermore, failure of reference of a term in a

---

[9] This is not quite accurate. Carnap does define L-truth in terms of the concept of state-description (*Meaning and Necessity*, pp. 9–10); but the definition of the latter concept is not very rigorous and is, at any rate, an essentially *syntactical* definition.

given context, e.g., in the context 'C(a)', prevents us from assuming that there exists any object of which the predicate (open
sentence) 'C(x)' is true, and therefore bars the inference, known
as existential generalization, from 'C(a)' to '($\exists$x)C(x)' ('There
exists an x such that C(x)'). Thus failure of substitutivity of a
term in a given context is also tantamount to failure of existential generalization with respect to that term. This explains
Quine's insistence that we cannot meaningfully "quantify *into*"
referentially opaque contexts, e.g., intentional contexts, or, as
Quine calls the latter, contexts expressing "propositional attitudes." For example, we cannot quantify with respect to the
term 'the author of Waverley' in 'George IV wished to know
whether Scott is the author of *Waverley*,' for that term is inside the propositional-attitude context. We can, of course, quantify (as well as apply the principle of substitutivity) with respect
to terms which are *outside* the propositional-attitude context,
e.g., with respect to 'George IV' in the same sentence.

The latter observation suggests an interpretation of propositional attitudes which Quine develops in the latter part of his
essay, according to which the singular terms inside the propositional attitude contexts are taken *outside* the context, so that
the propositional verb expresses not a dyadic relation between
a believer and a proposition, but a polyadic relation between a
believer, an attribute, and the references of each of the singular
terms taken outside the context. For example, the previous sentence about George IV is now interpreted as 'George IV wished
to know whether xy(x = y) with respect to Scott and the author
of *Waverley*,' where 'xy(x = y)' designates an attribute. This
interpretation specifies a *relational* as contrasted with the previous *notional* sense of propositional attitudes. All individual
terms being now outside the opaque context, they are now open
to quantification and substitutivity. Quine finally proposes a
way of interpreting attributes without reference to intensional
entities, so as to achieve a completely *extensional interpretation*
of propositional-attitude contexts. Whether Quine's interpretation of these contexts is philosophically adequate is for the
reader to decide.

Frege, Russell, and Quine had accepted the principle of substitutivity as the linguistic counterpart of the principle of indiscernibility of identicals. What L. Linsky seeks to show in the latter part of his essay "Substitutivity and Descriptions" is that these two principles are actually quite distinct, and that the latter implies the former only on the assumption that every predicate or open sentence *expresses a property* (p. 426). On this assumption, given the usual interpretation of statements as resulting from binding the free variables in the predicate or by replacing these by individual terms, it follows that every statement ascribes a property to (at least) an individual. (Recall our assumption on p. 19 above.) But given the principle of indiscernibility of identicals, that is, given that an individual has the properties it has by virtue of being the individual it is and not by virtue of being referred to by one name or another (cf. p. 18 above), it follows that a term referring to an individual in a true statement may be replaced by any other term referring to the same individual *salva veritate:* i.e., the principle of substitutivity holds.

However, is the assumption that every predicate or open sentence expresses a property *true?* According to Linsky, the existence of contexts guilty of "referential opacity" and therefore of failure of substitutivity, is evidence that not all predicates express properties. For example, the predicate 'George IV wished to know whether Scott was identical with x,' insofar as the variable 'x' cannot be replaced by any term having the same reference as 'Scott', cannot sensibly be said to express a property. For Linsky, then, if we wish to specify conditions on substitutivity and thereby *explain* why substitutivity fails in certain contexts, we must first answer this question: How can we tell (without of course appealing to failure of substitutivity) which predicates express properties and which do not? This is a question to which we do not know any satisfactory answer.

In the last essay included in this anthology, J. Hintikka develops a semantics for "propositional attitudes" which seeks to explain failure of substitutivity and quantification in intentional contexts without appealing to Carnap's intensions and without imputing this failure to a "referential opacity" of these contexts. Essentially, Hintikka develops the idea that in attributing a propositional attitude to a person we must consider the set of all the "possible worlds" which are compatible with the person's attitude. For example, to say that a believes that p is to say, on this interpretation, that "in all possible worlds compatible with what a believes, it is the case that p" (p. 436). A novelty of this approach is that an individual term occurring inside an intentional context does not specify a *unique* individual; instead, it specifies an individual in *each* of the possible worlds we have to consider. Thus, in order to determine the reference of such a term, we have to consider how the reference of the term may vary in the different possible worlds. Thus failure of substitutivity and quantification is not due to "referential opacity" but simply to "referential multiplicity,"[10] that is, to failure of *uniqueness* of reference, not to failure of reference altogether. The principle of substitutivity can then be applied in a propositional-attitude context on condition that the terms to be interchanged, e.g., 'b' and 'c', refer to the same individual in all the different worlds consistent with the person's attitude (that is, on condition that the person knows or believes that b is identical with c); and existential generalization can be applied with respect to a term 'b' occurring within such a context on condition that 'b' refers to the *same* individual in all those possible worlds (i.e., on condition that there be an individual known or believed by the person in question to be b).

For a more detailed discussion of these issues as well as of other relevant topics not explicitly covered in the present anthology the reader is invited to consult the bibliographical note at the end of this volume.

[10] Hintikka uses this expression in *Knowledge and Belief* (Ithaca, N.Y.: Cornell Univ. Press, 1962), p. 139.

PART I

Criteria of Intentionality

1

Sentences about Believing

RODERICK M. CHISHOLM

1. "I can look for him when he is not there, but not hang him when he is not there."[1] The first of these activities, Brentano would have said, is *intentional;* it may take as its object something which does not exist. But the second activity is "merely physical"; it cannot be performed unless its object is there to work with. "Intentionality," he thought, provides us with a mark of what is psychological.

I shall try to reformulate Brentano's suggestion by describing one of the ways in which we need to use language when we talk about certain psychological states and events. I shall refer to this use as the "intentional use" of language. It is a kind of use we can avoid when we talk about nonpsychological states and events.

In the interests of a philosophy contrary to that of Brentano, many philosophers and psychologists have tried to show, in effect, how we can avoid intentional language when we wish to talk about psychology. I shall discuss some of these attempts insofar as they relate to the sorts of things we wish to be able to say about *believing.* I believe that these attempts have been so far unsuccessful. And I think that this fact may provide

From *Minnesota Studies in the Philosophy of Science,* vol. II, eds. H. Feigl, M. Scriven, and G. Maxwell, pp. 510–520. Copyright, 1958, by the University of Minnesota. Reprinted by permission of the author, the University of Minnesota Press, and the editors of *Proceedings of the Aristotelian Society,* where an earlier version of this paper was first published [56 (1955–56), 125–148].

[1] L. Wittgenstein, *Philosophical Investigations* (London and New York: Macmillan, 1953), p. 133e.

some reason for saying, with Brentano, that "intentionality" is a mark of what is psychological.

2. In order to formulate criteria by means of which we can identify the "intentional" use of language, let us classify sentences as simple and compound. For our purposes I think it will be enough to say that a compound sentence is one compounded from two or more sentences by means of propositional connectives, such as "and," "or," "if-then," "although," and "because." A simple sentence is one which is not compound. Examples of simple sentences are "He is thinking of the Dnieper Dam," "She is looking for a suitable husband for her daughter," "Their car lacks a spare wheel," and "He believes that it will rain." I shall formulate three criteria for saying that simple declarative sentences are intentional, or are used intentionally.

(a) A simple declarative sentence is intentional if it uses a substantival expression—a name or a description—in such a way that neither the sentence nor its contradictory implies either that there is or that there isn't anything to which the substantival expression truly applies. The first two examples above are intentional by this criterion. When we say that a man is thinking of the Dnieper Dam, we do not imply either that there is or that there isn't such a dam; similarly when we deny that he is thinking of it. When we say that a lady is looking for a suitable husband for her daughter, we do not commit ourselves to saying that her daughter will, or that she will not, have a suitable husband; and similarly when we deny that the lady is looking for one. But the next sentence in our list of examples—"Their car lacks a spare wheel"—is not intentional. It is true that, if we affirm this sentence, we do not commit ourselves to saying either that there are or that there are not any spare wheels. But if we deny the sentence, affirming "Their car does not lack a spare wheel," then we imply that there is a spare wheel somewhere.

(b) We may describe a second type of intentional use by reference to simple sentences the principal verb of which takes as its object a phrase containing a subordinate verb. The subordinate verb may follow immediately upon the principal verb, as in "He is contemplating killing himself"; it may occur in a complete clause, as in "He believes it will rain"; it may occur in an infinitive, as in "He wishes to speak"; or it may occur in participial form, as in "He accused John of stealing the money" and "He asked John's brother to testify against him." I shall say that such a simple declarative sentence is intentional if neither the sentence nor its contradictory imply either that the phrase following the principal verb is true or that it is false.[2] "He is contemplating killing himself" is intentional, according to this second criterion, because neither it nor its denial implies either that he does or that he doesn't kill himself; similarly with our other examples. But "He prevented John from stealing the money" is not intentional, because it implies that John did not steal the money. And "He knows how to swim" is not intentional, because its denial implies that he isn't swimming.

Sometimes people use substantival expressions in place of the kind of phrases I have just been talking about. Instead of saying, "I want the strike to be called off," they may say, "The strike's being called off is what I want." The latter sentence could be said to be intentional according to our first criterion, for neither the sentence nor its contradictory implies either that "there is such a thing as" the strike's being called off, or that there isn't—that is to say, neither implies that the strike will be, or that it will not be, called off.

Many intentional sentences of our first type may be re-written in such a way that they become instances of our second

---

[2] This criterion must be so interpreted that it will apply to sentences wherein the verb phrases following the principal verb are infinitive, prepositional, or participial phrases; hence it must make sense to speak of such phrases as being true or false. When I say of the phrase, following the main verb of "He accused John of stealing the money," that it is true, I mean, of course, that John stole the money. More generally, when I say of such a sentence that the phrase following the principal verb is true, or that it is false, my statement may be interpreted as applying to that new sentence which is like the phrase in question, except that the verb appearing in infinitive or participial form in the phrase is the principal verb of the new sentence. I should add a qualification about tenses, but I do not believe that my failure to do so is serious. It should be noted that, in English, when the subject of an infinitive or of a participle is the same as that of the principal verb, we do not repeat the subject; although we say "I want John to go," we do not say "I want me to go" or "John wants himself to go." When I say, then, that the last two words of "I want to go" are true, my statement should be interpreted as applying to "I shall go."

type. Instead of saying "I would like a glass of water," one may say "I would like to have a glass of water." And instead of saying "He is looking for the Fountain of Youth," one may say "He is trying to find the Fountain of Youth." But some sentences of the first type seem to resist such transformation into the second type; for example, "I was thinking about you yesterday."

(c) If we make use of Frege's concept of "indirect reference," which is, of course, closely related to that of "intentionality," we can add another important class of sentence to our list of those which are intentional.[3] "Indirect reference" may be defined, without using the characteristic terms of Frege's theory of meaning, in the following way: a name (or description) of a certain thing has an indirect reference in a sentence if its replacement by a different name (or description) of that thing results in a sentence whose truth-value may differ from that of the original sentence.[4] It is useful to interpret this criterion in such a way that we can say of those names (or descriptions), such as "the Fountain of Youth" and "a building half again as tall as the Empire State," which don't apply to anything, that they are all names of the same thing. Let us add, then, that a simple declarative sentence is intentional if it contains a name (or description) which has an indirect reference in that sentence. We can now say of certain *cognitive* sentences—sentences which use words such as "know," "remember," "see," "perceive," in one familiar way—that they, too, are intentional. I may see that Albert is here and Albert may be the man who will win the prize; but I do not now *see that* the man who will win the prize is here. And we all remember that although George IV knew

[3] By adopting Frege's theory of meaning—or his terminology—we could make this criterion do the work of our first two. But I have made use of the first two in order that no one will be tempted to confuse what I want to say with what Frege had to say about meaning. The three criteria overlap to a considerable extent. [The concept of indirect reference is discussed by Frege in his paper "On Sense and Nominatum," reprinted in this volume. Ed.]

[4] If E is a sentence obtained merely by putting the identity sign between two names or descriptions of the same thing, if A is a sentence using one of these names or descriptions, if B is like A except that where A uses the one name or description B uses the other, then the one name or description may be said to have an *indirect reference* in A provided that the conjunction of A and E does not imply B.

that Scott was the author of Marmion he did not know that Scott was the author of Waverley.

(d) With respect to the intentionality of compound sentences—sentences constructed by means of propositional connectives from two or more sentences—it is enough to say this: a compound declarative sentence is intentional if and only if one or more of its component sentences is intentional. "I will be gratified if I learn that Albert wins the prize" is intentional, because the if-clause is intentional. But "The career of Ponce de Leon would have been most remarkable if he had found the Fountain of Youth" is not intentional, because neither of its components is intentional. (In order that this final criterion be applicable to sentences in the subjunctive, we should, of course, interpret it to mean a compound declarative sentence is intentional if and only if one or more of the component sentences of its indicative version is intentional.)

3. We may now formulate a thesis resembling that of Brentano by referring to intentional language. Let us say (1) that we do not need to use intentional language when we describe nonpsychological, or "physical," phenomena; we can express all that we know, or believe, about such phenomena in language which is not intentional.[5] And let us say (2) that, when we wish to describe certain psychological phenomena—in particular, when we wish to describe thinking, believing, perceiving, seeing, knowing, wanting, hoping, and the like—either (a) we must use language which is intentional or (b) we must use a vocabulary which we do not need to use when we describe nonpsychological, or "physical," phenomena.

I shall discuss this linguistic version of Brentano's thesis with reference to sentences about believing. I do not pretend to be able to show that it is true in its application to believing. But I think that there are serious difficulties, underestimated by

---

[5] Certain sentences describing relations of comparison (e.g., "Some lizards look like dragons") constitute exceptions to (1). Strictly speaking, then, (1) should read: "we do not need any intentional sentences, other than those describing relations of comparison, when we describe nonpsychological phenomena."

many philosophers, which stand in the way of showing that it is false.

I wish to emphasize that my question does not concern "subsistence" or "the being of objects which don't exist." Philosophers may ask whether it is possible to think about unicorns if there are no unicorns for us to think about. They may also ask whether you and I can believe "the same thing" if there is no proposition or objective toward which each of our beliefs is directed. But I am not raising these questions. Possibly the feeling that the intentional use of language commits us to the assumption that there are such entities is one motive for seeking to avoid such use. But I wish to ask only whether we *can* avoid such use and at the same time say all that we want to be able to say about believing.

4. The first part of our thesis states that we do not need to use intentional language when we describe nonpsychological, or "physical," phenomena. I do not believe that this statement presents any serious difficulty. It is true that we do sometimes use intentional sentences in nonpsychological contexts. The following sentences, for example, are all intentional, according to our criteria, but none of them describe anything we would want to call "psychological": "The patient will be immune from the effects of any new epidemics" and "It is difficult to assemble a prefabricated house." But these sentences are not examples counter to our thesis. Anyone who understands the language can readily transform them into conditionals which are not intentional. (A compound sentence, it should be recalled, is intentional only if it has a component which is intentional.) Instead of using intentional sentences, we could have said, "If there should be any new epidemics, the patient would not be affected by them" and "If anyone were to assemble a prefabricated house, he would have difficulties." (Perhaps the last sentence should be rendered as "If anyone were to *try* to assemble a prefabricated house, he would have difficulties." In this version the sentence is intentional, once again, but since it contains the verb "to try" it can no longer be said to be nonpsychological.)

I believe that any other ostensibly nonpsychological sentence which is intentional can be transformed, in an equally obvious way, into a sentence conforming to our version of Brentano's thesis. That is to say, it will become a sentence of one of two possible types: either (a) it will be no longer intentional or (b) it will be explicitly psychological. Sentences about probability may be intentional, but, depending upon one's conception of probability, they may be transformed either into the first or into the second type. If I say "It is probable that there is life on Venus," neither my sentence nor its denial implies either that there is life on Venus or that there is not. According to one familiar interpretation of probability, my sentence can be transformed into a nonintentional sentence about frequencies—sentences telling about places where there is life and places where there isn't and comparing Venus with such places, etc. According to another interpretation, my sentence can be transformed into a psychological statement about believing—e.g., "It is reasonable for us to believe that there is life on Venus." Intentional sentences about tendencies and purposes in nature may be treated similarly. If we say, nonintentionally, "The purpose of the liver is to secrete bile," we may mean, psychologically, that the Creator made the liver so that it would secrete bile, or we may mean, nonintentionally, that in most live animals having livers the liver does do this work and that when it does not the animal is unhealthy.

There are people who like to ascribe beliefs, perceptions, plans, desires, and the like to robots and computing machinery. A computing machine might be said to believe, truly, that 7 and 5 are 12; when it is out of order, it may be said to make mistakes and possibly to believe, falsely, that 7 and 5 are 11. But such sentences, once again, are readily transformed into other sentences, usually conditionals, which are no longer intentional. If a man says that the machine believes 7 and 5 to be 11, he may mean merely that, if the keys marked "7" and "5" are pressed, the machine will produce a slip on which "11" is marked. Other intentional sentences about the attitudes of machines may be more complex, but I'm sure that, if they have been given any meaning by those who use them, they can be readily transformed into sentences which are not intentional. Indeed the ease with which robot sentences may be made either intentional or nonintentional may be one ground, or cause, for believing that sentences about the attitudes of human beings may readily be transformed in ways counter to our version of Brentano's thesis.

It should be noted, with respect to those universal sentences of physics which have no "existential import," that they are not intentional. It is true that the sentence, "All moving bodies not acted upon by external forces continue in a state of uniform motion in a straight line," does not imply either that there are, or that there are not, such bodies. But its contradictory implies that there are such bodies.

5. The second part of our version of Brentano's thesis states that, when we wish to describe anyone's believing, seeing, knowing, wanting, and the like, either (a) we must use language which is intentional or (b) we must use a vocabulary we don't need when we talk about nonpsychological facts.

Perhaps the most instructive way of looking at our thesis is to contrast it with one which is slightly different. It has often been said, in recent years, that "the language of physical things" is adequate for the description of psychological phenomena—this language being any language whose vocabulary and rules are adequate for the description of nonpsychological phenomena. If we do not need intentional language for describing physical things, then this counterthesis—the thesis that the language of physical things is adequate for the description of psychological phenomena—would imply that we do not need intentional language for the description of psychological phenomena.

The easiest way to construct a nonintentional language for psychology is to telescope nouns and verbs. Finding a psychological verb, say "expects," and its grammatical object, say "food," we may manufacture a technical term by combining the two. We may say that the rat is "food-expectant" or that he "has a food-expectancy." Russell once proposed that, instead of saying "I perceive a cat," we say "I am cat-perceptive," and Professor Ryle has described a man seeing a thimble by saying that the man "is having a visual sensation in a thimble-seeing frame of mind."[6] Sentences about thinking, believing, desiring, and the like could readily be transformed in similar ways. But this way of avoiding intentional language has one serious limitation. If we wish to tell anyone what our technical terms mean, we must use intentional language again. Russell did not propose a definition of his technical term "cat-perceptive" in familiar nonintentional terms; he told us, in effect, that we should call a person "cat-perceptive" whenever the person takes something to be a cat. Our version of Brentano's thesis implies that, if we dispense with intentional language in talking about perceiving, believing, and expecting, we must use a vocabulary we don't need to use when we talk about nonpsychological facts. The terms "food-expectancy," "thimble-seeing frame of mind," and "cat-perceptive" illustrate such a vocabulary.

I shall comment upon three general methods philosophers and psychologists have used in their attempts to provide "physical" translations of belief sentences. The first of these methods makes use of the concepts of "specific response" and "appropriate behavior"; references to these concepts appeared in the writings of the American "New Realists" and can still be found in the works of some psychologists. The second method refers to "verbal behavior"; its clearest statement is to be found in Professor Ayer's *Thinking and Meaning*. The third refers to a peculiar type of "fulfillment" or "satisfaction"; its classic statement is William James' so-called pragmatic theory of truth. I shall try to show that, if we interpret these various methods as attempts to show that our version of Brentano's thesis is false, then we can say that they are inadequate. I believe that the last of these methods—the one which refers to "fulfilment" or "satisfaction" —is the one which has the best chance of success.

6. When psychologists talk about the behavior of animals, they sometimes find it convenient to describe certain types of response in terms of the stimuli with which such responses are

---

[6] See Russell's *Inquiry into Meaning and Truth*, American ed. (New York: Norton and Co., 1940), p. 142; and G. Ryle's *Concept of Mind* (London: Hutchinson's Univ. Library, 1949), p. 230.

usually associated. A bird's "nesting responses" might be defined by reference to what the bird does in the presence of its nest and on no other occasions. A man's "rain responses," similarly, might be defined in terms of what he does when and only when he is in the rain. I believe we may say that some of the American "New Realists" assumed that, for every object of which a man can be said ever to be conscious, there is some response he makes when and only when he is in the presence of that object —some response which is *specific* to that object.[7] And they felt that the specific response vocabulary—"rain response," "fire response," "cat response"—provided a way of describing belief and the other types of phenomena Brentano would have called "intentional." This "specific response theory" is presupposed in some recent accounts of "sign behavior."

I think Brentano would have said that, if smoke is a *sign* to me of fire, then my perception of smoke causes me to *believe* that there is a fire. But if we have a specific response vocabulary available, we might say this: smoke is a sign to me of fire provided smoke calls up my *fire responses*. We might then say, more generally, that S is a sign of E for O provided only S calls up O's E-responses. But what would O's E-responses be?

What would a man's fire responses be? If smoke alone can call up his fire responses—as it may when it serves as a sign of fire— we can no longer say that his fire responses are the ways he behaves when and *only* when he is stimulated by fire. For we want to be able to say that he can make these responses in the presence of smoke and not of fire. Should we modify our conception of "fire response," then, and say that a man's fire responses are responses which are *like* those—which are *similar* to those—he makes when stimulated by fire? This would be saying too much, for in some respects *every* response he makes is like those he makes in the presence of fire. *All* of his responses, for example, are alike in being the result of neural and physiological events. But we don't want to say that all of the man's responses are fire responses. It is not enough, therefore, to say that a man's fire

7 See ch. 9 of E.B. Holt, *The Concept of Consciousness* (London: G. Allen and Co., 1914).

responses are *similar* to those he makes, or would make, in the presence of fire; we must also specify the *respect* in which they are similar. But no one, I believe, has been able to do this.

The problem isn't altered if we say that a man's fire responses constitute some *part* of those responses he makes in the presence of fire. More generally, the problem isn't altered if we introduce this definition: S is a sign of E provided only that S calls up *part* of the behavior that E calls up. It is not enough to say that the sign and the object call up *some* of the same behavior. The books in this room are not a sign to me of the books in that room, but the books in the two rooms call up some of the same behavior. And it is too much to say that S calls up *all* of the behavior that E calls up—that the sign evokes *all* of the responses that the subject makes to the object. The bell is a sign of food to the dog, but the dog, as we know, needn't eat the bell.

We might try to avoid our difficulties by introducing qualifications of another sort in our definition of *sign*. Charles E. Osgood proposes the following definition in the chapter entitled "Language Behavior," in *Method and Theory in Experimental Psychology* (New York: Oxford Univ. Press, 1953): "A pattern of stimulation which is not the object is a sign of the object if it evokes in an organism a mediating reaction, this (a) being some fractional part of the total behavior elicited by the object and (b) producing distinctive self-stimulation that mediates responses which would not occur without the previous association of nonobject and object patterns of stimulation" (p. 696). The second qualification in this definition—the requirement that there must have been a "previous association of nonobject and object" and hence that the thing signified must at least once have been experienced by the subject provides a restriction we haven't yet considered. But this restriction introduces a new set of difficulties. I have never seen a tornado, an igloo, or the Queen of England. According to the present definition, therefore, nothing can signify to me that a tornado is approaching, that there are igloos somewhere, or that the Queen of England is about to arrive. Hence the definition leaves one of the principal functions of signs and language unprovided for.

We may summarize the difficulties such definitions involve by reference to our attempt to define what a man's "fire responses" might be—those responses which, according to the present type of definition, are evoked by anything that serves as a sign of fire, and by reference to which we had hoped to define *beliefs* about fires. No matter how we formulate our definition of "fire responses," we find that our definition has one or another of these three defects: (1) a man's fire responses become responses that *only* fire can call up—in which case the presence of smoke alone will *not* call them up; (2) his fire responses become responses he sometimes makes when he *doesn't* take anything to be a sign of fire, when he *doesn't* believe that anything is on fire; or (3) our definitions will make use of intentional language.[8]

The "appropriate action" terminology is a variant of the "specific response" terminology. Psychologists sometimes say that, if the bell is a sign of food, then the bell calls up responses *appropriate* to food. And one might say, more generally, that a man *believes* a proposition *p* provided only he behaves, or is disposed to behave, in a way that is "appropriate to *p*," or "appropriate to *p*'s being true." But unless we can find a way of defining "appropriate," this way of talking is intentional by our criteria. When we affirm, or when we deny, "The knight is acting in a way that is appropriate to the presence of dragons," we do not imply either that there are, or that there are not, any dragons.[9]

[8] If we say that smoke signifies fire to O provided only that, as a result of the smoke, "there is a fire in O's *behavioral environment*," or "there is a fire for O," and if we interpret the words in the quotations in the way in which psychologists have tended to interpret them, our language is intentional.

[9] R.B. Braithwaite, in "Belief and Action," *Proceedings of the Aristotelian Society*, supp. vol. 20, p. 10, suggests that a man may be said to believe a proposition *p* provided this condition obtains: "If at a time when an occasion arises relevant to *p*, his springs of action are *s*, he will perform an action which is such that, if *p* is true, it will tend to fulfill *s*, and which is such that, if *p* is false, it will not tend to satisfy *s*." But the definition needs qualifications in order to exclude those people who, believing the true proposition *p* that there are people who can reach the summit of Mt. Everest, and having the desire *s* to reach the summit themselves, have yet acted in a way which has not tended to satisfy *s*. Moreover, if we are to use such a definition to show that Brentano was wrong, we must provide a nonintentional definition of the present use of "wish," "desire," or "spring of action."

7. In the second type of definition we refer to the "verbal behavior" which we would ordinarily take to be symptomatic of belief. This time we try to describe a man's belief—his believing—in terms of his actual uses of words or of his dispositions to use words in various ways.

Let us consider a man who believes that the Missouri River has its source in the northern part of Montana. In saying that he believes this, we do not mean to imply that he is actually doing anything; we mean to say that, if the occasion arose, he would do certain things which he would not do if he did not believe that the Missouri had its source in northern Montana. This fact may be put briefly by saying that when we ascribe a belief to a man we are ascribing a certain set of dispositions to him. What, then, are these dispositions? According to the present suggestion, the man is disposed to use language in ways in which he wouldn't use it if he didn't have the belief. In its simplest form, the suggestion is this: if someone were to ask the man "Where is the source of the Missouri River?" the man would reply by uttering the words, "In the Northern part of Montana"; if someone were to ask him to name the rivers having their sources in the northern part of Montana, he would utter, among other things, the word "Missouri"; if someone were to ask "Does the Missouri arise in northern Montana?" he would say "Yes"; and so on.

We should note that this type of definition, unlike the others, is not obviously applicable to the beliefs of animals. Sometimes we like to say such things as "The dog believes he's going to be punished" and "Now the rat thinks he's going to be fed." But if we accept the present type of definition, we cannot say these things (unless we are prepared to countenance such conditions as "If the rat could speak English, he'd now say 'I am about to be fed' "). I do not know whether this limitation—the fact that the definition does not seem to allow us to ascribe beliefs to animals—should be counted as an advantage, or as a disadvantage, of the "verbal behavior" definition. In any case, the definition involves a number of difficulties of detail and a general difficulty of principle.

The if-then sentences I have used as illustrations describe the ways in which our believer would answer certain questions. But surely we must qualify these sentences by adding that the believer has no desire to deceive the man who is questioning him. To the question "Where is the source of the Missouri?" he will reply by saying "In northern Montana"—provided he wants to tell the truth. But this proviso brings us back to statements which are intentional. If we say "The man wants to tell the truth" we do not imply, of course, either that he does or that he does not tell the truth; similarly, if we assert the contradictory. And when we say "He wants to *tell* the *truth*"—or, what comes to the same thing, "He doesn't want to *lie*"—we mean, I suppose, he doesn't want to say anything he *believes* to be false. Perhaps we should also add that he has no objection to his questioner *knowing* what it is that he believes about the Missouri.

We should also add that the man speaks English and that he does not misunderstand the questions that are put to him. This means, among other things, that he should not *take* the other man to be saying something other than what he is saying. If he took the other man to be saying "Where is the source of the *Mississippi?*" instead of "Where is the source of the Missouri?" he might reply by saying "In Minnesota" and not by saying "In Montana." It would seem essential to add, then, that he must not *believe* the other man to be asking anything other than "Where is the source of the Missouri?"

Again, if the man does not speak English, it may be that he will not reply by uttering any of the words discussed above. To accommodate this possibility, we might qualify our if-then statements in some such way as this: "If someone were to ask the man a question which, for him, had the same meaning as 'Where is the source of the Missouri?' had for us, then he would reply by uttering an expression which, for him, has the same meaning as 'In the northern part of Montana' has for us."[10] Or we might qualify our original if-then statements by adding this

[10] See Alonzo Church's "On Carnap's Analysis of Statements of Assertion and Belief," *Analysis*, 10 (1950), 97–99.

provision to the antecedents: "and if the man speaks English."
When this qualification is spelled out, then, like the previous
one, it will contain some reference to the meanings of words—
some reference to the ways in which the man uses, applies, or
interprets words and sentences. These references to the meanings
of words and sentences—to their use, application, or interpretation—take us to the difficulty of principle involved in this linguistic interpretation of believing.

The sentences we use to describe the meanings and uses of
words are ordinarily intentional. If I say, "The German word
Riese means giant," I don't mean to imply, of course, either that
there are giants or that there aren't any giants; similarly, if I deny
the sentence. If we think of a word as a class of sounds or of designs, we may be tempted to say, at first consideration, that intentional sentences about the meanings and uses of words are
examples which run counter to our general thesis about intentional sentences. For here we have sentences which seem to be
concerned, not with anyone's thoughts, beliefs, or desires, but
rather with the properties of certain patterns of marks and
noises. But we must remind ourselves that such sentences are
elliptical.

If I say, of the noises and marks constituting the German
word *Riese*, that they mean giant, I mean something like this:
"When people in Germany talk about giants, they use the word
Riese to stand for giants, or to refer to giants." To avoid talking
about things which don't exist, we might use the expression
"gigantic" (interpreting it in its literal sense) and say: "People
in Germany would call a thing *ein Riese* if and only if the thing
were gigantic." And to make sure that the expression "to call a
thing *ein Riese*" does not suggest anything mentalistic, we might
replace it by a more complex expression about noises and marks.
"To say 'A man calls a thing *ein Riese*' is to say that, in the
presence of the thing, he would make the noise, or the mark, *ein
Riese*."

Let us ignore all of the difficulties of detail listed above and let
us assume, for simplicity, that our speakers have a childlike
desire to call things as frequently as possible by their con-

ventional names. Let us even assume that everything having a name is at hand waiting to be called. Is it true that people in Germany would call a thing *ein Riese*—in the present sense of "to call"—if and only if the thing were gigantic?

If a German were in the presence of a giant and took it to be something else—say, a tower or a monument—he would not call it *ein Riese*. Hence we cannot say that, if a thing were a giant, he would call it *ein Riese*. If he were in the presence of a tower or a monument and *took* the thing to be a giant, then he would call the tower or the monument *ein Riese*. And therefore we cannot say he would call a thing *ein Riese* only if the thing were a giant.

Our sentence "The German word *Riese* means giant" does not mean merely that people in Germany—however we may qualify them with respect to their desires—would call a thing *ein Riese* if and only if the thing were gigantic. It means at least this much more—that they would call a thing by this name if and only if they *took* the thing to be gigantic or *believed* it to be gigantic or *knew* it to be gigantic. And, in general, when we use the intentional locution, "People use such and such a word to mean so-and-so," part of what we mean to say is that people use that word when they wish to express or convey something they *know* or *believe*—or *perceive* or *take*—with respect to so-and-so.

I think we can say, then, that, even if we can describe a man's believing in terms of language, his actual use of language or his dispositions to use language in certain ways, we cannot describe his use of language, or his dispositions to use language in those ways, unless we refer to what he believes, or knows, or perceives.

The "verbal behavior" approach, then, involves difficulties essentially like those we encountered with the "specific response" theory. In trying to define "fire response," it will be recalled, we had to choose among definitions having at least one of three possible defects. We now seem to find that, no matter how we try to define that behavior which is to constitute "using the word *Riese* to mean giant," our definition will have one of these three undesirable consequences: (1) we will be unable to say that German-speaking people ever mistake anything for a giant and call something which is *not* a giant *ein Riese;* (2) we will be unable to say that German speaking people ever mistake a giant for something else and refuse to call a giant *ein Riese;* or (3) our definition will make use of intentional language.

The final approach I shall examine involves similar difficulties.

8. One of the basic points in the grammar of our talk about states of consciousness, as Professor Findlay has observed, is that such states always stand opposed to other states which will "carry them out" or "fulfill" them.[11] The final approach to belief sentences I would like to discuss is one based upon this conception of *fulfilment.* I believe that, if we are to succeed in showing that Brentano was wrong, our hope lies here.

Let us consider a lady who reaches for the teakettle, *expecting* to find it full. We can say of her that she has a "motor set" which would be *disrupted* or *frustrated* if the teakettle turns out to be empty and which would be *fulfilled* or *satisfied* if the teakettle turns out to be full. In saying that the empty teakettle would disrupt or frustrate a "motor set," I am thinking of the disequilibration which might result from her lifting it; at the very least, she would be startled or surprised. But in saying that her set would be fulfilled or satisfied if the teakettle turns out to be full, I am not thinking of a positive state which serves as the contrary of disruption or frustration. Russell has introduced the terms "yes-feeling" and "quite-so feeling" in this context and would say, I think, that if the teakettle were full the lady would have a quite-so feeling.[12] Perhaps she would have such a feeling if her expectation had just been challenged—if someone had said, just before she lifted the teakettle, "I think you're mistaken in thinking there's water in that thing." And perhaps expectation always involves a kind of tension, which is relieved, or consummated, by the presence of its object. But we will be on surer ground if we describe the requisite fulfilment or satis-

[11] "The Logic of *Bewusstseinslagen,*" *Philosophical Quarterly,* 5 (1955), 57–68.
[12] See *Human Knowledge,* American ed. (New York: Simon and Schuster, 1948), pp. 148, 125; compare *The Analysis of Matter* (New York: Harcourt, Brace, 1927), p. 184.

faction, in negative terms. To say that a full teakettle would
cause fulfilment, or satisfaction, is merely to say that, unlike an
empty teakettle, it would not cause disruption or frustration.
The kind of "satisfaction" we can attribute to successful expectation then, is quite different from the kind we can attribute
to successful strivings or "springs of action."

Our example suggests the possibility of this kind of definition:
"S *expects* that E will occur within a certain period" means
that S is in a bodily state which would be frustrated, or disrupted, if and only if E were not to occur within that period.
Or, if we prefer the term "fulfill," we may say that S is in a
bodily state which would be fulfilled if and only if E were to
occur within that period. And then we could define "believes"
in a similar way, or perhaps define "believes" in terms of "being-
disposed-to-expect."

I would like to remark, in passing, that in this type of definition we have what I am sure are the essentials of William
James' so-called pragmatic theory of truth—a conception which
has been seriously misunderstood, both in Great Britain and
in America. Although James used the terms "fulfill" and "fulfillment," he preferred "satisfy" and "satisfaction." In his terms,
our suggested definition of "believing" would read: "S believes
that E will occur within a certain period" means that S is in a
bodily state which would be *satisfied* if and only if E were to
occur within that period. If we say that S's belief is *true*, that he
is correct in thinking that E will occur within that period, then
we imply, as James well knew, that E is going to occur in that
period—and hence that S's belief will be satisfied. If we say
that S's belief is false, we imply that E is not going to occur—
and hence that S's belief will not be satisfied. And all of this
implies that the man's belief is true if and only if he is in a state
which is going to be satisfied. But unfortunately James' readers
interpreted "satisfy" in its more usual sense, in which it is applicable to strivings and desirings rather than to believings.

Our definitions, as they stand, are much too simple; they cannot be applied, in any plausible way, to those situations for
which we ordinarily use the words "believe," "take," and "ex-

pect." Let us consider, briefly, the difficulties involved in applying our definition of "believe" to one of James' own examples.

How should we re-express the statement "James believes there are tigers in India"? Obviously it would not be enough to say merely, "James is in a state which would be satisfied if and only if there are tigers in India, or which would be disrupted if and only if there are no tigers in India." We should say at least this much more: "James is in a state such that, if he were to go to India, the state would be satisfied if and only if there are tigers there." What if James went to India with no thought of tigers and with no desire to look for any? If his visit were brief and he happened not to run across any tigers, then the satisfaction, or disruption, would not occur in the manner required by the definition. More important, what if he came upon tigers and took them to be lions? Or if he were to go to Africa, *believing* himself to be in India—or to India, *believing* himself to be in Africa?

I think it is apparent that the definition cannot be applied to the example unless we introduce a number of intentional qualifications into the definiens. Comparable difficulties seem to stand in the way of applying the terms of this type of definition in any of those cases we would ordinarily call instances of believing. Yet this type of definition may have an advantage the others do not have. It may be that there are simple situations, ordinarily described as "beliefs" or "expectations," which can be adequately described, nonintentionally, by reference to fulfillment, or satisfaction, and disruption, or surprise. Perhaps the entire meaning of such a statement as "The dog expects to be beaten" or "The baby expects to be fed" can be conveyed in this manner. And perhaps "satisfaction" or "surprise" can be so interpreted that our ordinary beliefs can be defined in terms of "being disposed to have" a kind of expectation which is definable by reference to "satisfaction" or "surprise." And if all of these suppositions are true then we may yet be able to interpret belief sentences in a way which is contrary to the present version of Brentano's thesis. But, I believe, we aren't able to do so now.

9. The philosophers and psychologists I have been talking about seem to have felt that they were trying to do something

important—that it would be philosophically significant if they could show that belief sentences can be rewritten in an adequate language which is not intentional, or at least that it would be significant to show that Brentano was wrong. Let us suppose for a moment that we *cannot* rewrite belief sentences in a way which is contrary to our linguistic version of Brentano's thesis. What would be the significance of this fact? I feel that this question is itself philosophically significant, but I am not prepared to answer it. I do want to suggest, however, that the two answers which are most likely to suggest themselves are not satisfactory.

I think that, if our linguistic thesis about intentionality is true, then the followers of Brentano would have a right to take some comfort in this fact. But if someone were to say that this linguistic fact indicates that there is a ghost in the machine I would feel sure that his answer to our question is mistaken. (And it would be important to remind him that belief sentences, as well as other intentional sentences, seem to be applicable to animals.)

What if someone were to tell us, on the other hand, that intentional sentences about believing and the like don't really say anything and that, in consequence, the hypothetical fact we are considering may have no philosophical significance? He might say something like this to us: "The intentional sentences of ordinary language have many important tasks; we may use the ones about believing and the like to give vent to our feelings, to influence the behavior of other people, and to perform many other functions which psychiatrists can tell us about. But such sentences are not factual; they are not descriptive; they don't say things about the world in the way in which certain non-psychological sentences say things about the world." I do not feel that this answer, as it stands, would be very helpful. For we would not be able to evaluate it unless the man also (1) gave some meaning to his technical philosophical expressions, "factual," "descriptive," and "they don't say things about the world," and (2) had some way of showing that, although these expressions can be applied to the use of certain nonpsychological sentences they cannot be applied to the use of those psychological sentences which are intentional.

Or suppose something like this were suggested: "Intentional sentences do not say of the world what at first thought we tend to think they say of the world. They are, rather, to be grouped with such sentences as 'The average carpenter has 2.7 children,' 'Charity is an essential part of our obligations,' and 'Heaven forbid,' in that their uses, or performances, differ in very fundamental ways from other sentences having the same grammatical form. We need not assume, with respect to the words which make sentences intentional, such words as 'believe', 'desire', 'choose', 'mean', 'refer', and 'signify', that they stand for a peculiar kind of property, characteristic, or relation. For we need not assume that they stand for properties, characteristics, or relations at all." We could ask the philosopher taking such a stand to give us a positive account of the uses of these words which would be an adequate account and which would show us that Brentano was mistaken. But I do not believe that anyone has yet been able to provide such an account.

2

Intentionality and Intensionality

JAMES W. CORNMAN

Certain philosophers have held the thesis of the unity of science.[1] As often conceived, this thesis has two parts: the thesis of physicalism and the thesis of extensionality. For each of these two parts there is an outstanding problem, i.e., the problem of intentionality and the problem of intensionality respectively. The purpose of this paper is twofold: first, to make explicit the nature of these two problems, and second, to show to what extent they can be said to be the same problem. Thus I am not at all interested here either to defend or attack this thesis or either of its parts. I shall first define the two theses.

PHYSICALISM AND THE PROBLEM OF INTENTIONALITY

The thesis of physicalism has had several formulations. I shall interpret physicalism to be the thesis that the language of physics is adequate for a complete description of the world. This view, then, is that the language of physics can be made the universal language of science. An obvious problem for this view

From *Philosophical Quarterly*, 12 (1962), 44–52. Reprinted by permission of the author and the editors.

[1] Such a view in various forms has been held by men such as Neurath, Carnap, Schlick, and Hempel. For example, see R. Carnap, *The Unity of Science* (London: K. Paul, Trench, Trubner and Co., 1934). See also the articles by Schlick and Hempel in sec. VI of *Readings in Philosophical Analysis*, eds. H. Feigl and W. Sellars (New York: Appleton-Century-Crofts, Inc., 1949). [The article by Hempel here referred to is "The Logical Analysis of Psychology" and is reprinted in this volume with revisions by the author.]

is that one subdomain of science, psychology, seems to require terms such as 'believe', 'assume', 'desire', 'hope', 'think', and 'doubt' which are quite unnecessary for physics. The problem of intentionality is the problem of showing that we do not need such "intentional" terms in order fully to describe the world.

We can define the problem of intentionality more precisely by stating what I shall call the thesis of intentionality as it is presented by R.M. Chisholm.

> Let us say (1) that we do not need to use intentional language when we describe nonpsychological, or "physical," phenomena; we can express all that we know, or believe, about such phenomena in language which is not intentional. And let us say (2) that, when we wish to describe certain psychological phenomena—in particular, when we wish to describe thinking, believing, perceiving, seeing, knowing, wanting, hoping and the like—either (a) we must use language which is intentional or (b) we must use a vocabulary which we do not need to use when we describe nonpsychological or "physical" phenomena.[2]

If the view that Chisholm has expressed here is correct, then the thesis of physicalism is mistaken. The problem of showing that this view is wrong is the problem of intentionality.

We must now explain what is meant by intentional language. According to Chisholm there are three individually sufficient criteria of intentionality. I shall paraphrase them in order to use terminology more appropriate than Chisholm's for comparing intentionality and intensionality. The three criteria are as follows:

1. A simple declarative sentence is intentional if it uses a substantive expression (a name or description) as the direct object of an action verb and in such a way that neither the sentence nor its contradictory implies whether or not the expression has a non-null extension, i.e., whether or not the expression truly applies to anything.[3]

---

[2] R.M. Chisholm, "Sentences about Believing," this volume, p. 35. This is also discussed in Chisholm's book, *Perceiving* (Ithaca, N.Y.: Cornell Univ. Press, 1957), pp. 168–173.

[3] I have not only paraphrased this criterion but also added to it the phrase "as the direct object of an action verb" which makes explicit what seems to be the

2. A complex declarative sentence (a sentence containing a subordinate clause or an equivalent phrase) is intentional if neither the sentence nor its contradictory implies anything about the truth-value of the subordinate clause.

3. A declarative sentence is intentional if it contains a substantive expression which is such that its replacement in the sentence by an expression with the same extension, i.e., by an extensionally equivalent expression, results in a sentence, the truth-value of which will under certain conditions differ from that of the original sentence.[4]

Chisholm says two additional things. First, he says that a compound sentence is intentional "if and only if one or more of the component sentences of its indicative version is intentional."[5] Second, he suggests that by adopting Frege's terminology we can make the third criterion "do the work of the first two."[6] Since I have paraphrased Chisholm's criteria using a terminology similar to Frege's, this will enable us to examine Chisholm's suggestion during our discussion of the relationship between intentionality and intensionality.

At this point let me give three examples each of which proves to be intentional by at least one of the three criteria. First, the sentence "John is thinking of Pegasus" is intentional by the first criterion because neither the sentence nor its contradictory implies anything about the extension of the substantive expression 'Pegasus'. That is, neither the sentence nor its contradictory implies whether or not 'Pegasus' truly applies to anything. Second, the sentence "John believes that Pegasus once lived" is intentional by the second criterion because neither this complex sentence nor its contradictory implies anything about the truth-values of 'Pegasus once lived'. Third, the sentence "John knows that Pegasus is a fictitious creature" is intentional

---

point of such criteria, that is, that these are criteria of intentional activity. Thus, we are interested only in verbs expressing this kind of activity, and consequently only in those substantive expressions which occur as direct (grammatical) objects of these "intentional" verbs.

[4] For the way Chisholm expresses these three criteria, see "Sentences about Believing," pp. 32–34.

[5] *Ibid.*, p. 35.

[6] *Ibid.*, p. 34.

by the third criterion because if we replace the substantive expression 'Pegasus' by the extensionally equivalent expression 'the winged horse captured by Bellerophon', the truth-value of the resulting sentence would differ from the original if, for example, John knew that Pegasus is a creature from Greek mythology but had never heard of Bellerophon.

EXTENSIONALITY AND THE PROBLEM OF INTENSIONALITY

Let us now turn to the thesis of extensionality. This thesis states that a universal language of science may be extensional, or, as Carnap says, "for every given [nonextensional] language S_1 an extensional language S_2 may be constructed such that S_1 may be translated into S_2."[7] By an extensional language I mean a language such that each of its sentences is extensional. And I understand an extensional sentence to be a sentence the extension of which is a function of the extensions of its designative components.[8] Let me amplify this. First, let us take as the extension of a declarative sentence its truth-value. Thus the truth-value of a sentence is a function of the extensions of its components. Second, because the truth-value of such a sentence is a function of only the extensions of its components, then if a sentence is extensional, the replacement of one of those components by another expression with the same extension, i.e., by an extensionally equivalent expression, will not change the truth-value of the sentence. Third, if certain of the components are sentential elements, i.e., clauses, then the truth-value of the sentence is a function of the truth-values, i.e., extensions, of the simple sentential elements which are its components.

[7] R. Carnap, *The Logical Syntax of Language* (New York: Harcourt, Brace, 1937), p. 320.

[8] By a designative component of a sentence I mean a component which has independent meaning at least to some degree. Such components are to be distinguished from syncategorematic components. (See R. Carnap, *Meaning and Necessity* [Chicago: The Univ. of Chicago Press, 1958], pp. 6–7, for this distinction.) Consequently a designative component of a sentence has an extension which, however, may be null. Hereafter in this paper when I speak of components of sentences I shall be referring to designative components.

With this characterization of extensionality in mind, let me now state two jointly sufficient necessary conditions of the extensionality of a sentence. A sentence is extensional if and only if:

1. The truth-value of a sentence which results from the replacement of any expression contained in the original sentence by an extensionally equivalent expression will not differ from that of the original under any conditions.
2. The truth-value of the sentence, if it is compound or complex, i.e., if it contains coordinate main clauses or at least one subordinate clause, is a function of the truth-values of the simple sentential elements which make up the compound or complex sentence.

I shall say that any sentence which fails to meet at least one of these conditions is a nonextensional or an intensional sentence. The problem of intensionality, then, arises because contrary to the thesis of extensionality, it does not seem that all intensional sentences utilized by science can be translated into extensional sentences. As an example let us examine the sentence "John believes that all humans are mortal." This sentence is not extensional because it fails to meet the first condition of extensionality. Suppose that the sentence is true and that John has a false anthropomorphic view of God. If this belief-sentence were extensional, then by replacing 'humans' with the extensionally equivalent expression 'featherless bipeds' the truth-value would not change. But under the above-stated conditions it would be changed because John believes that there is a featherless biped who is immortal, i.e., God. Thus the new sentence is false and the original sentence intensional.

Another kind of intensional sentence is the kind which fails to meet the second condition. Contrary-to-fact or subjunctive conditionals are an important part of this class of sentences because of their relationship to scientific laws. For example, the sentence "If Socrates had pleaded guilty, he would not have had to drink hemlock" is a contrary-to-fact conditional, because it implies that its antecedent clause taken as a declarative sentence is false. If this conditional were extensional, and thus truth-functional, its truth-value would be a function of the truth-values of its antecedent and its consequent clauses. Thus, knowing the truth-value of these two clauses when taken as declarative sentences would be sufficient for calculating the truth-value of the conditional. However, for contrary-to-fact conditionals this is not sufficient. The two clauses are both false and, it seems, the conditional is true. Thus if the truth-value of this kind of conditional is a function of the truth-value of its two clauses, then all such conditionals should be true. But this is not the case. For example, the false conditional, "If Socrates had pleaded guilty, he would have been crowned queen of Sparta" has both clauses false. Thus such sentences are not truth-functional. Their truth-value seems to depend not merely on the truth-value of their component clauses, but also upon the "connection" among the expressions in the clauses. Contrary-to-fact conditionals, then, fail to meet the second condition of extensionality. They are intensional sentences.

INTENTIONALITY AND INTENSIONALITY

What we have done so far is to characterize the thesis of the unity of science as consisting of two subtheses, the thesis of physicalism and the thesis of extensionality. That is, the thesis of the unity of science is the claim that all expressions utilized by science can be translated into the physical language which itself can be made extensional. We have also described the two general problems for such a thesis, the problem of intentionality and the problem of intensionality. Let us now see to what extent these two problems overlap. To do so, let us examine the relationship of the three independently sufficient conditions of intentionality to the two jointly sufficient necessary conditions of extensionality. I shall begin with the third criterion of intentionality because it is the condition which Chisholm suggests can be made not only sufficient but necessary.

The third criterion of intentionality states that a sufficient condition of a sentence being intentional is that the truth-value of the sentence will change under certain conditions, when an expression in the sentence is replaced by an extensionally equivalent expression. Let us compare this with the first necessary condition of extensionality. This condition is that the truth-value of an extensional sentence will not change under any conditions when an expression in the sentence is replaced by an extensionally equivalent expression. Since this is a necessary condition of extensionality, then its denial is a sufficient condition for nonextensionality or intensionality. But this denial is the third criterion of intentionality. Thus any sentence intentional by the third criterion is also intensional because it fails to meet the first necessary condition of extensionality and any sentence which fails to meet this condition of extensionality is intensional. One consequence of this is that given a solution to the problem of intensionality, the problem of intentionality for those sentences intentional on the basis of the third criterion alone is also solved.

Another consequence is that if Chisholm's suggestion that the third criterion of intentionality can "do the work of the first two" is correct, then given the problem of intensionality there is no additional problem of intentionality. Intentional sentences would be a subclass, whether a proper subclass or not we shall have to see, of intensional sentences. This consequence, then, gives us an additional reason to examine Chisholm's suggestion. However, first let us consider the second criterion of intentionality.

The second criterion of intentionality states that a sufficient condition of a sentence being intentional is that neither the sentence nor its contradictory implies anything about the truth-value of a subordinate clause which is a component of the sentence. In other words, a sufficient condition of intentionality is that the truth-value of the sentence is completely independent of the truth-value of any subordinate clause in the sentence. This amounts to saying that a sufficient condition of intentionality is that the truth-value of the sentence is not a function of the truth-value of any subordinate clause in the sentence. If we look at the second necessary condition of extensionality and confine our attention to complex sentences, we can see that a necessary condition for the extensionality of a complex sentence is that the truth-value of the complex sentence is a function of the truth-values of the simple sentential elements of which it is composed. Thus, because a subordinate clause is a sentential element (either compound, complex, or simple), then a necessary condition of the extensionality of a complex sentence is that the truth-value of the sentence is a function of the truth-values of the subordinate clauses in it, which, if they are not simple themselves, are truth-functional with regard to the simple clauses of which they are composed.

The denial of this necessary condition of extensionality is a sufficient condition of intensionality. But this denial is that the truth-value of a complex sentence is not a function of the truth-value of the subordinate clauses in the sentence, which is the second sufficient condition of intentionality. Thus, as with the third criterion, any sentence intentional by the second criterion is also intensional because it fails to meet the second necessary condition of extensionality. Furthermore, any complex sentence which fails to meet the second necessary condition of extensionality is intentional. A consequence of this is that given a solution to the problem of intensionality the problem of intentionality for those sentences intentional on the basis of the second criterion alone is also solved.

We have so far found out that two of the three criteria of intentionality are also criteria of intensionality. Thus all sentences intentional by either the second or third criterion are also intensional. There are two questions left. First, are all sentences intentional by the first criterion also intensional? If they are, then we can conclude that all intentional sentences are intensional, and then proceed to examine Chisholm's suggestion which amounts to saying that the reason they are all intensional is that they fail to meet the first necessary condition of extensionality. Second, are all intensional sentences intentional? We have found that all simple and complex intensional sentences are. This leaves only compound sentences to be examined.

One way to approach the first question is to find out whether any sentences intentional by the first criterion are extensional. Unlike the second and third criteria there seems to be no necessary condition of extensionality comparable to this criterion. The first criterion concerns a certain kind of simple sentence which is not explicitly covered by either necessary condition of extensionality. It seems, then, that there should be an extensional sentence which is intentional by the first criterion. This is indeed the case. One example is the sentence "John is thinking of Alaska." This sentence is intentional by the first criterion because neither the sentence nor its contradictory implies whether or not the substantive expression 'Alaska' applies to anything, i.e., whether or not 'Alaska' has an extension.

However, the sentence is extensional because it is a simple sentence in which an expression may be replaced by an extensionally equivalent one under any conditions without change of truth-value. The only expression in the sentence which might seem to violate this condition is 'Alaska'. But when we say that John is thinking of Alaska we are claiming that he is thinking of a particular place and what name or description we use to express this claim is limited only by the condition that it indeed apply to the place. This is true of all expressions extensionally equivalent to 'Alaska'. Thus the sentence "John is thinking of Alaska" is true given and only given those conditions under which "John is thinking of the largest state of the United States" is true. In other words, substituting 'the largest state of the United States' for its extensionally equivalent 'Alaska' will not change the truth-value of the sentence under any conditions. The sentence will have the same truth-value as the original, for example, no matter what John is thinking about Alaska or what he thinks that the largest state is.

We have, therefore, found an example of an intentional sentence which is extensional. The class of intentional sentences and the class of intensional sentences are not co-extensive. This shows, incidentally, that Chisholm's suggestion about the applicability of the third criterion of intentionality is mistaken. If it were correct, then, as we have shown, all intentional sentences would be intensional. But we have just found an intentional sentence which is not intensional. Thus at least the first criterion of intentionality is not replaceable by the third.

Not all intentional sentences are intensional. Are all intensional sentences intentional? We have found all simple and complex intensional sentences to be intentional. However, certain intensional compound sentences do not seem to be intentional. Certain contrary-to-fact conditionals such as the example used previously are, as we have shown, intensional, but they are not all intentional. Thus the sentence "If Socrates had pleaded guilty, he would not have had to drink hemlock" is intensional because it violates the second necessary condition of extensionality. However, it is not intentional because it meets none of the three criteria of intentionality. It does not meet the third criterion because the two substantive expressions in the sentence, 'Socrates' and 'hemlock', can be replaced by extensionally equivalent expressions without change of truth-value. The second criterion does not apply because the sentence is not complex. It also fails to meet the first criterion. The only substantive expression which is the direct object of an action verb is 'hemlock'. Although this sentence does not imply whether or not 'hemlock' has a non-null extension, its contradictory, i.e., "He had to drink hemlock" implies that 'hemlock' does have an extension, i.e., does apply to something.

We could, however, make the sentence intentional by changing the second criterion of intentionality so that it applies not only to subordinate clauses and thus complex sentences, but also to coordinate clauses and thus compound sentences. By so doing the second and third criteria of intentionality would be the contradictories of the second and first necessary conditions of extensionality. Thus, all intensional sentences would be intentional. However, there is another way to interpret the relationship among the various criteria and conditions which seems more appropriate. Intentionality, as Chisholm points out, has to do with certain activities which can take as objects things which do not exist. Thus to say that someone is performing such an activity with regard to some object, e.g., thinking of it, dreaming of it, looking for it, or hoping for it, is to imply nothing about the existence or nonexistence of the object and, therefore, to imply nothing about the extension of the name or description of the object. Thus a criterion of intentionality should be one by which we can distinguish intentional from nonintentional activities, or, if it is a criterion relevant to language as those of Chisholm are, one by which we can distinguish verbs which express intentional activities and those which do not. It should seem, furthermore, that it should be a criterion which in some way considers the fact that intentional activities can take as objects things which do not exist.

This suggests two reasons why we might more profitably construe the first of Chisholm's three criteria as the one criterion of intentionality, and the other two as merely the contradictories of the two necessary conditions of extensionality, i.e., as the two sufficient conditions of intensionality. First, only the first criterion is concerned with the extensions of substantive expressions which are the direct objects of verbs expressing activities. Thus only the first criterion is directly concerned with matters related to "intentional inexistence." Second, the action verbs which are found to be intentional by the other two criteria, are, I believe, with the exception of one class of verbs also found intentional by the first criterion. Thus although the first criterion may not be sufficient to distinguish all intentional sentences, it seems to be sufficient to distinguish all intentional verbs, with the one possible exception. We must consider this apparent exception now.

There is a class of sentences which Chisholm calls "cognitive sentences," i.e., sentences using verbs such as 'know', 'perceive', and 'remember'. If we use just the first criterion, or even the second, such sentences and thereby such verbs will be excluded from the class of intentional expressions because they are intentional by the third criterion only. Let us take 'know' as an example. A simple sentence using 'know', such as "John knows the governor of Alaska," is not intentional by the first criterion because the sentence implies that the substantive expression 'the governor of Alaska' has an extension. Similarly a complex sentence such as "John knows that Alaska is the largest state" is not intentional by the second criterion because the sentence implies that the declarative sentence "Alaska is the largest state" is true. Thus if we merely use the first, or the first and second criteria, the verb 'know' would not be intentional. However, by the third criterion the sentence "John knows that Alaska is the largest state" is intentional and thereby so is 'know'. If we replace the substantive expression 'the largest state' by an extensionally equivalent expression, e.g., 'the forty-ninth state', then if John believes that Alaska is the fiftieth state the resulting sentence would be false although the original might well be true.

I have suggested that the first of Chisholm's criteria seems to be the appropriate criterion of intentionality. But, as we have just seen, cognitive sentences are not intentional by the criterion. This leads to the question of whether we should consider them intentional. There is, I believe, only one argument which would give us reason to think that we should. It goes as follows. All psychological or mental activities have the characteristic of intentionality. Cognitive verbs such as 'know', 'perceive', and 'remember' express mental activities. Therefore, such verbs are intentional. This argument can be attacked via either premise, depending upon the interpretation given to cognitive verbs. If we assume that such verbs express mental activities, then the first premise, i.e., all mental activities are intentional, seems to be false by the following argument. Intentional activities are those which can take as objects things which may never exist. Knowing, perceiving, and remembering take as objects things which must exist at least at some time or other. Therefore, knowing, perceiving, and remembering, and thus 'know', 'perceive', and 'remember', are not intentional.

The above argument is directed against the first premise only if cognitive verbs express mental activities, i.e., if the second premise is true. However, as certain philosophers have pointed out, there is good reason to believe that no matter what the truth-value of the first premise, the second premise is false. Cognitive verbs are not verbs which designate certain mental activities. In other words, there is no activity corresponding to 'know', for example, in the way the activity of desiring corresponds to the verb 'desire'. There is no unique activity (e.g., some certain infallible mental act which we can perform), such that its occurrence guarantees that we know and do not merely believe something.[9] Cognitive verbs are more like what Gilbert Ryle has called "achievement verbs"[10] because they express not that some activity has taken place but that over and above the occurrence of any activity some task, i.e., the achievement of knowledge, has been successfully completed. To know, to perceive, or to remember something are, as Ryle points out, analogous to winning rather than running a race. Winning a race, although it implies an activity such as running, is not itself an activity. It is an achievement. Furthermore, just as there is an activity essential to winning, there may be one essential to knowing (certainly believing seems to be necessary). However, just as 'win' does not express running or any other activity, neither does 'know' express any activity essential to it (e.g., 'believe' rather than 'know' expresses believing). Thus although cognitive verbs may imply and thus refer to certain mental activities, they do not express mental or any other kind of activity. They express achievements and achievements are not activities. Thus the second premise in the above argument is false.

In the preceding discussion I have tried to show why cognitive verbs do not seem to be intentional and thus why there is no need for the third of Chisholm's criteria. There is, however, a reason not only against the need of using this criterion, but against considering it at all related to intentionality. This reason is, incidentally, an additional reason against considering cognitive verbs intentional, because they are intentional only by the third criterion. If the third criterion is properly a criterion of sentences expressing intentional activity, whether that kind of activity be mental or not, then certain modal sentences such as "It is necessary that Alaska is Alaska", which do not seem to

[9] See J.L. Austin, "Other Minds," *Proceedings of the Aristotelian Society*, supp. vol. 20 (1946), for a discussion of this point.
[10] See G. Ryle, *The Concept of Mind* (London: Hutchinson, 1949), pp. 149–153.

express any activity, either mental or otherwise, would be intentional and thus express intentional activity.[11] For this reason as well as those above, I believe that the third of Chisholm's criteria should be interpreted merely as a criterion of intensionality, as should the second. Thus only the first criterion would be the criterion of intentionality. This interpretation, while maintaining the two problems distinct, as we have seen they are, not only allows us to reduce the number of criteria we must work with, but also, as I hope to have shown, pinpoints which of the criteria are relevant to the problem of intensionality and which to the problem of intentionality.

[11] It might be suggested that this objection could be avoided by adding to the third criterion a clause similar to the phrase which I added to the first criterion, e.g., "which is the direct object of an action verb." This would eliminate modal sentences from the class of intentional sentences. It would also, however, if the previous objections are sound, eliminate cognitive sentences because they would not contain action verbs. In such a case, I believe, the third criterion would become unnecessary because it would apply only to those action verbs found intentional by the first criterion.

3

Intentionality and Cognitive Sentences

AUSONIO MARRAS

I

In his book *Perceiving* Chisholm has proposed three criteria for intentionality:[1]

1. A simple declarative sentence is intentional if it uses a substantival expression—a name or a description—in such a way that neither the sentence nor its contradictory implies either that there is or that there isn't anything to which the substantival expression truly applies.

2. Any noncompound sentence which contains a propositional clause . . . is intentional provided that neither the sentence nor its contradictory implies either that the propositional clause is true or that it is false.

3. Suppose there are two names or descriptions which designate the same thing and that E is a sentence obtained merely by separating those two names or descriptions by means of "is identical with." . . . Suppose also that A is a sentence using one of those names or descriptions and that B is like A except that, where A uses the one, B uses the other. Let us say that A is intentional if the conjunction of A and E does not imply B.

From *Philosophy and Phenomenological Research*, 29 (1968), 257–263. Reprinted, with revisions by the author, by permission of the editors.
[1] R.M. Chisholm, *Perceiving: A Philosophical Study* (Ithaca, N.Y.: Cornell Univ. Press, 1957), pp. 170–171. Essentially the same criteria have been proposed by Chisholm in "Sentences about Believing," reprinted in the present volume.

The purpose of the third criterion is to show that cognitive sentences (i.e., sentences containing such verbs as 'know,' 'remember,' 'perceive,' etc.), which are not shown to be intentional by the first two criteria, are indeed intentional. Some have claimed[2] that this criterion is too restrictive: *simple* cognitive sentences (i.e., cognitive sentences whose grammatical object is a substantival expression—name or description) are not shown to be intentional by this criterion. On the other hand one may claim that Chisholm's third criterion is not sufficiently restrictive: sentences expressing logical modalities, though not intentional by the first two criteria, appear to be intentional by the third. If these claims are true, we seem to be faced with the consequence that Chisholm's criteria are not jointly necessary and sufficient conditions of intentionality, and that therefore some important revision of these criteria is called for. Since the burden of the apparent failure of Chisholm's criteria seems to lie primarily on the third, we may begin by taking a closer look at this criterion.

This criterion makes use of the well-known principle of substitutivity of identicals: failure of substitutivity of coreferential expressions in a context is taken by Chisholm to be sufficient evidence of the context's intentionality. Thus "John knows that the governor of Alaska is older than forty" is intentional because it does not imply "John knows that the governor of the largest state is older than forty," even though

(E)   The governor of Alaska = The governor of the largest
                                state in the Union

is true.

The third criterion seems indeed to be entirely satisfactory with respect to cognitive sentences of the form "A knows *that. . . .*" But what about simple cognitive sentences of the form "A knows X"? Does substitution still fail? There are some who take it to be obvious that substitution does not fail in such

[2] Anthony Kenny, *Action, Emotion and Will* (London: Routledge and Kegan Paul, 1963), p. 198; and J.W. Cornman, "Intentionality and Intensionality," this volume, p. 60 (see footnote 4 below).

simple sentences. Anthony Kenny, for example, has claimed that the sentences "Dr. Jekyll = Mr. Hyde" and "James knows Dr. Jekyll" together imply that James knows Mr. Hyde.[3] Similarly, J.W. Cornman seems to believe that "John knows the governor of Alaska" and (3) together imply that John knows the governor of the largest state in the Union.[4] If Kenny and Cornman are right,[5] it seems that at least some cognitive sentences are not shown to be intentional by Chisholm's third criterion. And since cognitive sentences are not shown to be intentional by the first and second criterion either, it seems that Chisholm's criteria are not jointly necessary conditions of intentionality (on the assumption, of course, that cognitive sentences *are* intentional).

On the other hand by Chisholm's third criterion some modal sentences appear to be intentional. For example, "It is necessary that the governor of Alaska is identical with the governor of Alaska" appears to be intentional because it does not presum-

---

[3] *Action, Emotion and Will*, p. 198.

[4] At least, I take Cornman to be committed to this belief. For, on p. 60 of the above-cited article, he argues that the sentence "John is thinking of Alaska" (call it *p*) is extensional by a condition of extensionality whose contradictory corresponds to Chisholm's third criterion of intentionality (cf. pp. 56–58). It follows, on Cornman's own account, that *p* must not only be extensional but also not intentional by Chisholm's third criterion. Now if *p* is not intentional by this criterion, neither should a sentence that is like *p* in all relevant respects except for employing a cognitive verb where *p* employs 'is thinking.'

[5] My own feeling is that the conditions for substitutivity in intentional contexts are less clear than either Kenny or Cornman make it appear. It seems to me that Quine's distinction between the "notional" (opaque) and "relational" (transparent) sense of intentional verbs (cf. "Quantifiers and Propositional Attitudes," this volume, pp. 402–403) can be applied not only to contexts of the *propositional* form "A knows that . . . ," but also to *simple* intentional contexts of the form "A knows X." One reason is that, if philosophers such as Quine and W. Sellars are right, intentional verbs have explicitly or implicitly a propositional form, and therefore it may be possible to "recast" simple intentional sentences into intentional sentences having a propositional form; and if the latter are subject to Quine's distinction, one would expect the former should too. Another (obviously related) reason is that there is a sense in which the simple sentences "A knows the governor of Alaska" and "A knows the governor of the largest state in the Union" may be so interpreted as to entail respectively "A knows of someone that he is the governor of Alaska" and "A knows of someone that he is the governor of the largest state in the Union"; but since it is possible that one of the latter pair of sentences be true and the other be false even when (E) is true, it must likewise be possible that one of the former pair of sentences be true and the other be false even when (E) is true.

ably imply, together with (E), "It is necessary that the governor of Alaska is identical with the governor of the largest state in the Union." And "It is possible that the governor of Alaska is not identical with the governor of the largest state in the Union" appears to be intentional because it does not presumably imply, together with (E), "It is possible that the governor of Alaska is not identical with the governor of Alaska." In general, as Quine has argued, modal contexts are "referentially opaque," that is, substitutivity of coreferential expressions may fail in such contexts. But failure of substitutivity, according to Chisholm's third criterion, is sufficient evidence of intentionality; hence, acceptance of this criterion compels us to regard modal sentences as intentional. There is no intuitively acceptable sense, however, in which modal sentences may be said to be intentional, that is, about psychological acts or states. Chisholm's third criterion, therefore, does not seem to be a sufficient condition of intentionality.

Notice, however, that modal sentences are not shown to be intentional by Chisholm's first two criteria. For example, "It is possible that p" is not shown to be intentional because its contradictory implies that p is false; and "It is necessary that p" is not shown to be intentional because it implies that p is true.[6]

---

[6] Contingency sentences seem to be intentional by Chisholm's second criterion. For neither "It is contingent that p" (where 'contingent' means 'neither necessary nor impossible') nor its contradictory implies either that p is true or that p is false. However, since such contingency sentences are really abbreviations for *compound* sentences (i.e., sentences containing two or more sentences conjoined by means of such sentence connectors as 'and,' 'or,' 'because,' etc.), they are as such beyond the scope of Chisholm's second criterion, which applies only to noncompound sentences. Chisholm also provides a criterion of intentionality for compound sentences, according to which a compound (declarative) sentence is intentional if and only if at least one of its component sentences is intentional (*Perceiving*, p. 172). By this criterion, the compound sentence "It is neither necessary that p nor impossible that p" (which is the unabbreviated form of "It is contingent that p") is not intentional since neither of its component sentences is intentional.

Similar considerations apply to sentences about probability, dispositions, tendencies, abilities, etc., which may be analyzed into nonintentional sentences about frequencies and/or nonintentional counterfactual sentences (cf. Chisholm, "Sentences about Believing," p. 37).

Deontic sentences like "It is obligatory that p," "It is morally wrong that p," etc., seem to pose a more serious problem since they appear to be intentional by Chisholm's second criterion and do not seem to be exponible into nonintentional

II

We have noted that the purpose of Chisholm's third criterion was to show that cognitive sentences are intentional. We have found, however, that by this criterion some cognitive sentences do not appear to be intentional, whereas some obviously non-intentional sentences, such as those expressing logical modalities, would seem to be intentional. In rejecting Chisholm's criterion while unable to formulate a better one, one is naturally led to the question whether cognitive sentences *should* be considered intentional. The obvious advantage of not considering them such is that Chisholm's first and second criteria would be two jointly necessary and sufficient conditions of intentionality.

J.W. Cornman has recently argued in favor of considering cognitive sentences as nonintentional. His argument rests on the supposition that cognitive verbs do not express or designate any mental *activities*:

> Cognitive verbs are not verbs which designate certain mental activities. In other words, there is no activity corresponding to 'know,' for example, in the way the activity of desiring corresponds to the verb 'desire.' There is no unique activity . . . such that its occurrence guarantees that we know and do not merely believe something. Cognitive verbs are more like what Gilbert Ryle has called "achievement verbs." . . . Winning a race, although it implies an activity such as running, is not itself an activity. . . . Furthermore, just as there is an activity essential to winning, there may be one essential to knowing (certainly believing seems to be necessary). However, just as 'win' does not express running or any other activity, neither does 'know' express any activity essential to it (e.g., 'believe' rather than 'know' expresses believing). Thus, although cognitive verbs may imply and thus refer to certain mental activities, they do not express mental or any other kind of activity.[7]

---

sentences. But then it may be that deontic sentences, if they presuppose a context of agents giving "commands" (to others or to themselves), do have an intentional component.

    [7] "Intentionality and Intensionality," pp. 63–64.

However, one may grant the correctness of Cornman's claim that cognitive verbs are ill-conceived as verbs expressing kinds of activity, and still wonder why the kinds of "achievement" expressed by cognitive verbs should not also be considered, in some important sense, essentially intentional. It is not clear, in other words, why intentionality should be an exclusive characteristic of certain kinds of activity and not, for example, of achievement.[8] Cornman does not explain why what is generally considered as the essential feature of intentionality, namely, a certain *aboutness* or intrinsic *reference* to something as an object (which may or may not exist), should be shared only by certain activities and not, for example, by certain states or by certain achievements.

It is surprising, at any rate, that Cornman should experience no perplexity in deciding which kinds of things (phenomena) are to be called *activities* and which not. He has apparently no hesitation in calling *believing* an activity. But then one wonders what activity is really expressed by such sentence as "I firmly believe in Science." Or one may wonder why the sentence "At last, I believe there is God" expresses less of an achievement than the sentence "At last, I know there is God," if there is God. My feeling is that the distinction between activities and nonactivities (achievements), while perhaps useful for distinguishing kinds of physical phenomena, is too fuzzy to be of any help for sorting out intentional from nonintentional phenomena. Whether or not we should want to class as an activity the phenomenon expressed by the main verb in the sentence "I infer the thief probably escaped through the window," I think there should be no hesitation about classing it as intentional. If an intentional phenomenon does not "look like" an activity, that only shows how misleading it is to assimilate intentional activity to physical activity.[9]

[8] Cornman seems to take as obvious that intentionality has to do only with kinds of activity. Thus he rephrases Chisholm's three criteria so as to "make explicit what seems to be the point of such criteria, that is, that these are criteria of intentional activity" (p. 53, footnote 3).

[9] It is unfortunate that philosophers should often tend to explicate the notion of a mental *act* in terms of the notion of *activity* (on the analogy with

But one may also challenge the claim that a cognitive verb does not designate any activity, even in the minimal sense in which 'believe' or 'desire' may be said to designate activities. It is interesting that Cornman should begin the passage quoted above with the assertion that cognitive verbs do not *designate* mental activities and end it by admitting that, nevertheless, cognitive verbs may *refer* to mental activities. The difference between 'designate' and 'refer to' corresponds, for Cornman, to the difference between 'express' and 'imply'. Thus, for example, 'believe' designates and thus expresses the activity of believing, while 'know' designates and expresses the achievement of knowing but refers to and implies the activity of believing.

This distinction, however, fails to support Cornman's claim that "there is *no* activity corresponding to 'know'." Instead, it supports the view that there is an activity (whether implied or expressed) *common to both* 'know' and 'believe' (namely the activity of believing), though, to be sure, the occurrence of this activity does not guarantee that one knows and does not merely believe something. The claim that no activity corresponds to 'know' is defensible only on the presupposition that knowing is not essentially a *kind* of believing—e.g., believing that meets certain logical and epistemological requirements. That is, Cornman can defend his claim only on a view of knowledge and belief such that knowledge is essentially different from something like *true and adequately evidenced belief*. Suppose we take knowledge to be just that. Then we can no longer say, as Cornman does, that 'know' designates *no* activity (if by this is meant that there is no activity corresponding to 'know'). We must say, instead, that 'know' does not designate *merely* an activity, or that it designates *more* than an activity, namely an

---

physical activity). What they seem to forget is that the notion of act as applied to the mental traditionally derived its sense from the Aristotelian distinction between *actuality* and *potentiality*. To say that one performs a mental act was for the Scholastics to say that one's mind is *in act*, that is, in a state of actuality with respect to a certain (mental) capacity. On this account, it is clear that the question whether a certain verb expresses an activity or not is totally irrelevant to the question whether it expresses a mental act.

activity of believing that meets certain special requirements—e.g., the requirements that what is believed be true and that the believer have adequate evidence for his belief.

III

We may now return to the main question at issue, that is, whether cognitive sentences should be considered as intentional. Despite Cornman's objections, I think we may answer this question in the affirmative on the ground that cognitive sentences (or the cognitive verbs contained in them) imply or presuppose or refer to certain intentional activities. "A knows that *p*" is intentional because an essential part of what is *meant* (and thus *expressed*) by it is that A *believes* that *p*. Similarly, "A perceives [sees] X [or an X]" is intentional because it implies or presupposes some such intentional sentence as "A has a visual experience of [an] X" (which does not imply that there is [an] X), or perhaps "A takes [an] X to be so-and-so" (which does not imply that [an] X is so-and-so).[10]

The question that now remains to be answered is whether Chisholm's criteria of intentionality can be so emended or supplemented as to cover also cognitive sentences. The important consequence of my account is that all that is required for this purpose is that Chisholm's *first two* criteria be supplemented by a condition like the following: "Any sentence which entails[11] a sentence that is intentional by either of these criteria is itself intentional."

---

[10] Cf. Chisholm, *Perceiving*, ch. 10.

[11] I am obviously using 'entail' in a stronger sense than the logician's 'imply': e.g., in a sense which excludes the paradoxes of strict implication (thus on my account an intentional sentence would not be entailed by a contradiction). Though no one, to my knowledge, has yet been able to give a satisfactory *formal* analysis of entailment, I think it would be unfair to reject, on this account, my (unanalyzed) use of this concept. It would also be unfair to reject, as D.J. O'Connor does in a somewhat unsympathetic critique of Chisholm ("Tests for Intentionality," *American Philosophical Quarterly*, 4 [1967], 173–178), Chisholm's criteria of intentionality on the ground that Chisholm does not specify his use of the word 'imply.'

We are now, I think, in a position to say that Chisholm's first two criteria, together with the added condition, are a jointly necessary and sufficient condition of intentionality, without being compelled to exclude cognitive sentences from the class of intentional sentences.

4

Notes on the Logic of Believing

RODERICK M. CHISHOLM

1. I shall set forth what appears to be a fruitful method of investigating certain philosophical concepts and illustrate it by applying it to the concept of *believing*. The illustration will show, if I am not mistaken, that the logic of belief-statements differs in certain fundamental respects from that of all the familiar nonpsychological modalities. And this fact, in turn, enables us to formulate at least one logical mark of the psychological.

2. Let us consider statements which are like those of ordinary English except for making use of the apparatus of quantification theory and its variables, and let us single out two fairly simple types of such statements. The first comprises those statements which begin with a universal quantifier, which have no other quantifier, and which are such that the scope of the quantifier extends to the end of the statement. For example:

> For every *x*, *x* is material.

The second type comprises those statements which are like the first, except for having an existential quantifier where the first has a universal quantifier. Hence:

> There exists an *x* such that *x* is material.

We may think of a modal operator (somewhat broadly) as a phrase which, when prefixed to a complete statement, yields a

From *Philosophy and Phenomenological Research*, 24 (1963–64), 195–201. Reprinted by permission of the author and the editors.

new complete statement. Examples are: "It is possible," "It is permissible," and "Jones believes that." The two types of statement just singled out are such that a modal operator may be inserted either at the beginning of the statement or immediately after the quantifier. Hence for any modal operator—e.g., "it is possible that"—we may distinguish four statement forms, exemplified by the following:

(UC)   It is possible that, for every x, x is material;
(UD)   For every x, it is possible that x is material;
(EC)   It is possible that there exists an x such that x is material;
(ED)   There exists an x such that it is possible that x is material.

The letters "U" and "E" refer to universal and existential quantifications, respectively, and the letters "C" and "D" are short for the medieval terms *in sensu composito* and *in sensu diviso*.

One way of determining the logical properties of any modal operator, then, is to note the relations among its occurrences in these four forms. Thus, if we were studying the logic of "it is possible that," we could ask whether a statement of the form UC, in which "it is possible that" is the modal operator, implies the corresponding statement of the UD; or, as I shall put it, "For possibility, does UC imply UD?" And we may ask whether UC implies EC, or ED; whether UD implies UC, or EC, or ED; whether EC implies UC, or UD, or ED; and whether ED implies UC, or UD, or EC.

These twelves possibilities, for any modality, may be set forth in a table. Thus the following three tables indicate the relevant properties of *truth*, *falsehood*, and *moral indifference*, respectively. The top line of each table indicates what is being considered as implicans, and the first vertical column indicates what is being considered as implicate; the occurrence of a minus-sign in any particular place indicates that the implication which that place represents does not hold; the nonoccurrence of a minus-sign in any place indicates that the relevant implication does hold.

	UC	UD	EC	ED
UC			—	—
UD			—	—
EC				
ED				

Truth

	UC	UD	EC	ED
UC				
UD	—			—
EC	—			—
ED				

Falsehood

	UC	UD	EC	ED
UC		—	—	—
UD	—		—	—
EC	—	—		—
ED	—		—	

Indifference

The first table tells us, with respect to "It is true that," the forms UC and UD each imply each of the other three forms, that the forms EC and ED are implied by each of the other forms, and that neither EC nor ED implies either UC or UD. For example, "There exists an *x* such that it is true that *x* is material" (ED) implies "It is true that there exists an *x* such that *x* is material" (EC), but it does not imply "It is true that, for every *x*, *x* is material" (UC) and it does not imply "For every *x*, it is true that *x* is material" (UD).

The second table tells us, with respect to "It is false that," that the forms UC and ED are each implied by each of the other three forms, that UD and EC each imply each of the other three forms, and that neither UD nor EC is implied by UC or by ED.

And the third table tells us, with respect to one sense, at least, of "it is morally indifferent whether," that UD implies ED, and that, otherwise, no one of the forms implies any of the others. The implication from UD to ED—from "For every *x*, it is indifferent whether *x* smoke(s)" to "There exists an *x* such that it is indifferent whether *x* smoke(s)"—is valid on the general assumption that a statement beginning with "For every *x*" implies the corresponding statement which begins instead with "There exists an *x*."

3. Before considering the pattern for believing—for "S believes that" or "It is believed that"—I shall make three general observations about the nature of belief. If these observations are accepted, as I believe they should be, then the pattern for believing will be as obvious as are those for truth and falsehood.

(i) We should note first that people—i.e., believers—are capable of error: they may believe propositions which are false, including those which are logically false; they may withhold belief from propositions which are true; they may have beliefs which contradict each other; and they are capable of two rather special types of error, which I shall call "defective instantiation" and "defective generalization."

A *defective* instantiation is to be distinguished from a merely mistaken instantiation. We make a mistake of the latter sort when we suppose, with respect to something which does not in fact exemplify a certain property, that it does exemplify that property. Defective instantiation, on the other hand, is illustrated by the following: Suppose that materialism is true—i.e., that everything is a material object—and that there is a "complete physicist," a man who has informed himself with respect to each particular object. He would make the mistake of defective instantiation if he were to believe that, in addition to these physical objects, there is still another thing—say, God. Knowing each particular thing as he does, he has not made the mistake of supposing any of them to be nonmaterial, much less to be divine. Without contradicting himself, he believes that there exists an x such that x is *non*material and he also believes, with respect to every object, that it *is* material. Defective instantiation might thus be described, paradoxically, as the mistake of believing, with respect to something that does not exist, that it does exist; or, somewhat less paradoxically, as being a mistake which is not a mistake with respect to anything at all.

"Defective generalization," on the other hand, may be described as the mistake of believing, with respect to some nonuniversal set of things, that those things comprise everything that there is. Suppose that the universe is exactly what our "complete physicist" thought it to be: there are many material objects, $a, b, c \ldots n$, and just one other thing, namely, God. Consider now a second "complete physicist": he, too, is well-informed with respect to each and every particular physical object, but he has the mistaken belief that they are everything that there is, thus committing the error of "defective generalization." He then reasons: "$a, b, c \ldots n$ comprise everything that there is; each of these things is a material object; therefore everything is a material object." His conclusion, then, is a mistake which is based upon a "defective generalization." The mistake which it embodies does not consist in identifying God with any material thing, or in supposing God to be material; like our first physicist, he knows each material thing too well to identify it with God, and (we may also suppose) he has no heretical views as to the material nature of God. He has made a defective generalization—a generalization which is false, not because of the nature of the things to which it does refer, but simply because of the nature of one of the things to which it does not refer.

(ii) Having noted that believers are capable of various types of error, we should also remind ourselves that they are capable of avoiding error—at least, that they may occasionally have beliefs which do not involve any of the mistakes just described.

(iii) And, finally, I shall make the following general assumption about the nature of belief (an assumption which most philosophers now take for granted, but which has been denied by Meinong and others): To believe with respect to any particular thing *x*, that *x* has a certain property F is, in part at least, to believe that there *exists* an *x* such that *x* is F; to believe that some S are P is to believe that there exists an *x* such that *x* is S and *x* is P; and, more generally, to believe anything at all, with respect to any particular thing, is to believe, with respect to some property, that there exists something exemplifying that property.

This general assumption implies that no one who is consistent can be said to believe, with respect to any particular thing, that that thing does not exist.[1] The consistent atheist, though he may be said to believe that God does not exist, cannot be said to believe, with respect to God, that he does not exist; and this is true whether God exists or not. That is to say, from the sup-

---

[1] Hence I resisted the temptation to say that, if defective instantiation is the mistake of believing, with respect to something that does not exist, that it does exist, then defective generalization is the mistake of believing, with respect to something that does exist, that it does not exist.

position that God does exist, we would not be justified in inferring that there exists an x such that the atheist believes that x does not exist. The atheist believes, with respect to the things that do exist, that none of them has the properties which are peculiar to being God and hence that none of them is God.

According to Meinong, most of us believe that all centaurs have hooves, thus believing, with respect to certain nonexistent things, that *they* have hooves. But it has often been pointed out that what we believe, in such cases, might be expressed in one or the other of these two ways: (a) we believe, with respect to the property of being a centaur and the property of having hooves, that the former includes the latter; or (b) we believe, with respect to the things that do exist, that if any of them were a centaur it would have hooves.[2] In short: from the supposition that, unknown to us, there exists a centaur somewhere, it does not follow, with respect to those of us whom Meinong describes as believing that all centaurs have hooves, that there exists an x such that we believe that x has hooves.

Let us say, then, that to believe anything, with respect to any particular thing, is to have a belief with "existential import": it is, in part at least, to believe with respect to some property, that there *exists* something having that property.

We may now turn to the modal table for believing.

4. To make our examples uniform and relatively clear, let us restrict our universe of discourse to the people who are in a given room at a certain time; but each example may be readily revised so as to hold of a universe which is not thus restricted. Let us suppose that there are exactly three such people—Jones, Smith, and Robinson; these will be the things which our variables will designate (or, at least, which the variables of our true sentences will designate). Consider, then, the following statements:

(UC)   S believes that, for every x, x is sitting;
(UD)   For every x, S believes that x is sitting;

---

[2] Another possibility, of course, would be: we believe that there exists nothing which is a centaur and which lacks hooves. But I think Meinong was right in saying that this is not plausible as an account of what it is to believe that all centaurs have hooves.

(EC) S believes that there exists an x such that x is sitting;
(ED) There exists an x such that S believes that x is sitting.

We may now turn to the columns of our diagram.

(i) UC does not imply UD. S may commit the error of "defective generalization," believing that only Jones and Smith are in the room, and he may believe that each of them is sitting. If he is consistent, he will conclude that, for every x, x is sitting (UC), thus basing a conclusion upon his defective generalization. But UD in this instance will be false; for there exists an x, namely, Robinson—such that it is false that S believes that x is sitting.

(ii) UC does not imply EC. S may believe, mistakenly, that there is no one in the room; hence he may also believe it to be false that there exists an x such that x is sitting; and therefore he will believe that, for every x, it is false that x is sitting (UC).[3] But believing that there is no one in the room, S does not believe that there exists an x such that it is false that x is sitting.

(iii) UC does not imply ED. In the situation just described, S believes that, for every x, it is false that x is sitting (UC); but there exists no x such that S believes it is false that x is sitting. He believes that there is *no one* there, not that the people who *are* there are not sitting.

(iv) UD does not imply UC. S may believe that Smith, Jones, and Robinson are all sitting; hence, for every x, S believes that x is sitting (UD). But he may believe, mistakenly, that there is a fourth person and that this fourth person is standing; in which case he will not believe that, for every x, x is sitting. He will have made the mistake of "defective instantiation." Or he may simply be agnostic and reluctant to conclude that the three people, whom he believes to be sitting, comprise everyone in the room.

(v) But UD implies EC. If each person in the room is such

---

[3] If the reader is inclined to reject the "therefore," he may replace the "for every x," in its various occurrences throughout our discussion, by "it is false that there exists an x such that it is false that," and the inference will then be unobjectionable.

that S believes him to be sitting (UD), then S believes that there exists an x such that x is sitting (EC). For, as we have noted, to believe with respect to any particular thing (and hence also to believe with respect to each particular thing) that that thing has a certain property is to believe that there exists something having that property.

(vi) UD implies ED. This is a consequence of our assumption that there is something—i.e., that there are people in the room. If S believes something with respect to each thing, then there exists something with respect to which he believes it.

(vii) EC does not imply UC. In believing that there exists an x such that x is sitting (EC), S need not believe that, for every x, x is sitting; for, as we have noted, he need not always be in error.

(viii) EC does not imply UD. If S believes that only Robinson is sitting, then it is not true that, for every x, S believes that x is sitting. But it is true (EC) that there exists an x such that x is sitting.

(ix) EC does not imply ED. S may believe that some one of the three persons is sitting and yet have no belief as to which one of the three it happens to be; hence there will be no x such that S believes that x is sitting, but S believes that there exists an x such that x is sitting (EC).

(x) ED does not imply UC. One of the people, say Robinson, may be such that S believes him to be sitting (ED), without S also believing that for every x, x is sitting.

(xi) Nor does ED imply UD. One of the people, again, may be such that S believes him to be sitting (ED), without every x being such that S believes that x is sitting.

(xii) But ED implies EC. If S believes, with respect to Robinson, that he is sitting, i.e., if there exists an x such that S believes that x is sitting (ED), then, as we have noted, S will also believe that there exists an x such that x is sitting.

We have, then, completed our table for believing.

	UC	UD	EC	ED
UC		—	—	—
UD	—		—	—

EC   —

ED   —   —

Believing

The pattern has the failures of both truth and falsehood; but, unlike that for moral indifference, it is such that UD implies EC, and ED implies EC.

5. The patterns for various "alethic," "deontic," and causal modalities differ, in some respect or other, from the pattern here attributed to believing.[4]

Thus the "alethic" and causal modalities differ from believing in the following respects, among others: for *necessity* and *possibility*, UC implies EC; for *impossibility*, ED implies UC; and for *nonnecessity, contingency,* and *noncontingency*, ED does not imply EC.

The "deontic" modalities differ in these respects, among others: for *obligatory* and *not wrong* ("permitted"), UC implies EC; for *wrong* ("forbidden"), ED implies UC; and for *nonobligatory, indifference,* and *nonindifference*, ED does not imply EC.

The "epistemic" modalities—*evidence, probability, acceptability,* and the like—are more difficult to interpret. On some interpretations they pertain to believing—to what people are justified in believing—and it is possible that, on some one of these interpretations, they will conform to the pattern I have attributed to believing.

But I know of no modalities which are obviously nonpsychological and which conform to this pattern. It may be, therefore, that the pattern gives us a mark of intentionality. Perhaps we can say that a modal prefix is intentional if its occurrence in our four statement-forms is such that: UC does not imply UD, EC, or ED; UD implies EC and ED, but does not imply UC; EC does not imply UC, UD, or ED; and ED implies

---

[4] I have discussed the patterns for these and for certain other psychological modalities in greater detail in "On Some Psychological Concepts and the 'Logic' of Intentionality," read at a Symposium on the Philosophy of Mind held at Wayne State University on December 6, 1962, and published in H.-N. Castañeda, ed., *Intentionality, Minds, and Perception* (Detroit: Wayne State Univ. Press, 1967), pp. 11–35.

EC, but does not imply UC or UD. Or, more aphoristically: "A modal prefix is intentional if it may say that some or say that all when it doesn't say of any, and say of each and hence that some when it doesn't say that all."[5]

[5] I am indebted to Robert Sleigh, Herbert Heidelberger, and Hector-Neri Castañeda.

5

Discussion on "Notes on the Logic of Believing"

On the Logic of Belief
DAVID R. LUCE

It would seem to be a harmless and noncontroversial assumption, that in the case of anyone who believes anything at all, there is something about which he believes something or other. A logic of belief predicated upon that assumption ought not to suffer any noticeable lack of generality. The logic of belief which Roderick M. Chisholm presents in a recent issue of *Philosophy and Phenomenological Research*, however, rules out that assumption.

In that article Professor Chisholm seeks to specify the logical relationships holding between four types of statement called "UC," "UD," "EC," and "ED." They are represented by the following:

UC: S believes that for all x, Fx.
UD: For all x, S believes that Fx.
EC: S believes that there is an x such that Fx.
ED: There is an x such that S believes that Fx.

The following table summarizes Professor Chisholm's findings. A "1" at the intersection of a column and a row means that the statement at the head of the column entails the statement at the left-hand side of the row; a "0" in that place means that the statement at the head of the column does *not* entail the

From *Philosophy and Phenomenological Research*, 25 (1964–65), 259–269. Reprinted by permission of the authors and the editors.

statement at the left-hand side of the row. (In the four trivial
cases I have enclosed the "1" in parentheses.)

	UC	UD	EC	ED
UC	(1)	o	o	o
UD	o	(1)	o	o
EC	o	1	(1)	1
ED	o	1	o	(1)

One of Professor Chisholm's conclusions, that UC does not
entail EC, holds up only if we suppose either that S is not
capable of the simplest sort of inference, or that S believes that
nothing at all exists. For in any nonempty domain of discourse,
"$(x)Fx$" entails "$(\exists x)Fx$." Hence if S acknowledges that logical
principle, the only condition under which "S believes that
$(x)Fx$" can be true and "S believes that $(\exists x)Fx$" false is that
S does not acknowledge the existence of anything at all. This
view is so odd that contemporary logicians do not treat it as a
serious possibility, contenting themselves with principles of
inference valid in nonempty domains.

A second conclusion, that UC does not entail ED, holds up
only if we suppose that none of the things which S believes to
exist actually does exist. (The condition is satisfied if S believes
that nothing at all exists.) This means, among other things,
that S cannot believe even in his own existence. Again we must
ascribe a rather incredible view to S.

I suggest that the table of entailment-relations that Professor
Chisholm gives us is based on a logic of belief that is needlessly
general. If we assume that for every S, there is at least one
thing which S admits into his ontology, then the table must be
modified at the two indicated places. Stated more precisely, the
assumption is:

> For every subject S, there is some entity x such that S believes
> that if $(x)Fx$ then Fx.

The assumption, as I say, seems to be harmless and noncontroversial.

Notes on Chisholm on the Logic
of Believing
ROBERT C. SLEIGH

I

In an article entitled "Notes on the Logic of Believing,"[1] R.M. Chisholm has proposed an ingenious method of investigating certain concepts and has, apparently, derived an important result from this method, i.e., the formulation of a logical mark of the intentional, i.e., the psychological.

Briefly, the method consists in studying the logical behavior of modal operators[2] in contexts of quantification. Let M be some modal operator; we may distinguish the following forms:

(UC)	$M(x) (\phi x)$;
(UD)	$(x) M (\phi x)$;
(EC)	$M(\exists x) (\phi x)$;
(ED)	$(\exists x) M (\phi x)$.[3]

Having distinguished these forms, Chisholm points out that "one way of determining the logical properties of any modal operator . . . is to note the relations among its occurrences in these four forms."[4] This Chisholm does for various prefixes by setting forth a table or pattern for that prefix. An example of such a table, taken from Chisholm, is the table for the prefix 'it is morally indifferent that', which goes as follows.[5]

---

[1] "Notes on the Logic of Believing" (included in this volume; page references are to this volume).

[2] Chisholm says: "We may think of a modal operator (somewhat broadly) as a phrase which when prefixed to a complete sentence, yields a new statement," pp. 75–76.

[3] Chisholm says: "The letters 'U' and 'E' refer to universal and existential quantifications, respectively, and the letters 'C' and 'D' are short for the medieval terms 'in sensu composite' and 'in sensu diviso,' " p. 76.

[4] p. 76.

[5] p. 77.

	UC	UD	EC	ED
UC		—	—	—
UD	—		—	—
EC	—	—		—
ED	—		—	

Chisholm explains how this and other tables are to be read as follows: "The top line of each table indicates what is being considered as implicans, and the first vertical column indicates what is being considered as implicate; the occurrence of a minus-sign in any particular place indicates that the implication which that place represents does not hold; the nonoccurrence of a minus-sign in any place indicates that the relevant implication does hold."[6]

Chisholm proceeds to formulate and defend certain principles concerning *believing* in the light of which he concludes that the table for the prefix 'S believes that' is as follows:[7]

	UC	UD	EC	ED
UC		—	—	—
UD	—		—	—
EC	—			
ED	—		—	

He concludes: "I know of no modalities which are obviously nonpsychological and which conform to this pattern. It may be, therefore, the pattern gives us a mark of intentionality."[8]

II

Like Chisholm, I have been unable to discover any modal prefix which is obviously nonpsychological and which conforms to this pattern. What I am not convinced of is that *believing* conforms to it. Although, in certain cases, I find Chisholm's use of the notion of a restricted universe of discourse not altogether clear, it seems to me that Chisholm is right in all of the as-

[6] p. 76.
[7] p. 82 f.
[8] p. 83.

signments he makes concerning *believing* except for two, i.e., his claims that, for *believing* UD implies EC and ED implies EC. Presumably, these two claims are on a par; that is, if we accept one we will accept the other and if we reject one, we will reject the other. Hence, we may concentrate on the claim that for *believing* ED implies EC. That ED implies EC for *believing* follows from a general principle about the nature of belief which Chisholm formulates as follows: "To believe with respect to any particular thing X, that X has a certain property F is, in part at least, to believe that there exists an X such that X is F; to believe that some S are P is to believe that there exists an X such that X is S and X is P; and, more generally, to believe anything at all, with respect to any particular thing, is to believe, with respect to some property, that there exists something exemplifying that property."[9]

Consider the following two examples which I propose as counterexamples to Chisholm's principle:

(a) Suppose that Jones believes that every positive integer is either zero or one of its successors. Jones, perhaps overly immersed in his study of arithmetic, believes that there is nothing but the various positive integers. Jones being otherwise rational, we can imagine that, in these circumstances, the following is true:

(i) It is false that Jones believes that there is an X such that it is possible that X is neither zero nor any of its successors. Jones thinks that only numbers exist; yet he believes that things might have been otherwise. Thus Jones believes that it is possible that there exists an X, e.g., the tallest man in Boston, such that X is neither zero nor any of its successors. Contrary to Jones's views, there is such a thing as the tallest man in Boston. Hence we may conclude:

(ii) There is an X such that Jones believes that it is possible that X is neither zero nor any of its successors.

(b) Suppose Ed believes the following: that there is no God, but that, nonetheless, Al thinks incessantly of God, and that God is the only thing of which Al incessantly thinks. Hence we have:

[9] p. 79.

(iii) It is false that Ed believes that there is an X such that Al incessantly thinks of X.

Suppose that, contrary to Ed's atheism, there is a God. We may, then, conclude:

(iv) There is an X such that Ed believes that Al incessantly thinks of X.[10]

Note that both of the proposed counterexamples make use of intensional predicates. It might be thought that any counterexample to Chisholm's principle would employ intensional predicates. Suppose this were so; might we use this information to preserve Chisholm's claim with respect to the table to which *believing* conforms? Note that, in claiming that a given modal operator M is such that, for M, ED implies EC, we are claiming that:

$$(v) \ (\phi) \ [(\exists x) \ M \ (\phi x) \supset M \ (\exists x) \ (\phi x)]$$

We might restrict the range of 'ϕ' to nonintensional properties, thus restricting its substituends to nonintensional predicates. This strategy would be promising provided that there were available a criterion for an intensional predicate, and that only examples employing intensional predicates provide counterexamples to Chisholm's principle. But the latter claim seems dubious to me. Counterexamples seem most clear cut when intensional predicates are employed. But there appear to be counterexamples which do not employ intensional predicates.

Consider the following:

(c) Imagine a philosopher who is persuaded by Meinong's arguments and comes to believe that, although there is no golden mountain, nonetheless, the golden mountain is golden. Suppose he also believes that anything golden is slippery. Hence, he concludes that the golden mountain is slippery. Perhaps this last inference would be rejected even by Meinong. But we can imagine a philosopher making it. Clearly, we can fill in the remaining details so as to produce a case in which it is possible that ED is true for *believing* and EC is false.

[10] I owe this point to Edmund Gettier.

Chisholm has foreseen this type of purported counterexample and considered two ways of handling it, i.e., two ways of expressing beliefs like those ascribed to our Meinongian philosopher which avoid violating Chisholm's principle. We are assuming the following:

(vi) Jones believes that the golden mountain is slippery.

We might try putting this as:

(vii) Jones believes the property of being the golden mountain includes the property of being slippery.[11]

But (vii), clearly, attributes to Jones a belief in a necessary connection between properties not at all involved in (vi).

We might try:

(viii) Jones believes that, for any X, if X were a golden mountain then X would be slippery.[12]

But (viii) asserts of Jones a belief quite different from that asserted of Jones in (vi). A non-Meinongian might accept the proposition that (viii) says Jones believes without accepting the proposition that (vi) says Jones believes.

Hence. I conclude, tentatively, that neither UD nor ED imply EC for *believing*. If these tentative conclusions are correct, then *believing* has the same table as *moral indifference*. If this were so, then we would not have found a logical mark of the psychological.[13]

Believing and Intentionality:
A Reply to Mr. Luce and Mr. Sleigh
RODERICK M. CHISHOLM

1. In the paper discussed by Mr. Luce and Mr. Sleigh, I had argued that, of the four types of sentence, "S believes that, for every x, x is F" (UC), "For every x, S believes that x is

[11] p. 80.
[12] p. 80.
[13] I am indebted in these matters to Roderick M. Chisholm, Edmund Gettier, and Hector-Neri Castañeda.

F" (UD), "S believes that there exists an x such that x is F"
(EC), and "There exists an x such that S believes that x is F"
(ED), neither the first nor the third implies any of the others,

	UC	UD	EC	ED
UC		—	—	—
UD	—		—	—
EC	—			
ED	—		—	

the second implies all but the first, and the fourth implies the
third but does not imply either the first or the second. Thus I
said that "S believes" follows the pattern exhibited on the accompanying diagram, and that no nonpsychological prefix follows that pattern. Mr. Luce argues that the diagram has two too
many minus-signs and Mr. Sleigh that it has two too few.

2. According to Mr. Luce, if we assume that "there is an x
such that x believes that there is some property F such that x is
F, or S believes that there is some property F such that x is not
F," then we may say that, for believing, UC implies both EC
and ED. The result is exhibited on the accompanying diagram.
If we do make use of Mr. Luce's assumption, then the "S"
in our formulas will have to be restricted to persons and indeed

	UC	UD	EC	ED
UC		—	—	—
UD	—		—	—
EC				
ED			—	

only to those persons who are not irrational, and even then it
will be logically possible for there to be a person S concerning
whom the assumption is false. Since I was concerned only with
the nature of *believing,* and with what kinds of belief-sentences
entail what, I did not make any contingent assumption about
the nature of believers and thus did not restrict the discussion in
the way in which Mr. Luce proposes. I would say, therefore, that
he is concerned with a slightly different problem, but that what
he says, in connection with that problem, is correct.

If we do proceed as Mr. Luce does, we can still say, if I am not mistaken, that believing exemplifies a logical pattern that is exemplified only by psychological prefixes; i.e., that no non-psychological prefix satisfies the second of the accompanying diagrams.[14] Mr. Sleigh has noted, independently, that this second pattern is also satisfied by certain semantical prefixes, such as "With respect to the propositions expressed by the sentences on page 124 of Volume One of the *Psychologie vom empirischen Standpunkt,* it is more probable than not that. . . ." But to say of a sentence that it "expresses" a certain proposition is to say that there are people who use, or may use, the sentence in order to *assert, express,* or *convey* the proposition, and to say any of the latter things is to say something that is psychological or intentional.

3. Mr. Sleigh attempts to show that, for believing, ED does

	UC	UD	EC	ED
UC		—	—	—
UD	—		—	—
EC	—	—		—
ED	—		—	

not imply EC (and also, by implication, that UD does not imply EC). If he is right, then the logical pattern for believing is exhibited in the accompanying diagram. Since this pattern is also satisfied by "it is morally indifferent whether," and by certain other nonpsychological prefixes, it does not yield any characteristic that is peculiar to what is intentional.

To show that, for believing, ED does not imply EC, Mr. Sleigh offers three sets of premises, the premises being plausible and mutually consistent and each set implying, according to Mr. Sleigh, (i) that a certain belief-sentence in the form EC is false and (ii) that the corresponding belief-sentence in the form ED is true. But in each case, I suggest, the derivation of the

[14] In a Symposium on the Philosophy of Mind held at Wayne State University in December, 1962, I defended the pattern which Mr. Luce proposes; but as a result of criticisms by Mr. Sleigh, I then proposed the first of these patterns instead. See my paper "On Some Psychological Concepts and the 'Logic' of Intentionality" in H-N. Castañeda, ed., *Intentionality, Minds, and Perception* (Detroit: Wayne State University Press, 1967).

ED sentence is invalid. For in each case, the argument may be put in the following form:

(1) S believes that the G is F;

(2) There exists an x such that x is identical with the G;

(3) Hence there exists an x such that S believes that x is F.[15]

The following example seems to me to indicate that this type of inference is invalid. S believes, patriotically, that some American astronaut (he doesn't know who) is the person who will be first to land on the moon; he also believes that the person who will be first to land on the moon is a future President of the United States (this second sentence may be put in the form of (1) above); but, undreamed of by S, the person who will be first to land on the moon is a Russian lady astronaut from Minsk (and therefore we may assert a sentence in the form of (2) above). May we conclude, in accordance with (3), that there exists an x such that S believes that x is a future President of the United States—i.e., that the lady astronaut from Minsk is a person whom our patriotic American believes to be a future President of the United States? It seems to me that we cannot draw this conclusion, and therefore, since Mr. Sleigh's argument presupposes that we can, that his examples do not falsify what I had said.

4. Mr. Sleigh's discussion leads us to ask the following question, which, I believe, is of considerable importance: Can we find any consistent set of premises which contain no psychological sentence in *sensu diviso* (i.e., in the form UD or ED) but which imply a belief sentence in *sensu diviso*?[16]

When we say, of a man and some property F, not only that the man believes that there is something which is F, but also

---

[15] In Mr. Sleigh's second example "the G" would be "God," and "F" would be "thought of incessantly by A"; in his third example, as I interpolate it, "the G" would be "the golden mountain," and "F" would be "slippery." In his first example, the sentence corresponding to (1) and (3) above is slightly more complex, but the added complexity does not affect the question of the validity of the inference.

[16] Compare the discussion of this question (in a terminology somewhat different from that used here) in Jaakko Hintikka, *Knowledge and Belief* (Ithaca, N.Y.: Cornell Univ. Press, 1962), ch. 6, and in Roderick M. Chisholm, "The Logic of Knowing," *The Journal of Philosophy*, 60 (1963), 773–795.

that there is something which the man believes to be F, we imply that, to an extent at least, he has gotten "beyond the circle of his own ideas" and that he has not shot entirely beside the mark.[17] An answer to our logical question, therefore, will describe conditions under which it can be said, of some particular entity, that a man's belief is directed upon *that* particular entity.

I suggest, as a partial answer, that a belief-sentence, in the form "There exists an *x* such that S believes that *x* is F," in *sensu diviso*, is implied by any sentence of the following forms, in *sensu composito*: "S *perceives* (e.g., *sees*) something to be the G and he believes that the G is F"; "S *remembers perceiving* something to be the G and he believes that the G is F"; and "S *believes himself* to be F." Presumably other types of sentences should be added to the list.

5. I have suggested, then, that if we consider the behavior of "S believes that" in certain quantificational contexts, we will find a kind of logical mark of the psychological or intentional. It should not be supposed, however, that this is the only way of finding such a mark. The following is another possibility.

Let us refine upon ordinary written English in the following way: instead of writing propositional clauses as "that"-clauses, we will eliminate the initial "that" and put the remainder of the clause in parentheses. Thus, instead of "It is possible that there are angels" and "Socrates believes that all men are mortal," we will have "It is possible (all men are angels)" and "Socrates believes (all men are mortal)." Let us define a (simple) sentence-prefix as an expression which is such that: (1) the result of prefixing it to such a sentence in parentheses is another sentence, and (2) it contains no proper part that is logically equivalent to a sentence or to a sentence-function. We may say that a (simple) prefix M is *intentional* if, for every sentence *p*, M(p) is logically contingent. And we may now say that a sentence is intentional if it implies the result of prefixing some sentence by an intentional prefix.

Thus "it is necessary" is not intentional since, when pre-

[17] Compare *Theatetus,* 194a.

fixed to "(all squares are rectangles)," it yields a sentence which is necessary and therefore not contingent. "It is wrong" is not intentional since, when prefixed to "(there is not anything of which it can be truly said that it is wrong)," it yields a sentence which is contradictory and therefore not logically contingent. But "S believes," "S desires," and certain other psychological prefixes, are such that, no matter what sentence we prefix them to, the result will be logically contingent. We might put the matter somewhat loosely in this way: in dealing with propositions, persons, unlike nonpersons, need not be hampered by any of the logical properties of the propositions.

The "thesis of intentionality" would now become: the psychological, unlike the nonpsychological, can be adequately described only by using sentences that are intentional.

6

On "Intentionality" and the Psychological

W. GREGORY LYCAN

Several philosophers have recently tried, following Brentano, to establish the claims that (1) a logical property called "intentionality" is an ineluctable characteristic of all sentences containing "psychological" verbs, and that (2) no nonpsychological verb is "intentional." Statements about propositional attitudes receive special consideration in these investigations.

What initially renders evaluation of their results difficult is that no less than nine differently specified concepts of "intentionality" are being appealed to, some of which are restricted to sentences of certain logical types. If we are to make sense of the above two claims, we must sort these out as rigorously as possible. But first let me set out the exact nature of the enterprise more explicitly, so that the grounds upon which some of the definitions are to be rejected will be absolutely clear: We want to find a definition of "intentionality" *such that* it will be a contingent (or in some sense nontrivial) fact that (1) and (2) above obtain. A criterion put forth as logically necessary and sufficient of intentionality may be rejected, therefore, if some psychological verb fails to count as intentional according to it, or if it is fulfilled by some obviously nonpsychological verb. In such a case, the recalcitrant verb will be referred to as a *counterexample*. But it must be emphasized that rejections are to be made entirely on the basis of *usefulness* vis à vis

From *American Philosophical Quarterly*, 6 (1969), 305–311. Reprinted, with revisions by the author, by permission of the author and editors.

the vindication of what I shall hereafter call Brentano's thesis;[1] I shall never mean to suggest that a proposed *definition* admits of a truth-value, although I shall insist that a definition whose adoption would lead to the falsification of Brentano's claims be discarded. (My purpose is to see *if there is* any logical property for which (1) and (2) are correct.)

I think it is most fruitful to stipulate at the outset that the adjective "intentional" shall apply in its primary usage to *verbs*, and then to say that a sentence is intentional if and only if one of its verbs is intentional. (Later on I shall liberalize this recommendation slightly.) Now we can make the preliminary distinction between verbs which take propositions as their objects and those which take entities given by names or by descriptions. Clearly there are psychological verbs of each of these types; so a useful concept of intentionality must apply to both, although I shall occupy myself primarily with investigating the logical properties of the former sort. Let us hereafter speak of "proposition-verbs" ("*p*-verbs") and "object-verbs" ("*o*-verbs"). Some verbs can, of course, be both *p*-verbs and *o*-verbs.

I can now list the most interesting of the sets of defining conditions offered. (Some of the nine, like (B), are not "defining" conditions in the most precise sense, since they are not put forth as both necessary and sufficient, and hence are not intended to be strictly constitutive of intentionality. But they are at least fragmentarily suggestive ways of marking off the concept.) I shall restrict my discussion in this paper to those sets of criteria to which there are no *obvious* counterexamples, or at least to those sets which have yet to be controverted in the literature. In recording them as suggestions, however, I do not mean to say that their original proponents still adhere to them.

(A) In "The Intentionality of Sensation: A Grammatical

---

[1] Brentano, of course, did not formulate (1) and (2) in terms of "logical" or "grammatical" properties (see his *Psychologie vom empirischen Standpunkt* [Vienna, 1874]). The connection between his "psychological theory of intentionality" and the view that I am evaluating in this paper is best expressed by Roderick Chisholm in his article "Intentionality," in Paul Edwards, ed., *The Encyclopedia of Philosophy* (New York: Macmillan, 1967).

Feature,"[2] G.E.M. Anscombe loosely describes an intentional o-verb as one which satisfies three conditions; they are jointly sufficient, although she does not say that the first, at least, is necessary:

(i) The actual existence of the referent of the direct object of the verb is not a necessary condition of the truth of the clause in which the verb occurs.

(ii) The substitution of a different description of the object can affect the truth-value of the original clause.

(iii) The description of the object may be "indeterminate," in that certain questions which might sensibly be asked of an object failing to meet (i) and (ii) need not be applicable in the intentional case. ("I can think of a man without thinking of a man of any particular height [intentional o-verb]; I cannot hit a man without hitting a man of some particular height [nonintentional o-verb], because there is no such thing as a man of no particular height.")

I talk of "clauses" rather than of sentences in my formulation to prevent us from saying, e.g., that "I kicked a rock or I punched Fred" meets (i) because I may have kicked a rock even if there is no such thing as Fred.

This definition is the only one of the nine that applies exclusively to o-verbs. It can, I think, easily be translated into p-verb language; I shall attempt to do so below.

(B) In "Sentences About Believing"[3] Roderick Chisholm offers three conditions of intentionality for whole sentences (the first two appear in *Perceiving*[4] as well). He apparently means them to be individually sufficient, but none of them taken singly is necessary. Thus he gives us no way of saying that a particular sentence is *not* intentional, unless we assume that the disjunction of the sufficient conditions is actually a necessary condition. The first of the three is similar to (A–i),

[2] G.E.M. Anscombe, "The Intentionality of Sensation: A Grammatical Feature," in R.J. Butler, ed., *Analytical Philosophy: Second Series* (Oxford: Basil Blackwell, 1965).

[3] Chisholm, "Sentences about Believing" (included in this volume).

[4] Chisholm, *Perceiving* (Ithaca, N.Y.: Cornell Univ. Press, 1957).

Anscombe's nonexistence condition for o-verbs. The other two apply to sentences containing p-verbs: When a sentence S contains a p-verb having q as its object, S is intentional if

> (ii) neither S nor its contradictory implies either the truth or the falsity of q;

or

> (iii) q contains a name or description which (in Frege's terminology) has an "indirect reference."

The remaining definitions presuppose a notation devised by Chisholm in "On Some Psychological Concepts and the 'Logic' of Intentionality."[5] He introduces the idea of a "modal prefix," which is "a phrase ending with 'that,' or at the end of which the word 'that' may be inserted"; it serves as a sentence-forming operator, applicable just to sentences. He now specifies four kinds of sentences which can be formed by combining a given modal prefix "It is M that" with a universal or an existential quantifier. They are

$$\text{(UC)}\quad \text{It is } M \text{ that } (x)\ (Fx);$$
$$\text{(UD)}\quad (x)\ (\text{It is } M \text{ that } Fx);$$
$$\text{(EC)}\quad \text{It is } M \text{ that } (\exists x)\ (Fx);$$
$$\text{(ED)}\quad (\exists x)\ (\text{It is } M \text{ that } Fx).$$

For any modal prefix, Chisholm says, a set of four sentences of this form can be constructed, and various entailment-relations can be found to hold between its members. We can represent the pattern of entailments by a 4 x 4 matrix as shown:

	UC	UD	EC	ED
UC				
UD				
EC		b		
ED				

[5] Chisholm, "On Some Psychological Concepts and the 'Logic' of Intentionality," in H.-N. Castañeda, ed., *Intentionality, Minds, and Perception* (Detroit: Wayne State Univ. Press, 1966); includes comments by Robert Sleigh and a rejoinder by R.M. Chisholm.

(A dash is placed in box *b*, for example, if UD does *not* entail EC.) Every modal prefix "It is *M* that" determines such a matrix, which holds for any *Fx* whatsoever.

Let us now speak of *entire prefixes* as being intentional, since any *p*-verb can be made into a prefix without much trouble (but some prefixes cannot easily be paraphrased into verb phrases).

(C) Now Chisholm defines an intentional modal prefix as one whose matrix is

	UC	UD	EC	ED
UC		—	—	—
UD	—		—	—
EC				
ED		—		

PATTERN I

or one which conforms to Pattern I when prefixed by "it is false that."

(D) For reasons which I shall discuss below, Robert Sleigh counters (C) with the suggestion that a better set of defining conditions might be:

	UC	UD	EC	ED
UC		—	—	—
UD	—		—	—
EC	—	—		—
ED	—		—	

PATTERN II (THE MINIMUM PATTERN)

(E) Finally, Chisholm proposes that a sentence prefix might be called intentional if and only if "for every sentence *q*, the result of modifying *q* by *M* is logically contingent."[6]

---

[6] A sixth definition, first proposed by Russell, is discussed and discarded by Anthony Kenny in *Action, Emotion and Will* (London: Routledge and Kegan Paul, 1963), pp. 199–201. A seventh and eighth are to be found in the aforementioned Chisholm-Sleigh symposium: On p. 49 Chisholm advances a new matrix-pattern (III) which includes the condition that UD and ED both entail EC. But it is easy to see that few, if any, psychological verbs meet this requirement, whether or not conformity to Pattern III is a sufficient criterion. The

I should like to examine (C) and (D) first, since the Chisholm-Sleigh dispute centers around these despite its shotgun-like mode of attack. Chisholm obtains Pattern I for "S believes that," "S hopes that," "S fears that," etc., by presupposing that S is a "rational being," i.e., a being who believes everything entailed by each of his beliefs, hopes everything entailed by each of his hopes, etc. Sleigh claims that this assumption, when formulated explicitly (we can change "S believes that" to "S is rational with respect to his beliefs and he believes that"), is "utterly *ad hoc* and, perhaps, circular," since such a change does not constitute an analysis or paraphrase. Certainly to alter the prefix in this way would be to render Chisholm's claims about *belief* much less interesting; it might be said that they would not really be about our garden variety of belief at all, but rather about "rational opinion" or some such, since it is true and importantly so that one *can* fail to believe a deductive consequence of one's beliefs. (In fact, it is not senseless to suppose that some one of my beliefs might happen to be *incompatible* with a distant consequence of some other set of beliefs held by me. One need grant only that at least one of the beliefs in question be false.) If we are to represent the facts about our ordinary use of psychological verbs in filling out Chisholm's matrices, then we must agree that Sleigh is correct. His "rationality" criticism holds for belief, desire, hope, fear, and wondering, to name a few; I conclude that definition (C) can be dispensed with.

In the light of this, it seems that most psychological verbs (all, I think, save perhaps "knows," "is aware," and the like) tend more to conform to the Minimum Pattern (II), as Sleigh says, and not to Pattern I or (see note 6) to Pattern III. Let us then see if we can fare any better by adopting recommendation (D).

Sleigh himself claims to show that Pattern II is not manifested only by psychological prefixes; he offers "there is a sentence derivable in the system explained on p. 32 of *Perceiving*

---

eighth proposal appears on pp. 50–51; it is clearly violated by "S knows that" and "S is having the occurrent thought that," and fulfilled by "It is legally obligatory somewhere in the world that."

which expresses the proposition that" as a counterexample. Chisholm replies, in essence, that "Sentence S expresses the proposition that s" can be analyzed only as "Certain people use S, or may use S, in order to assert, express, or otherwise convey the proposition that s." To be as fair as possible to him, we might further paraphrase this latter sentence as "S is or may be used by a person when this person is thinking that s" or as "S is or may be used by a person when this person wishes to cause someone to entertain the proposition that s." Now it may be a fact about the world that, if persons did not occasionally have thoughts and express them by creating physical (public) sentence-tokens, we would not have the idea of what it is for sentence S to express 's'. But it is not certain that we must for this reason restrict ourselves to the analyses I attribute to Chisholm above. Sleigh's counterexample may materially imply some intentional sentence as a matter of contingent fact, but I think we should require a stronger logical connection than this if we intend to use it as grounds for saying that "semantic statements, as such, are intentional" or psychological. If we are to make a case for semantical sentences being psychological, then it would be nice to have a connection which could rule out the possibility that computers and robots may use S to express 's' just as persons do. One might want to say that, if something P "uses S to express 's'" (as opposed to merely uttering S), then P is surely a person, no matter what sort of electronic devices it may have for viscera. But this would be only a stipulation, not an argument.

There is some question, moreover, as to whether or not conformity to the Minimum Pattern could be a necessary criterion, even if it were a sufficient one, since "knowing" appears to violate it. Chisholm may be right in saying that UD entails EC for "S knows that," if we accept the view that "S knows that p" entails p. It seems justifiable to hold that, if (x) $(S$ knows that $Fx)$, then S must know that Fa and hence that $(\exists x)$ (Fx). Since Chisholm assumes his universe to be nonempty, there is no problem about fictional existence here.

But it might be objected that to maintain this is to presuppose that S is a "rational" knower. Even though the inference from "Fa" to "$(\exists x) (Fx)$" seems so automatic as to be called an "inference" only by courtesy, it is still (technically speaking) an inference; and there is no difference in principle between permitting S the leap from "Fa" to "$(\exists x) (Fx)$" and insisting that, if S knows the Peano axioms, he knows every theorem of arithmetic. So I shall leave it an open question whether or not our concept of "knowing" precludes our conceiving of someone's knowing that a, b, c, and everything else are F but not knowing that there are any F things.

What, then, does (D) leave us with? At most it is of dubious usefulness, since it is not sufficient and only *may* be necessary. (C) and (D) comprehend, in any case, only small sets of psychological prefixes which Chisholm and Sleigh have selected themselves. The Chisholm matrix thus seems at best only the beginning of what might, if radically expanded, prove to be a useful definition of "intentionality." For now it is more fruitful to explore some of the other more general proposals. Eventually we hope to find an idea of intentionality which applies, not only to all psychological p-verbs, but to all psychological o-verbs as well.

(E) seems at the outset to fail for most psychological p-verbs: "If S is not having any occurrent thought, then S is not having the occurrent thought that" forms a logically necessary statement when completed by *any* value of p. Another family of counterexamples is provided by sentences such as "If S knows that snow is white, then it is true that p," which is logically necessary for at least one value of p, and logically contradictory for several others, so long as we treat the incompatibility of colors as a logical matter. Nor can (E) serve as a sufficient condition: "It is legally obligatory that p" is logically contingent for all values of p (unless one holds the unpalatable view that "It is legally obligatory that q and not-q" is not merely false but nonsensical).

It will be noticed, on the other hand, that all my objections to calling (E) a *necessary* condition of intentionality turn on counterexamples which contain truth-functional particles. In one of his other essays[7] Chisholm offers a way of undercutting these criticisms. He suggests restricting (E) by allowing it to apply only to "simple" modal prefixes, a "simple" prefix being one which "contains no proper part that is logically equivalent to a sentence or to a sentence-function." Now even with this modification, (E) is still violated by "S knows that," "S notices that," etc., since "S knows that q and not-q" entails a contradiction. But if Chisholm were, further, definitionally to prohibit p's being logically incoherent or containing a psychological verb, I think he would have a useful necessary (but not a sufficient) criterion of intentionality such that Brentano's claim (1) is vindicated. At any rate, I can think of no counterexamples to this.

It is now time to examine (A) and (B). I have earlier stated my hope that (A) can be generalized to apply to p-verbs as well as to o-verbs; I shall now put forward analogues which should accomplish this.

(A′) A p-verb or a modal prefix "It is M that" is intentional if and only if the following three conditions are satisfied:

(i) For every argument p which is not logically incoherent and contains no psychological verb,[8] p entails neither "It is M that p" nor "It is false that it is M that p."

(ii) Where p and q are different contingent sentences which contain no psychological verbs, the conjunction of "It is M that p" and "p and q are logically equivalent" does not entail "It is M that q."

((ii) is closely related to but is not the same as Chisholm's condition (B–iii).)[9]

(iii) EC does not entail ED and UC does not entail UD.

[7] Chisholm, "Believing and Intentionality: A Reply to Mr. Luce and Mr. Sleigh" (included in this volume, pp. 91–96.

[8] I add the latter restriction in response to a valuable criticism of David Sanford's.

[9] The ninth (traditional) criterion of intentionality is simply failure of substitutivity, or "referential opacity." But, as Quine says, alethic contexts are referentially opaque; and further objections to the sufficiency of (A′–ii) alone are raised in this paper.

This definition is composed of a variant of proposal (B) plus the condition (A′–iii). It is easy to show that (A′–iii) is needed, since "It is legally obligatory somewhere in the world that" fulfills (B) but is nonpsychological. D.J. O'Connor has discussed (B) at some length,[10] so I shall not belabor his points. But I do not think he has touched upon this particular difficulty. I find it important chiefly because, in that it centers around (B–iii), it may threaten (A′–ii) as well. Shortly I shall investigate this in detail; but for now, I think, the valuable aspects of (B) may be left to stand or fall with (A′).

And (A′) seems to be met by all psychological *p*-verbs, including "knows"; despite the fact that "S knows that *p*" entails *p*, not-*p* does not entail that S either does or does not know that not-*p*. So, until someone provides a counterexample, it seems that we have three adequate necessary conditions of intentionality. Are they jointly sufficient? Let us test them with Chisholm's examples of nonpsychological prefixes, starting with the alethic modalities "It is possible that" violates condition (A′–i), because *p* entails "It is possible that *p*," and (A′–iii), because "It is possible that (x) (Fx)" entails "(x) (It is possible that Fx)." "It is necessary that" violates (A′–iii) similarly; and "It is logically impossible that" violates (A′–i), since *p* entails "It is not logically impossible that *p*."

What of causal modalities? Neither "S brings it about that" nor "S does something causally sufficient to prevent that" meets (A′–iii). Both satisfy (A′–ii), moreover, only when a psychological verb appears in the argument *p*; in this case we merely have a *new* prefix "It is M that it is M′ that," and it should not surprise us that this latter prefix should be intentional when M′ contains a psychological verb.

The aforementioned difficulty for (B–iii) arises when we consider prefixes of obligation. "It is legally obligatory that" does not violate (B–iii): Examine the case in which *p* takes the

<hr />

[10] D.J. O'Connor, "Tests for Intentionality," *American Philosophical Quarterly*, 4 (1967), 173–178. O'Connor reads Chisholm very unsympathetically, but he does raise a number of real and formidable difficulties which prompt me to seek a better set of criteria than (B).

value "The number of tons of food produced next year shall be greater than four billion." Suppose we produce five billion tons of food next year. Then, in Frege's sense, "the number of tons of food produced next year" and "five billion" denote the same (arithmetical) object. But "It is legally obligatory that the number of tons of food produced next year shall be greater than four billion" does not entail "It is legally obligatory that five billion be greater than four billion," which is (I presume) false; (B–iii) is met, and (B) is therefore not sufficient.

Is (A′–ii) susceptible to this objection? Clearly not, for these values of p and q. (A′–ii) requires that p and q be logically equivalent, i.e., that they strictly imply each other; and "Five billion is greater than four billion" does not strictly imply anything about tons of produce save perhaps for "Five billion tons is greater than four billion tons." Perhaps a problem will appear if we use an example involving undisputably real proper names. Does "Cicero should be provided with an armed guard at all times" entail "Tully should be provided with an armed guard at all times"? By presupposing Leibniz' Law and a certain theory of naming, we can answer in the affirmative. If we do so, then it seems reasonable to say that substitutivity holds in the "legal obligation" context—what is obligatory for Cicero is obligatory for Tully, so far as the law is concerned. But I think we are still allowed to claim, denying parity of reasoning, that substitutivity fails for "S knows that," "S hopes that," etc., since S may (obviously) not have heard of Tully, though he knows some facts about Cicero. The difference is that a law book is not the sort of thing that is generally said to have "heard of" anyone; by its very nature it presupposes that all the empirical facts of a case (e.g., that "Cicero" and "Tully" name the same person) are known. Even if the judge who *applies* the legal canons of obligation does not know such truths, this is no concern of the impersonal law book.

I realize that what I have just offered is a somewhat artificial, naive, and dogmatic theoretical model of how legal obligation works. But it does not seem outrageous as a tentative paradigm; it gives us reasonable grounds for saying, in my opinion, that "It is legally obligatory that" violates (A'–ii), even if we permit the aforementioned Cicero-Tully entailment to go through. And if we deny Leibniz' Law or the theory of naming on which the entailment-claim rests, then (A'–ii) has clearly not been met, since it requires mutual entailment between p and q. So I conclude that legal obligation does not provide a counterexample to (A').

I take it that (A'–ii) also rules out semantical prefixes like the one invoked by Sleigh in his rejection of the Minimum Pattern. If "There is a sentence derivable in the system explained on p. 32 of *Perceiving* which expresses the proposition that p" is true, and p and q entail each other, then there is a similarly derivable sentence which expresses the proposition that q. In general, semantical prefixes of the form " '*S*' means that" violate (A'–ii); if '*S*' means that p, and p and q are logically equivalent, then surely '*S*' means that q.[11] Perhaps my claims here trade on a certain view of the relation between sentences and propositions which I have neither the space nor the inclination to defend; but, again, they do not seem *prima facie* implausible.

I conclude at this point that (A') is our leading candidate for the characteristic sought by Brentano. The reason I have taken such pains to establish this, perhaps at the cost of being slightly unfair to some of the other suggested definitions, is that (A) and (A') together cover *both* p-verbs and o-verbs in a coherent manner, as no other proposal does. This, as I have said earlier, is an enormous advantage, since many psychological verbs do not express propositional attitudes at all: "S is thinking of . . . ," "S is dreaming of . . . ," "S loves . . . ," "S worships. . . ." All of these are ruled intentional according to (A).

It may be difficult to give (A) a precise interpretation in some cases. For example, we ought to establish exactly how

[11] Some philosopher might be squeamish about admitting this, if he holds that (p) "Snow is white" and (q) "Snow is white and $2 + 2 = 4$" are logically equivalent in that they strictly imply each other, but that they are not equivalent in *meaning*. If this view is correct, " '*S*' means that" fulfills (A'–ii). Let us therefore read "logically equivalent" in (A'–ii) as "synonymous" or as "equivalent in meaning"; on this interpretation, " '*S*' means that" no longer meets (A'–ii), and the possible objection is forestalled.

(A–ii) is to be used. Certainly there is a sense in which it would
be true for Marius to say "I admire Cicero but I don't admire
Tully (who's Tully?)," while at the same time it would be true
for me to say that Marius admired Tully. But in this sense most
nonpsychological verbs satisfy (A–ii) as well: Marius could with
equal sincerity deny that he had punched Tully in the nose. So
mere first-person–third-person disparity will not serve as an adequate interpretation of (A–ii). Perhaps if (A–i) and (A–ii) were
regarded as jointly rather than as individually necessary, (A–ii)
would acquire some force.

In any case, I think that it is along the lines of (A′) that we
should assess Brentano's position. So far as I can tell, (A) and
(A′) is fulfilled by all of these psychological statements:

S thinks that p	S realizes that p
S notices that p	S knows that p
S surmises that p	S requests that p
S believes that p	S fears that p
S doubts that p	S remembers that p
S wonders whether p	S thinks of X
S intends that p	S dreams of X
S affirms that p	S worships X
S states that p	S loves X
S wishes that p	S desires X
S hopes that p	S fears X
S dreams that p	S perceives X

and by no clearly nonpsychological ones.

I shall not stop to argue for the intentionality of perception-
verbs according to (A). Discussion of this is most clearly set
forth in Anscombe's own article. It will be noticed, however,
that the above list fails to include any of an important category
of mental concepts: those involving bodily sensations, such as
pain, itches, tickles, etc. Sentences like "My arm hurts" or "My
ear itches" do not contain objects at all, at least in the superficial grammatical sense. We have our choice of (a) holding
that bodily sensations are too firmly rooted in the world of
physiology to be (strictly) called "psychological" (a dubious
assertion, since the mental states associated with them are at

least not *logically* identical with bodily states); (b) hoping that at some future time we shall be able to incorporate some intentionality criterion, independent of the idea of a direct object, into our program which will cover these sentences; or (c) claiming that bodily-sensation sentences entail other sentences which are unequivocally intentional, e.g., belief-sentences. The last option seems the most promising.

Supposing that we have indeed proved (1) and (2), what are we entitled to infer from them about the "mental" and the "physical"? "Absolutely nothing," says O'Connor:

> The most that we could say would be that the structure of language does reflect our unsophisticated beliefs on this point. But it would have no tendency whatever to show that the distinction between mental and physical was in any way basic or that the mental could not be reduced to or explained in terms of the physical. So even if the distinction could be shown to be well founded, it would be no more than an interesting fact about language. . . . Perhaps our language merely embodies a discredited Cartesianism.[12]

But surely the issue is not quite so simple as this. No doubt the plain man's language does embody many false dualistic presuppositions ("An idea popped into his mind"; "His body was in the classroom, but his thoughts were miles away"; "Everyone was charmed by his outer shell, but no one could tell what he was really like inside," etc.) But these are examples of how the plain man is commonly misled by a picture or a metaphor, viz., by the "inner"–"outer" model rightly rejected by Wittgenstein. The logical property given by (A) and (A') does not have this same character; it does not betray a "diseased" adherence to any kind of metaphor. Rather, the necessity of its presence suggests that psychological statements perform a *function*, play a linguistic role, enable us to talk about something—which task cannot be fulfilled by physicalistic statements. Brentano expressed this (but in the material mode) by saying that mental phenomena were distinguished by their essential "reference to a

---

[12] O'Connor, "Tests for Intentionality," pp. 174–175.

content" (and that physical items cannot "refer" to anything extrinsic to themselves in this same way). All this is terribly unclear, and it may not in the end serve to point up any irreconcilable difference between the "mental" and the "physical"; but it should be clarified and then assessed, not dismissed out of hand.[13]

[13] I am indebted to Professors G.E.M. Anscombe, V.C. Chappell, Roderick Chisholm, and Robert Coburn for having pointed out several errors in an earlier version of this paper. They share no responsibility for any remaining mistakes.

PART II

Intentionality and Physicalism

7

Logical Analysis of Psychology

CARL G. HEMPEL

Author's preamble, 1970. By the time this article appeared in English translation, I had abandoned the narrow translationist version of physicalism here set forth in favor of a more liberal reductionism, as is stated in note 1. Since then, however, I have come to consider even that version as inadequate, for reasons I have mentioned in two more recent essays: "Logical Positivism and the Social Sciences," in P. Achinstein and S.F. Barker, eds., *The Legacy of Logical Positivism* (Baltimore: The Johns Hopkins Press, 1969); and "Reduction: Ontological and Linguistic Facets," in S. Morgenbesser, P. Suppes, M. White, eds., *Philosophy, Science, and Method. Essays in Honor of Ernest Nagel* (New York: St. Martin's Press, 1969).

My—somewhat hesitant—consent to this republication of the article was prompted by the editor's plea that it offers a concise statement of a version of physicalism which has played an influential role in the philosophical analysis of psychology, and which should therefore, at least for didactic purposes, be represented in this volume.

I have made a number of small changes in the original English version; these are meant to enhance the closeness of

From *Readings in Philosophical Analysis*, eds. H. Feigl and W. Sellars, pp. 373–384. Copyright 1949 by Appleton-Century-Crofts, Inc. Translated from the French by W. Sellars and reprinted here, with minor revisions by the author, by permission of the author, the publisher, and the editors of *Revue de Synthèse*, where the original French version of this essay was first published.

translation and the simplicity of formulation; none of them changes in any way the substance of the original article.

I

One of the most important and most discussed problems of contemporary philosophy is that of determining how psychology should be characterized in the theory of science. This problem, which reaches beyond the limits of epistemological analysis and has engendered heated controversy in metaphysics itself, is brought to a focus by the familiar alternative, "Is psychology a natural science, or is it one of the sciences of mind and culture (*Geisteswissenschaften*)?"

The present article attempts to sketch the general lines of a new analysis of psychology, one which makes use of rigorous logical tools, and which has made possible decisive advances towards the solution of the above problem.[1] This analysis was carried out by the "Vienna Circle" (*Wiener Kreis*), the members of which (M. Schlick, R. Carnap, P. Frank, O. Neurath, F. Waismann, H. Feigl, etc.) have, during the past ten years, developed an extremely fruitful method for the epistemological examination and critique of the various sciences, based in part on the work of L. Wittgenstein.[2] We shall limit ourselves essentially to the examination of psychology as carried out by Carnap and Neurath.

The method characteristic of the studies of the Vienna

----

[1] I now (1947) consider the type of physicalism outlined in this paper as too restrictive; the thesis that all statements of empirical science are *translatable*, without loss of theoretical content, into the language of physics, should be replaced by the weaker assertion that all statements of empirical science are *reducible* to sentences in the language of physics, in the sense that for every empirical hypothesis, including, of course, those of psychology, it is possible to formulate certain test conditions in terms of physical concepts which refer to more or less directly observable physical attributes. But those test conditions are not asserted to exhaust the theoretical content of the given hypothesis in all cases.

For a more detailed development of this thesis, cf. R. Carnap, "Logical Foundations of the Unity of Science" (included in this volume).

[2] *Tractatus Logico-Philosophicus* (London, 1922).

Circle can be briefly defined as a *logical analysis of the language of science*. This method became possible only with the development of a subtle logical apparatus which makes use, in particular, of all the formal procedures of modern symbolic logic.[3] However, in the following account, which does not pretend to give more than a broad orientation, we shall limit ourselves to setting out the general principles of this new method, without making use of strictly formal procedures.

II

Perhaps the best way to characterize the position of the Vienna Circle as it relates to psychology, is to say that it is the exact antithesis of the current epistemological thesis that there is a fundamental difference between experimental psychology, a natural science, and introspective psychology; and in general, between the natural sciences on the one hand, and the sciences of mind and culture on the other.[4] The common content of the widely different formulations used to express this contention, which we reject, can be set down as follows. Apart from certain aspects clearly related to physiology, psychology is radically dif-

[3] A recent presentation of symbolic logic, based on the fundamental work of Whitehead and Russell, *Principia Mathematica*, is to be found in R. Carnap, *Abriss der Logistik* (Vienna: Springer, 1929; vol. II of the series *Schriften zur Wissenschaftlichen Weltauffassung*). It includes an extensive bibliography, as well as references to other logistic systems.

[4] The following are some of the principal publications of the Vienna Circle on the nature of psychology as a science: R. Carnap, *Scheinprobleme in der Philosophie. Das Fremdpsychische und des Realismusstreit* (Leipzig: Meiner, 1928); *Der Logische Aufbau der Welt* (Leipzig: Meiner, 1928) [English trans.: *Logical Structure of the World* (Berkeley: Univ. of California Press, 1967)]; "Die Physikalische Sprache als Universalsprache der Wissenschaft," *Erkenntnis*, 2 (1931–32), 432–465 [English trans.: *The Unity of Science* (London: Kegan Paul, 1934)]; "Psychologie in physikalischer Sprache," *Erkenntnis*, 3 (1932–33), 107–142 [English trans.: "Psychology in Physical Language," in A.J. Ayer, ed., *Logical Positivism* (New York: Free Press, 1959)]; "Ueber Protokollsaetze," *Erkenntnis*, 3 (1932–33), 215–228; O. Neurath, "Protokollsaetze," *Erkenntnis*, 3 (1932–33), 204–214 [English trans.: "Protocol Sentences," in *Logical Positivism*]; *Einheitswissenschaft und Psychologie* (Vienna: Springer, 1933; vol. I of the series *Einheitswissenschaft*). See also the publications mentioned in the notes below.

ferent, both as to subject-matter and as to method, from physics in the broad sense of the term. In particular, it is impossible
to deal adequately with the subject-matter of psychology by
means of physical methods. The subject-matter of physics includes such concepts as mass, wave length, temperature, field
intensity, etc. In developing these, physics employs its distinctive method which makes a combined use of description and
causal explanation. Psychology, on the other hand, has for its
subject-matter notions which are, in a broad sense, mental. They
are *toto genere* different from the concepts of physics, and the
appropriate method for dealing with them scientifically is that
of sympathetic insight, called "introspection," a method which
is peculiar to psychology.

One of the principal differences between the two kinds of
subject-matter is generally believed to consist in the fact that
the objects investigated by psychology—in contradistinction to
those of physics—are specifically endowed with meaning. Indeed, several proponents of this idea state that the distinctive
method of psychology consists in "understanding the sense of
meaningful structures" (*sinnvolle Gebilde verstehend zu
erfassen*). Take, for example, the case of a man who speaks.
Within the framework of physics, this process is considered to be
completely explained once the movements which make up the
utterance have been traced to their causes, that is to say, to certain physiological processes in the organism, and, in particular,
in the central nervous system. But, it is said, this does not even
broach the psychological problem. The latter begins with understanding the sense of what was said, and proceeds to integrate it
into a wider context of meaning.

It is usually this latter idea which serves as a principle for
the fundamental dichotomy that is introduced into the classification of the sciences. There is taken to be an *absolutely impassable gulf* between the *natural sciences* which have a subject-
matter devoid of meaning and the *sciences of mind and culture*,
which have an intrinsically meaningful subject-matter, the appropriate methodological instrument for the scientific study of
which is "comprehension of meaning."

III

The position in the theory of science which we have just sketched has been attacked from several different points of view.[5] As far as psychology is concerned, one of the principal counter-theses is that formulated by behaviorism, a theory born in America shortly before the war. (In Russia, Pavlov has developed similar ideas.) Its principal methodological postulate is that a scientific psychology should limit itself to the study of the bodily behavior with which man and the animals respond to changes in their physical environment, and should proscribe as nonscientific any descriptive or explanatory step which makes use of terms from introspective or "understanding" psychology, such as 'feeling', 'lived experience', 'idea', 'will', 'intention', 'goal', 'disposition', 'repression'.[6] We find in behaviorism, consequently, an attempt to construct a scientific psychology which would show by its success that even in psychology we have to do with purely physical processes, and that therefore there can be no impassable barrier between psychology and physics. However, this manner of undertaking the critique of a scientific thesis is not completely satisfactory. It seems, indeed, that the soundness of the behavioristic thesis expounded above depends on the possibility of fulfilling the program of behavioristic psychology. But one cannot expect the question as to the scientific status of psychology to be settled by empirical research in psychology itself. To achieve this is rather an undertaking in epistemology. We turn, therefore, to the considerations advanced by members of the Vienna Circle concerning this problem.

[5] P. Oppenheim, for example, in his book *Die Natuerliche Ordnung der Wissenschaften* (Jena: Fischer, 1926), opposes the view that there are fundamental differences between any of the different areas of science. On the analysis of "understanding," cf. M. Schlick, "Erleben, Erkennen, Metaphysik," *Kantstudien*, 31 (1926), 146.

[6] For further details see the statement of one of the founders of behaviorism: J.B. Watson, *Behaviorism* (New York: Norton, 1930); also A.A. Roback, *Behaviorism and Psychology* (Cambridge, Mass.: Univ. Bookstore, 1923); and A.P. Weiss, *A Theoretical Basis of Human Behavior*, 2nd ed. rev. (Columbus, Ohio: Adams, 1929); see also the work by Koehler cited in note 11 below.

IV

Before attacking the question whether the subject-matters of physics and psychology are essentially the same or different in nature, it is necessary first to clarify the very concept of the subject-matter of a science. The theoretical content of a science is to be found in statements. It is necessary, therefore, to determine whether there is a fundamental difference between the statements of psychology and those of physics. Let us therefore ask what it is that determines the content—one can equally well say the "meaning"—of a statement. When, for example, do we know the meaning of the following statement: "Today at one o'clock, the temperature of such and such a place in the physics laboratory was 23.4° centigrade"? Clearly when, and only when, we know under what conditions we would characterize the statement as true, and under what circumstances we would characterize it as false. (Needless to say, it is not necessary to know whether or not the statement is true.) Thus, we understand the meaning of the above statement since we know that it is true when a tube of a certain kind filled with mercury (in short, a thermometer with a centigrade scale), placed at the indicated time at the location in question, exhibits a coincidence between the level of the mercury and the mark of the scale numbered 23.4. It is also true if in the same circumstances one can observe certain coincidences on another instrument called an "alcohol thermometer"; and, again, if a galvanometer connected with a thermopile shows a certain deviation when the thermopile is placed there at the indicated time. Finally, there is a long series of other possibilities which make the statement true, each of which is defined by a "physical test sentence," as we should like to call it. The statement itself clearly affirms nothing other than this: all these physical test sentences obtain. (However, one verifies only some of these physical test sentences, and then "concludes by induction" that the others obtain as well.) The statement, therefore, is nothing but an abbreviated formulation of all those test sentences.

Before continuing the discussion, let us sum up this result as follows:

1. A statement that specifies the temperature at a selected point in space-time can be "retranslated" without change of meaning into another statement—doubtlessly longer—in which the word "temperature" no longer appears. This term functions solely as an abbreviation, making possible the concise and complete description of a state of affairs the expression of which would otherwise be very complicated.

2. The example equally shows that *two statements which differ in formulation* can nevertheless have the *same meaning.* A trivial example of a statement having the same meaning as the above would be: "Today at one o'clock, at such and such a location in the laboratory, the temperature was 19.44° Réaumur."

As a matter of fact, the preceding considerations show—and let us set it down as another result—that *the meaning of a statement is established by the conditions of its verification.* In particular, two differently formulated statements have the same meaning or the same effective content when, and only when, they are both true or both false in the same conditions. Furthermore, a statement for which one can indicate absolutely no conditions which would verify it, which is in principle incapable of confrontation with test conditions, is wholly devoid of content and without meaning. In such a case we have to do, not with a statement properly speaking, but with a "pseudo-statement," that is to say, a sequence of words correctly constructed from the point of view of grammar, but without content.[7]

In view of these considerations, our problem reduces to one concerning the difference between the circumstances which verify psychological statements and those which verify the statements of physics. Let us therefore examine a statement which involves a psychological concept, for example: "Paul has a

---

[7] Space is lacking for further discussion of the logical form of test sentences (recently called "protocol sentences" by Neurath and Carnap). On this question see Wittgenstein, *Tractatus Logico-Philosophicus,* as well as the articles by Neurath and Carnap which have appeared in *Erkenntnis* (above, note 4).

toothache." What is the specific content of this statement, that is to say, what are the circumstances in which it would be verified? It will be sufficient to indicate some test sentences which describe these circumstances.

    a. Paul weeps and makes gestures of such and such kinds.
    b. At the question "What is the matter?," Paul utters the words "I have a toothache."
    c. Closer examination reveals a decayed tooth with exposed pulp.
    d. Paul's blood pressure, digestive processes, the speed of his reactions, show such and such changes.
    e. Such and such processes occur in Paul's central nervous system.

This list could be expanded considerably, but it is already sufficient to bring out the fundamental and essential point, namely, that all the circumstances which verify this psychological statement are expressed by physical test sentences. [This is true even of test condition *b*, which merely expresses the fact that in specified physical circumstances (the propagation of vibrations produced in the air by the enunciation of the words, "What is the matter?") there occurs in the body of the subject a certain physical process (speech behavior of such and such a kind).]

The statement in question, which is about someone's "pain," is therefore, just like that concerning the temperature, simply an abbreviated expression of the fact that all its test sentences are verified.[8] (Here, too, one verifies only some of the test sentences and then infers by way of induction that the others obtain as well.) It can be retranslated without loss of content into a statement which no longer contains the term "pain," but only physical concepts. Our analysis has consequently established that a certain statement belonging to psychology has the same content as a statement belonging to physics; a result which is

---

[8] (Added in 1970.) This reference to verification involves a conceptual confusion. The thesis which the preceding considerations were meant to establish was clearly that the statement about Paul's pain is, in effect, an abbreviated expression of all its test sentences; not that it is an expression of the claim (let alone "fact") that all those test sentences have in fact been tested and verified.

in direct contradiction to the thesis that there is an impassable gulf between the statements of psychology and those of physics.

The above reasoning can be applied to *any psychological statement*, even to those which concern, as is said, "deeper psychological strata" than that of our example. Thus, the assertion that Mr. Jones suffers from intense inferiority feelings of such and such kinds can be confirmed or falsified only by observing Mr. Jones' behavior in various circumstances. To this behavior belong all the bodily processes of Mr. Jones, and, in particular, his gestures, the flushing and paling of his skin, his utterances, his blood pressure, the events that occur in his central nervous system, etc. In practice, when one wishes to test statements concerning what are called the deeper layers of the psyche, one limits oneself to the observation of external bodily behavior, and, particularly, to speech movements aroused by certain physical stimuli (the asking of questions). But it is well known that experimental psychology has also developed techniques for making use of the subtler bodily states referred to above in order to confirm the psychological discoveries made by cruder methods. The statement concerning the inferiority feelings of Mr. Jones— whether true or false—means only this: such and such happenings take place in Mr. Jones' body in such and such circumstances.

We shall call a statement which can be translated without change of meaning into the language of physics, a "physicalistic statement," whereas we shall reserve the expression "statement of physics" to those which are already formulated in the terminology of physical science. (Since every statement is in respect of content equivalent to itself, every statement of physics is also a physicalistic statement.) The result of the preceding considerations can now be summed up as follows: *All psychological statements which are meaningful, that is to say, which are in principle verifiable, are translatable into statements which do not involve psychological concepts, but only the concepts of physics. The statements of psychology are consequently physicalistic statements. Psychology is an integral part of physics.* If a distinction is drawn between psychology and the other areas of physics, it is only from the point of view of the practical aspects of research and the direction of interest, rather than a matter of principle. This logical analysis, of which the result shows a certain affinity with the fundamental ideas of behaviorism, constitutes the physicalistic conception of psychology.

v

It is customary to raise against the above conception the following fundamental objection: The physical test sentences of which you speak are absolutely incapable of formulating the intrinsic nature of a mental process; they merely describe the physical *symptoms* from which one infers, by purely psychological methods—notably that of understanding—the presence of a certain mental process.

But it is not difficult to see that the use of the method of understanding or of other psychological procedures is bound up with the existence of certain observable physical data concerning the subject undergoing examination. There is no psychological understanding that is not tied up physically in one way or another with the person to be understood. Let us add that, for example, in the case of the statement about the inferiority complex, even the "introspective" psychologist, the psychologist who "understands," can confirm his conjecture only if the body of Mr. Jones, when placed in certain circumstances (most frequently, subjected to questioning), reacts in a specified manner (usually, by giving certain answers). Consequently, even if the statement in question had to be arrived at, *discovered*, by "sympathetic understanding," the only *information* it gives us is nothing more nor less than the following: under certain circumstances, certain specific events take place in the body of Mr. Jones. It is this which constitutes the meaning of the psychological statement.

The further objection will perhaps be raised that men can feign. Thus, though a criminal at the bar may show physical symptoms of mental disorder, one would nevertheless be justified in wondering whether his mental confusion was "real" or only simulated. One must note that in the case of the simulator, only some of the conditions are fulfilled which verify the statement "This man is mentally unbalanced," those, namely, which are most accessible to direct observation. A more penetrating examination—which should in principle take into account events occurring in the central nervous system—would give a decisive answer; and this answer would in turn clearly rest on a physicalistic basis. If, at this point, one wished to push the objection to the point of admitting that a man could show *all the "symptoms"* of a mental disease without being "really" ill, we reply that it would be absurd to characterize such a man as "really normal"; for it is obvious that by the very nature of the hypothesis we should possess no criterion in terms of which to distinguish this man from another who, while exhibiting the same bodily behavior down to the last detail, would "in addition" be "really ill." (To put the point more precisely, one can say that this hypothesis contains a *logical contradiction,* since it amounts to saying, "It is possible that a statement should be false even when the necessary and sufficient conditions of its truth are fulfilled.")

Once again we see clearly that the meaning of a psychological statement consists solely in the function of abbreviating the description of certain modes of physical response characteristic of the bodies of men or animals. An analogy suggested by O. Neurath may be of further assistance in clarifying the logical function of psychological statements.[9] The complicated statements that would describe the movements of the hands of a watch in relation to one another, and relatively to the stars, are ordinarily summed up in an assertion of the following form: "This watch runs well (runs badly, etc.)." The term "runs" is introduced here as an auxiliary defined expression which makes it possible to formulate briefly a relatively complicated system of statements. It would thus be absurd to say, for example, that the movement of the hands is only a "physical symptom" which

---

[9] "Soziologie im Physikalismus," *Erkenntnis,* 2 (1931–32), 393–431, particularly p. 411 [English trans.: "Sociology and Physicalism," in *Logical Positivism*].

reveals the presence of a running which is intrinsically incapable of being grasped by physical means, or to ask, if the watch should stop, what has become of the running of the watch.

It is in exactly the same way that abbreviating symbols are introduced into the language of physics, the concept of temperature discussed above being an example. The system of physical test sentences *exhausts* the meaning of the statement concerning the temperature at a place, and one should not say that these sentences merely have to do with "symptoms" of the existence of a certain temperature.

Our argument has shown that it is necessary to attribute to the characteristic concepts of psychology the same logical function as that performed by the concepts of "running" and of "temperature." They do nothing more than make possible the succinct formulation of propositions concerning the states or processes of animal or human bodies.

The introduction of new psychological concepts can contribute greatly to the progress of scientific knowledge. But it is accompanied by a danger, that, namely, of making an excessive and, consequently, improper use of new concepts, which may result in questions and answers devoid of sense. This is frequently the case in metaphysics, notably with respect to the notions which we formulated in section II. Terms which are abbreviating symbols are imagined to designate a special class of "psychological objects," and thus one is led to ask questions about the "essence" of these objects, and how they differ from "physical objects." The time-worn problem concerning the relation between mental and physical events is also based on this confusion concerning the logical function of psychological concepts. Our argument, therefore, enables us to see that *the psycho-physical problem is a pseudo-problem*, the formulation of which is based on an inadmissible use of scientific concepts; it is of the same logical nature as the question, suggested by the example above, concerning the relation of the running of the watch to the movement of the hands.[10]

[10] Carnap, *Der Logische Aufbau der Welt*, pp. 231–236; id. *Scheinprobleme in der Philosophie*. See also note 4 above.

VI

In order to bring out the exact status of the fundamental idea of the physicalistic conception of psychology (or logical behaviorism), we shall contrast it with certain theses of psychological behaviorism and of classical materialism, which give the appearance of being closely related to it.[11]

1. Logical behaviorism claims neither that minds, feelings, inferiority complexes, voluntary actions, etc., do not exist, nor that their existence is in the least doubtful. It insists that the very question as to whether these psychological constructs really exist is already a pseudo-problem, since these notions in their "legitimate use" appear only as abbreviations in physicalistic statements. Above all, one should not interpret the position sketched in this paper as amounting to the view that we can know only the "physical side" of psychological processes, and that the question whether there are mental phenomena behind the physical processes falls beyond the scope of science and must be left either to faith or to the conviction of each individual. On the contrary, the logical analyses originating in the Vienna Circle, one of whose consequences is the physicalistic conception of psychology, teach us that every meaningful question is, in principle, capable of a scientific answer. Furthermore, these analyses show that what, in the case of the mind-body problem, is considered as an object of belief, is absolutely incapable of being expressed by a factual proposition. In other words, there can be no question here of an "article of faith." Nothing can be an object of faith which cannot, in principle, be known.

2. The thesis here developed, though related in certain ways to the fundamental idea of behaviorism, does not demand, as does the latter, that psychological research restrict itself meth-

---

[11] A careful discussion of the ideas of so-called "internal" behaviorism is to be found in *Psychologische Probleme* by W. Koehler (Berlin: Springer, 1933). See particularly the first two chapters.

odologically to the study of the responses made by organisms to certain stimuli. It by no means offers a theory belonging to the domain of psychology, but rather a logical theory about the statements of scientific psychology. Its position is that the latter are without exception physicalistic statements, by whatever means they may have been obtained. Consequently, it seeks to show that if in psychology only physicalistic statements are made, this is not a limitation because it is logically *impossible* to do otherwise.

3. In order for logical behaviorism to be valid, it is not necessary that we be able to describe the physical state of a human body which is referred to by a certain psychological statement— for example, one dealing with someone's feeling of pain— down to the most minute details of the phenomena of the central nervous system. No more does it presuppose a knowledge of all the physical laws governing human or animal bodily processes; nor *a fortiori* is the existence of rigorously deterministic laws relating to these processes a necessary condition of the truth of the behavioristic thesis. At no point does the above argument rest on such a concrete presupposition.

VII

In concluding, I should like to indicate briefly the clarification brought to the problem of the division of the sciences into totally different areas, by the method of the logical analysis of scientific statements, applied above to the special case of the place of psychology among the sciences. The considerations we have advanced can be extended to the domain of sociology, taken in the broad sense as the science of historical, cultural, and economic processes. In this way one arrives at the result that every sociological assertion which is meaningful, that is to say, in principle verifiable, "has as its subject-matter nothing else than the states, processes and behavior of groups or of individuals (human or animal), and their responses to one another and to their environment,"[12] and consequently that every socio-logical statement is a physicalistic statement. This view is characterized by Neurath as the thesis of "social behaviorism," which he adds to that of "individual behaviorism" which we have expounded above. Furthermore, it can be shown[13] that every statement of what are called the "sciences of mind and culture" is a sociological statement in the above sense, provided it has genuine content. Thus one arrives at the "thesis of the unity of science":

The division of science into different areas rests exclusively on differences in research procedures and direction of interest; *one must not regard it as a matter of principle. On the contrary, all the branches of science are in principle of one and the same nature; they are branches of the unitary science, physics.*

VIII

The method of logical analysis which we have attempted to explicate by clarifying, as an example, the statements of psychology, leads, as we have been able to show only too briefly for the sciences of mind and culture, to a "physicalism" based on logic (Neurath): *Every statement of the above-mentioned disciplines, and, in general, of empirical science as a whole,* which is not merely a meaningless sequence of words, *is translatable, without change of content, into a statement in which appear only physicalistic terms, and consequently is a physicalistic statement.*

This thesis frequently encounters strong opposition arising from the idea that such analyses violently and considerably reduce the richness of the life of mind or spirit, as though the aim of the discussion were purely and simply to eliminate vast and

[12] R. Carnap, "Die Physikalische Sprache als Universalsprache," p. 451. See also: O. Neurath, *Empirische Soziologie* (Vienna: Springer, 1931; the fourth monograph in the series *Schriften zur wissenschaftlichen Weltauffassung*).

[13] See R. Carnap, *Der Logische Aufbau der Welt*, pp. 22–34 and 185–211, as well as the works cited in the preceding note.

important areas of experience. Such a conception comes from a false interpretation of physicalism, the main elements of which we have already examined in section VII above. As a matter of fact, nothing can be more remote from a philosophy which has the methodological attitude we have characterized than the making of decisions, on its own authority, concerning the truth or falsity of particular scientific statements, or the desire to eliminate any matters of fact whatsoever. *The subject-matter of this philosophy is limited to the form of scientific statements, and the deductive relationships obtaining between them.* It is led by its analyses to the thesis of physicalism, and establishes on purely logical grounds that a certain class of venerable philosophical "problems" consists of pseudo-problems. It is certainly to the advantage of the progress of scientific knowledge that these imitation jewels in the coffer of scientific problems be known for what they are, and that the intellectual powers which have till now been devoted to a class of senseless questions which are by their very nature insoluble, become available for the formulation and study of new and fruitful problems. That the method of logical analysis stimulates research along these lines is shown by the numerous publications of the Vienna Circle and those who sympathize with its general point of view (H. Reichenbach, W. Dubislav, and others).

In the attitude of those who are so bitterly opposed to physicalism, an essential role is played by certain psychological factors relating to individuals and groups. Thus the contrast between the concepts (*Gebilde*) developed by the psychologist, and those developed by the physicist, or, again, the question as to the nature of the specific subject-matter of psychology and the cultural sciences (which present the appearance of a search for the essence and unique laws of "objective spirit") is usually accompanied by a strong emotional coloring which has come into being during the long historical development of the "philosophical conception of the world," which was considerably less scientific than normative and intuitive. These emotional factors are still deeply rooted in the picture by which our epoch represents the world to itself. They are protected by certain affective dispositions which surround them like a rampart, and for all these reasons appear to us to have a verifiable content—something which a more penetrating analysis shows to be impossible.

A psychological and sociological study of the causes for the appearance of these "concomitant factors" of the metaphysical type would take us beyond the limits of this study,[14] but without tracing it back to its origins, it is possible to say that if the logical analyses sketched above are correct, the fact that they necessitate at least a partial break with traditional philosophical ideas which are deeply dyed with emotion can certainly not justify an opposition to physicalism—at least if one admits that philosophy is to be something more than the expression of an individual vision of the world, that it aims at being a science.

[14] O. Neurath has made interesting contributions along these lines in *Empirische Soziologie* and in "Soziologie im Physikalismus" (see above, note 9), as has R. Carnap in his article "Ueberwindung der Metaphysik durch logische Analyse der Sprache," *Erkenntnis*, 2 (1931–32), 219–241 [English trans.: "The Elimination of Metaphysics through Logical Analysis of Language," in *Logical Positivism*].

8

Logical Foundations of the Unity of Science

RUDOLF CARNAP

I. WHAT IS LOGICAL ANALYSIS OF SCIENCE?

The task of analyzing science may be approached from various angles. The analysis of the subject matter of the sciences is carried out by science itself. Biology, for example, analyzes organisms and processes in organisms, and in a similar way every branch of science analyzes its subject matter. Mostly, however, by 'analysis of science' or 'theory of science' is meant an investigation which differs from the branch of science to which it is applied. We may, for instance, think of an investigation of scientific *activity*. We may study the historical development of this activity. Or we may try to find out in which way scientific work depends upon the individual conditions of the men working in science, and upon the status of the society surrounding them. Or we may describe procedures and appliances used in scientific work. These investigations of scientific activity may be called history, psychology, sociology, and methodology of science. The subject matter of such studies is science as a body of actions carried out by certain persons under certain circumstances. Theory of science in this sense will be dealt with at various other places. . . .

From *International Encyclopedia of Unified Science*, vol. I, no. 1, ed. Otto Neurath. Copyright 1938 by the University of Chicago. Reprinted by permission of the University of Chicago Press.

We come to a theory of science in another sense if we study not the actions of scientists but their results, namely, science as a body of ordered knowledge. Here, by 'results' we do not mean beliefs, images, etc., and the behavior influenced by them. That would lead us again to psychology of science. We mean by 'results' certain linguistic expressions, viz., the statements asserted by scientists. The task of the theory of science in this sense will be to analyze such statements, study their kinds and relations, and analyze terms as components of those statements and theories as ordered systems of those statements. A statement is a kind of sequence of spoken sounds, written marks, or the like, produced by human beings for specific purposes. But it is possible to abstract in an analysis of the statements of science from the persons asserting the statements and from the psychological and sociological conditions of such assertions. The analysis of the linguistic expressions of science under such an abstraction is *logic of science.*

Within the logic of science we may distinguish between two chief parts. The investigation may be restricted to the forms of the linguistic expressions involved, i.e., to the way in which they are constructed out of elementary parts (e.g., words) without referring to anything outside of language. Or the investigation goes beyond this boundary and studies linguistic expressions in their relation to objects outside of language. A study restricted in the first-mentioned way is called *formal;* the field of such formal studies is called formal logic or *logical syntax.* Such a formal or syntactical analysis of the language of science as a whole or in its various branches will lead to results of the following kinds. A certain term (e.g., a word) is defined within a certain theory on the basis of certain other terms, or it is definable in such a way. A certain term, although not definable by certain other terms, is reducible to them (in a sense to be explained later). A certain statement is a logical consequence of (or logically deducible from) certain other statements; and a deduction of it, given within a certain theory, is, or is not, logically correct. A certain statement is incompatible with certain other statements, i.e., its negation is a logical consequence of them. A certain statement is independent of certain other statements, i.e., neither a logical consequence of them nor incompatible with them. A certain theory is inconsistent, i.e., some of its statements are incompatible with the other ones. The last sections of this essay will deal with the question of the unity of science from the logical point of view, studying the logical relations between the terms of the chief branches of science and between the laws stated in these branches; thus it will give an example of a syntactical analysis of the language of science.

In the second part of the logic of science, a given language and the expressions in it are analyzed in another way. Here also, as in logical syntax, abstraction is made from the psychological and sociological side of the language. This investigation, however, is not restricted to formal analysis but takes into consideration one important relation between linguistic expressions and other objects—that of designation. An investigation of this kind is called *semantics*. Results of a semantical analysis of the language of science may, for instance, have the following forms. A certain term designates a certain particular object (e.g., the sun), or a certain property of things (e.g., iron), or a certain relation between things (e.g., fathership), or a certain physical function (e.g., temperature); two terms in different branches of science (e.g., 'homo sapiens' in biology and 'person' in economics, or, in another way, 'man' in both cases) designate (or: do not designate) the same. What is designated by a certain expression may be called its *designatum*. Two expressions designating the same are called *synonymous*. The term 'true,' as it is used in science and in everyday life, can also be defined within semantics. We see that the chief subject matter of a semantical analysis of the language of science are such properties and relations of expressions, and especially of statements, as are based on the relation of designation. (Where we say 'the designatum of an expression,' the customary phrase is 'the meaning of an expression.' It seems, however, preferable to avoid the word 'meaning' wherever possible because of its ambiguity, i.e., the multiplicity of its designata. Above all, it is important to distinguish between the semantical and the psychological use of the word 'meaning.')

It is a question of terminological convention whether to use the term 'logic' in the wider sense, including the semantical analysis of the designata of expressions, or in the narrower sense of logical syntax, restricted to formal analysis, abstracting from designation. And accordingly we may distinguish between logic of science in the narrower sense, as the syntax of the language of science, and logic of science in the wider sense, comprehending both syntax and semantics.

II. THE MAIN BRANCHES OF SCIENCE

We use the word 'science' here in its widest sense, including all theoretical knowledge, no matter whether in the field of natural sciences or in the field of the social sciences and the so-called humanities, and no matter whether it is knowledge found by the application of special scientific procedures, or knowledge based on common sense in everyday life. In the same way the term 'language of science' is meant here to refer to the language which contains all statements (i.e., theoretical sentences as distinguished from emotional expressions, commands, lyrics, etc.) used for scientific purposes or in everyday life. What usually is called science is merely a more systematic continuation of those activities which we carry out in everyday life in order to know something.

The first distinction which we have to make is that between *formal science* and *empirical science*. Formal science consists of the analytic statements established by logic and mathematics; empirical science consists of the synthetic statements established in the different fields of factual knowledge. The relation of formal to empirical science will be dealt with at another place; here we have to do with empirical science, its language, and the problem of its unity.

Let us take 'physics' as a common name for the nonbiological field of science, comprehending both systematic and historical investigations within this field, thus including chemistry, mineralogy, astronomy, geology (which is historical), meteorology, etc. How, then, are we to draw the boundary line between physics and biology? It is obvious that the distinction between these two branches has to be based on the distinction between two kinds of things which we find in nature: organisms and nonorganisms. Let us take this latter distinction as granted; it is the task of biologists to lay down a suitable definition for the term 'organism,' in other words, to tell us the features of a thing which we take as characteristic for its being an organism. How, then, are we to define 'biology' on the basis of 'organism'? We could perhaps think of trying to do it in this way: biology is the branch of science which investigates organisms and the processes occurring in organisms, and physics is the study of nonorganisms. But these definitions would not draw the distinction as it is usually intended. A law stated in physics is intended to be valid universally, without any restriction. For example, the law stating the electrostatic force as a function of electric charges and their distance, or the law determining the pressure of a gas as a function of temperature, or the law determining the angle of refraction as a function of the coefficients of refraction of the two media involved, are intended to apply to the processes in organisms no less than to those in inorganic nature. The biologist has to know these laws of physics in studying the processes in organisms. He needs them for the explanation of these processes. But since they do not suffice, he adds some other laws, not known by the physicist, viz., the specifically biological laws. Biology presupposes physics, but not vice versa.

These reflections lead us to the following definitions. Let us call those terms which we need—in addition to logico-mathematical terms—for the description of processes in inorganic nature *physical terms*, no matter whether, in a given instance, they are applied to such processes or to processes in organisms. That sublanguage of the language of science, which contains—besides logico-mathematical terms—all and only physical terms, may be called *physical language*. The system of those statements which are formulated in the physical language and are acknowledged by a certain group at a certain time is called the physics of that group at that time. Such of these statements as have a specific universal form are called *physical laws*. The physical laws are needed for the explanation of processes in inorganic nature; but, as mentioned before, they apply to processes in organisms also.

The whole of the rest of science may be called *biology (in the wider sense)*. It seems desirable, at least for practical purposes, e.g., for the division of labor in research work, to subdivide this wide field. But it seems questionable whether any distinctions can be found here which, although not of a fundamental nature, are at least clear to about the same degree as the distinction between physics and biology. At present, it is scarcely possible to predict which subdivisions will be made in the future. The traditional distinction between bodily (or material) and mental (or psychical) processes had its origin in the old magical and later metaphysical mind-body dualism. The distinction as a practical device for the classification of branches of science still plays an important role, even for those scientists who reject that metaphysical dualism; and it will probably continue to do so for some time in the future. But when the aftereffect of such prescientific issues upon science becomes weaker and weaker, it may be that new boundary lines for subdivisions will turn out to be more satisfactory.

One possibility of dividing biology in the wider sense into two fields is such that the first corresponds roughly to what is usually called biology, and the second comprehends among other parts those which usually are called psychology and social science. The second field deals with the behavior of individual organisms and groups of organisms within their environment, with the dispositions to such behavior, with such features of processes in organisms as are relevant to the behavior, and with certain features of the environment which are characteristic of and relevant to the behavior, e.g., objects observed and work done by organisms.

The first of the two fields of biology in the wider sense may

be called biology in the narrower sense, or, for the following
discussions, simply *biology*. This use of the term 'biology' seems
justified by the fact that, in terms of the customary classification, this part contains most of what is usually called biology,
namely, general biology, botany, and the greater part of zoology. The terms which are used in this field in addition to
logico-mathematical and physical terms may be called biological terms in the narrower sense, or simply *biological terms*.
Since many statements of biology contain physical terms besides
biological ones, the *biological language* cannot be restricted to
biological terms; it contains the physical language as a sublanguage and, in addition, the biological terms. Statements and
laws belonging to this language but not to physical language
will be called *biological statements* and *biological laws*.

The distinction between the two fields of biology in the wider
sense has been indicated only in a very vague way. At the
present time it is not yet clear as to how the boundary line may
best be drawn. Which processes in an organism are to be assigned to the second field? Perhaps the connection of a process
with the processes in the nervous system might be taken as
characteristic, or, to restrict it more, the connection with speaking activities, or, more generally, with activities involving signs.
Another way of characterization might come from the other
direction, from outside, namely, selecting the processes in an
organism from the point of view of their relevance to achievements in the environment (see Brunswik and Ness). There is
no name in common use for this second field. (The term 'mental sciences' suggests too narrow a field and is connected too
closely with the metaphysical dualism mentioned before.) The
term 'behavioristics' has been proposed. If it is used, it must be
made clear that the word 'behavior' has here a greater extension
than it had with the earlier behaviorists. Here it is intended to
designate not only the overt behavior which can be observed
from outside but also internal behavior (i.e., processes within
the organism); further, dispositions to behavior which may not
be manifest in a special case; and, finally, certain effects upon
the environment. Within this second field we may distinguish

roughly between two parts dealing with individual organisms and with groups of organisms. But it seems doubtful whether any sharp line can be drawn between these two parts. Compared with the customary classification of science, the first part would include chiefly psychology, but also some parts of physiology and the humanities. The second part would chiefly include social science and, further, the greater part of the humanities and history, but it has not only to deal with groups of human beings but also to deal with groups of other organisms. For the following discussion, the terms 'psychology' and 'social science' will be used as names of the two parts because of lack of better terms. It is clear that both the question of boundary lines and the question of suitable terms for the sections is still in need of much more discussion.

III. REDUCIBILITY

The question of the unity of science is meant here as a problem of the logic of science, not of ontology. We do not ask: "Is the world one?" "Are all events fundamentally of one kind?" "Are the so-called mental processes really physical processes or not?" "Are the so-called physical processes really spiritual or not?" It seems doubtful whether we can find any theoretical content in such philosophical questions as discussed by monism, dualism, and pluralism. In any case, when we ask whether there is a unity in science, we mean this as a question of logic, concerning the logical relationships between the terms and the laws of the various branches of science. Since it belongs to the logic of science, the question concerns scientists and logicians alike.

Let us first deal with the question of terms. (Instead of the word 'term' the word 'concept' could be taken, which is more frequently used by logicians. But the word 'term' is more clear, since it shows that we mean signs, e.g., words, expressions consisting of words, artificial symbols, etc., of course with the meaning they have in the language in question. We do not mean 'concept' in its psychological sense, i.e., images or thoughts somehow connected with a word; that would not belong to logic.) We know the meaning (designatum) of a term if we know under what conditions we are permitted to apply it in a concrete case and under what conditions not. Such a knowledge of the conditions of application can be of two different kinds. In some cases we may have a merely practical knowledge, i.e., we are able to use the term in question correctly without giving a theoretical account of the rules for its use. In other cases we may be able to give an explicit formulation of the conditions for the application of the term. If now a certain term x is such that the conditions for its application (as used in the language of science) can be formulated with the help of the terms y, z, etc., we call such a formulation a *reduction statement* for x in terms of y, z, etc., and we call x *reducible* to y, z, etc. There may be several sets of conditions for the application of x; hence x may be reducible to y, z, etc., and also to u, v, etc., and perhaps to other sets. There may even be cases of mutual reducibility, e.g., each term of the set x_1, x_2, etc., is reducible to y_1, y_2, etc.; and, on the other hand, each term of the set y_1, y_2, etc., is reducible to x_1, x_2, etc.

A *definition* is the simplest form of a reduction statement. For the formulation of examples, let us use '$\equiv$' (called the symbol of equivalence) as abbreviation for 'if and only if.' Example of a definition for 'ox': 'x is an ox $\equiv$ x is a quadruped and horned and cloven-footed and ruminant, etc.' This is also a reduction statement because it states the conditions for the application of the term 'ox,' saying that this term can be applied to a thing if and only if that thing is a quadruped and horned, etc. By that definition the term 'ox' is shown to be reducible to—moreover definable by—the set of terms 'quadruped,' 'horned,' etc.

A reduction statement sometimes cannot be formulated in the simple form of a definition, i.e., of an equivalence statement, '$\ldots \equiv \ldots$,' but only in the somewhat more complex form 'If $\ldots$, then: $\ldots \equiv \ldots$.' Thus a reduction statement is either a simple (i.e., explicit) definition or, so to speak, a conditional definition. (The term 'reduction statement' is generally used in the narrower sense, referring to the second, conditional form.) For instance, the following statement is a reduction statement for the term 'electric charge' (taken here for the sake of simplicity as a nonquantitative term), i.e., for the statement form 'the body x has an electric charge at the time t': 'If a light body y is placed near x at t, then: x has an electric charge at $t \equiv y$ is attracted by x at t.' A general way of procedure which enables us to find out whether or not a certain term can be applied in concrete cases may be called a *method of determination* for the term in question. The method of determination for a quantitative term (e.g., 'temperature') is the method of measurement for that term. Whenever we know an experimental method of determination for a term, we are in a position to formulate a reduction statement for it. To know an experimental method of determination for a term, say 'Q_3,' means to know two things. First, we must know an experimental situation which we have to create, say the state Q_1, e.g., the arrangement of measuring apparatuses and of suitable conditions for their use. Second, we must know the possible experimental result, say Q_2, which, if it occurs, will confirm the presence of the property Q_3. In the simplest case—let us leave aside the more complex cases—Q_2 is also such that its nonoccurrence shows that the thing in question does not have the property Q_3. Then a reduction statement for 'Q_3,' i.e., for the statement form 'the thing (or space-time-point) x is Q_3 (i.e., has the property Q_3) at the time t,' can be formulated in this way: 'If x is Q_1 (i.e., x and the surroundings of x are in the state Q_1) at time t, then: x is Q_3 at $t \equiv x$ is Q_2 at t.' On the basis of this reduction statement, the term 'Q_3' is reducible to 'Q_1,' 'Q_2,' and spatio-temporal terms. Whenever a term 'Q_3' expresses the disposition of a thing to behave in a certain way (Q_2) to certain conditions (Q_1), we have a reduction statement of the form given above. If there is a connection of such a kind between Q_1, Q_2, and Q_3, then in biology and psychology in certain cases the following terminology is applied: 'To the stimulus Q_1 we find the reaction Q_2 as a symptom for Q_3.' But the situation is not essentially different from the analogous one in physics, where we usually do not apply that terminology.

Sometimes we know several methods of determination for a certain term. For example, we can determine the presence of an electric current by observing either the heat produced in the conductor, or the deviation of a magnetic needle, or the quantity of a substance separated from an electrolyte, etc. Thus the term 'electric current' is reducible to each of many sets of other terms. Since not only can an electric current be measured by measuring a temperature but also, conversely a temperature can be measured by measuring the electric current produced by a thermo-electric element, there is mutual reducibility between the terms of the theory of electricity, on the one hand, and those of the theory of heat, on the other. The same holds for the terms of the theory of electricity and those of the theory of magnetism.

Let us suppose that the persons of a certain group have a certain set of terms in common, either on account of a merely practical agreement about the conditions of their application or with an explicit stipulation of such conditions for a part of the terms. Then a reduction statement reducing a new term to the terms of that original set may be used as a way of introducing the new term into the language of the group. This way of introduction assures conformity as to the use of the new term. If a certain language (e.g., a sublanguage of the language of science, covering a certain branch of science) is such that every term of it is reducible to a certain set of terms, then this language can be constructed on the basis of that set by introducing one new term after the other by reduction statements. In this case we call the basic set of terms a *sufficient reduction basis* for that language.

IV. THE UNITY OF THE LANGUAGE OF SCIENCE

Now we will analyze the logical relations among the terms of different parts of the language of science with respect to reducibility. We have indicated a division of the whole language of science into some parts. Now we may make another division cutting across the first, by distinguishing in a rough way, without any claims to exactness, between those terms which we use on a prescientific level in our everyday language, and for whose application no scientific procedure is necessary, and scientific terms in the narrower sense. That sublanguage which is the common part of this prescientific language and the physical language may be called physical thing-language or briefly *thing-language*. It is this language that we use in speaking about the properties of the observable (inorganic) things surrounding us. Terms like 'hot' and 'cold' may be regarded as belonging to the thing-language, but not 'temperature' because its determination requires the application of a technical instrument; further, 'heavy' and 'light' (but not 'weight'); 'red,' 'blue,' etc.; 'large,' 'small,' 'thick,' 'thin,' etc.

The terms so far mentioned designate what we may call observable properties, i.e., such as can be determined by a direct observation. We will call them *observable thing-predicates*. Besides such terms the thing-language contains other ones, e.g., those expressing the disposition of a thing to a certain behavior under certain conditions, e.g., 'elastic,' 'soluble,' 'flexible,' 'transparent,' 'fragile,' 'plastic,' etc. These terms—they might be called disposition-predicates—are reducible to observable thing-predicates because we can describe the experimental conditions and the reactions characteristic of such disposition-predicates in terms of observable thing-predicates. Example of a reduction statement for 'elastic': 'If the body x is stretched and then released at the time t, then: x is elastic at the time $t = x$ contracts at t,' where the terms 'stretched,' 'released,' and 'contracting' can be defined by observable thing-predicates. If these predicates are taken as a basis, we can moreover introduce, by iterated application of definition and (conditional) reduction, every other term of the *thing-language*, e.g., designations of substances, e.g., 'stone,' 'water,' 'sugar,' or of processes, e.g., 'rain,' 'fire,' etc. For every term of that language is such that we can apply it either on the basis of direct observation or with the help of an experiment for which we know the conditions and the possible result determining the application of the term in question.

Now we can easily see that every term of the *physical language* is reducible to those of the thing-language and hence finally to observable thing-predicates. On the scientific level, we have the quantitative coefficient of elasticity instead of the qualitative term 'elastic' of the thing-language; we have the quantitative term 'temperature' instead of the qualitative ones 'hot' and 'cold'; and we have all the terms by means of which physicists describe the temporary or permanent states of things or processes. For any such term the physicist knows at least one method of determination. Physicists would not admit into their language any term for which no method of determination by observations were given. The formulation of such a method, i.e., the description of the experimental arrangement to be carried out and of the possible result determining the application of the term in question, is a reduction statement for that term. Sometimes the term will not be directly reduced by the reduction statement to thing-predicates, but first to other scientific terms, and these by their reduction statements again to other scientific terms, etc.; but such a reduction chain must in any case finally lead to predicates of the thing-language and, moreover, to observable thing-predicates because otherwise there would be no way of determining whether or not the physical term in question can be applied in special cases, on the basis of given observation statements.

If we come to *biology* (this term now always understood in the narrower sense), we find again the same situation. For any biological term the biologist who introduces or uses it must know empirical criteria for its application. This applies, of course, only to biological terms in the sense explained before, including all terms used in scientific biology proper, but not to certain terms used sometimes in the philosophy of biology—'a whole,' 'entelechy,' etc. It may happen that for the description of the criterion, i.e., the method of determination of a term, other biological terms are needed. In this case the term in question is first reducible to them. But at least indirectly it must be reducible to terms of the thing-language and finally to observable thing-predicates, because the determination of the term in question in a concrete case must finally be based upon observations of concrete things, i.e., upon observation statements formulated in the thing-language.

Let us take as an example the term 'muscle.' Certainly biologists know the conditions for a part of an organism to be a muscle; otherwise the term could not be used in concrete cases. The problem is: Which other terms are needed for the formulation of those conditions? It will be necessary to describe the functions within the organism which are characteristic of muscles, in other words, to formulate certain laws connecting the processes in muscles with those in their environment, or, again in still other words, to describe the reactions to certain stimuli characteristic of muscles. Both the processes in the environment and those in the muscle (in the customary terminology: stimuli and reactions) must be described in such a way that we can determine them by observations. Hence the term 'muscle,' although not definable in terms of the thing-language, is reducible to them. Similar considerations easily show the reducibility of any other biological term—whether it be a designation of a kind of organism, or of a kind of part of organisms, or of a kind of process in organisms.

The result found so far may be formulated in this way: The terms of the thing-language, and even the narrower class of the observable thing-predicates, supply a sufficient basis for the languages both of physics and of biology. (There are, by the way, many reduction bases for these languages, each of which is much more restricted than the classes mentioned.) Now the question may be raised whether a basis of the kind mentioned is sufficient even for the whole language of science. The affirmative answer to this question is sometimes called *physicalism* (because it was first formulated not with respect to the thing-language but to the wider physical language as a sufficient basis). If the thesis of physicalism is applied to biology only, it scarcely meets any serious objections. The situation is somewhat changed, however, when it is applied to psychology and social science (individual and social behavioristics). Since many of the objections raised against it are based on misinterpretations, it is necessary to make clear what the thesis is intended to assert and what not.

The question of the reducibility of the terms of psychology to those of the biological language and thereby to those of the thing-language is closely connected with the problem of the various methods used in psychology. As chief examples of methods used in this field in its present state, the physiological, the behavioristic, and the introspective methods may be considered. The *physiological approach* consists in an investigation of the functions of certain organs in the organism, above all, of the nervous system. Here, the terms used are either those of biology or those so closely related to them that there will scarcely be any doubt with respect to their reducibility to the terms of the biological language and the thing-language. For the *behavioristic approach* different ways are possible. The investigation may be restricted to the external behavior of an organism, i.e., to such movements, sounds, etc., as can be observed by other organisms in the neighborhood of the first. Or processes within the organism may also be taken into account so that this approach overlaps with the physiological one. Or, finally, objects in the environment of the organism, either observed or worked on or produced by it, may also be studied. Now it is easy to see that a term for whose determination a behavioristic method—of one of the kinds mentioned or of a related kind—is known, is reducible to the terms of the biological language, including the thing-language. As we have seen before, the formulation of the method of determination for a term is a reduction statement for that term, either in the form of a simple definition or in the conditional form. By that statement the term is shown to be reducible to the terms applied in describing the method, namely, the experimental arrangement and the characteristic result. Now, conditions and results consist in the behavioristic method either of physiological processes in the organism or of observable processes in the organism and in its environment. Hence they can be described in terms of the biological language. If we have to do with a behavioristic approach in its pure form, i.e., leaving aside physiological investigations, then the description of the conditions and results characteristic for a term can in most cases be given directly in terms of the thing-language. Hence the behavioristic reduction of psychological terms is often simpler than the physiological reduction of the same term.

Let us take as an example the term 'angry.' If for anger we knew a sufficient and necessary criterion to be found by a physiological analysis of the nervous system or other organs, then we could define 'angry' in terms of the biological language. The same holds if we knew such a criterion to be determined by the observation of the overt, external behavior. But a physiological criterion is not yet known. And the peripheral symptoms known are presumably not necessary criteria because it might be that a person of strong self-control is able to suppress these symptoms. If this is the case, the term 'angry' is, at least at the present time, not definable in terms of the biological language. But, nevertheless, it is reducible to such terms. It is sufficient for the formulation of a reduction sentence to know a behavioristic procedure which enables us—if not always, at least under suitable circumstances—to determine whether the organism in question is angry or not. And we know indeed such procedures; otherwise we should never be able to apply the term 'angry' to another person on the basis of our observations of his behavior, as we constantly do in everyday life and in scientific investigation. A reduction of the term 'angry' or similar terms by the formulation of such procedures is indeed less useful than a definition would be, because a definition supplies a complete (i.e., unconditional) criterion for the term in question, while a reduction statement of the conditional form gives only an incomplete one. But a criterion, conditional or not, is all we need for ascertaining reducibility. Thus the result is the following: If for any psychological term we know either a physiological or a behavioristic method of determination, then that term is reducible to those terms of the thing-language.

In psychology, as we find it today, there is, besides the physiological and the behavioristic approach, the so-called *introspective method*. The questions as to its validity, limits, and necessity
are still more unclear and in need of further discussion than
the analogous questions with respect to the two other methods.
Much of what has been said about it, especially by philosophers,
may be looked at with some suspicion. But the facts themselves
to which the term 'introspection' is meant to refer will scarcely
be denied by anybody, e.g., the fact that a person sometimes
knows that he is angry without applying any of those procedures
which another person would have to apply, i.e., without looking
with the help of a physiological instrument at his nervous
system or looking at the play of his facial muscles. The problems
of the practical reliability and theoretical validity of the introspective method may here be left aside. For the discussion of
reducibility an answer to these problems is not needed. It will
suffice to show that in every case, no matter whether the introspective method is applicable or not, the behavioristic method
can be applied at any rate. But we must be careful in the interpretation of this assertion. It is not meant as saying: "Every
psychological process can be ascertained by the behavioristic
method." Here we have to do not with the single processes
themselves (e.g., Peter's anger yesterday morning) but with
kinds of processes (e.g., anger). If Robinson Crusoe is angry
and then dies before anybody comes to his island, nobody except himself ever knows of this single occurrence of anger. But
anger of the same kind, occurring with other persons, may be
studied and ascertained by a behavioristic method, if circumstances are favorable. (Analogy: if an electrically charged rain-
drop falls into the ocean without an observer or suitable
recording instrument in the neighborhood, nobody will ever
know of that charge. But a charge of the same kind can be
found out under suitable circumstances by certain observations.)
Further, in order to come to a correct formulation of the thesis,
we have to apply it not to the kinds of processes (e.g., anger)
but rather to the terms designating such kinds of processes (e.g.,
'anger'). The difference might seem trivial but is, in fact, es-

sential. We do not at all enter a discussion about the question whether or not there are kinds of events which can never have any behavioristic symptoms, and hence are knowable only by introspection. We have to do with psychological terms, not with kinds of events. For any such term, say, 'Q,' the psychological language contains a statement form applying that term, e.g., "The person . . . is at the time . . . in the state Q." Then the utterance by speaking or writing of the statement "I am now (or: I was yesterday) in the state Q," is (under suitable circumstances, e.g., as to reliability, etc.) an observable symptom for the state Q. Hence there cannot be a term in the psychological language, taken as an intersubjective language for mutual communication, which designates a kind of state or event without any behavioristic symptom. Therefore, there is a behavioristic method of determination for any term of the psychological language. Hence every such term is reducible to those of the thing-language.

The logical nature of the psychological terms becomes clear by an analogy with those physical terms which are introduced by reduction statements of the conditional form. Terms of both kinds designate a state characterized by the disposition to certain reactions. In both cases the state is not the same as those reactions. Anger is not the same as the movements by which an angry organism reacts to the conditions in his environment, just as the state of being electrically charged is not the same as the process of attracting other bodies. In both cases that state sometimes occurs without these events which are observable from outside; they are consequences of the state according to certain laws and may therefore under suitable circumstances be taken as symptoms for it; but they are not identical with it.

The last field to be dealt with is *social science* (in the wide sense indicated before; also called social behavioristics). Here we need no detailed analysis because it is easy to see that every term of this field is reducible to terms of the other fields. The result of any investigation of a group of men or other organisms can be described in terms of the members, their relations to one another and to their environment. Therefore, the conditions for the application of any term can be formulated in terms of psychology, biology, and physics, including the thing-language. Many terms can even be defined on that basis, and the rest is certainly reducible to it.

It is true that some terms which are used in psychology are such that they designate a certain behavior (or disposition to behavior) within a group of a certain kind or a certain attitude toward a group, e.g., "desirous of ruling," "shy," and others. It may be that for the definition or reduction of a term of this kind some terms of social science describing the group involved are needed. This shows that there is not a clear-cut line between psychology and social science and that in some cases it is not clear whether a term is better assigned to one or to the other field. But such terms are also certainly reducible to those of the thing-language because every term referring to a group of organisms is reducible to terms referring to individual organisms.

The result of our analysis is that the class of observable thing-predicates is a sufficient reduction basis for the whole of the language of science, including the cognitive part of the everyday language.

V. THE PROBLEM OF THE UNITY OF LAWS

The relations between the terms of the various branches of science have been considered. There remains the task of analyzing the relations between the laws. According to our previous consideration, a biological law contains only terms which are reducible to physical terms. Hence there is a common language to which both the biological and the physical laws belong so that they can be logically compared and connected. We can ask whether or not a certain biological law is compatible with the system of physical laws, and whether or not it is derivable from them. But the answer to these questions cannot be inferred from the reducibility of the terms. At the present state of the development of science, it is certainly not possible to derive the biological laws from the physical ones. Some philosophers believe that such a derivation is forever impossible because of the very nature of the two fields. But the proofs attempted so far for this thesis are certainly insufficient. This question is, it seems, the scientific kernel of the problem of vitalism; some recent discussions of this problem are, however, entangled with rather questionable metaphysical issues. The question of derivability itself is, of course, a very serious scientific problem. But it will scarcely be possible to find a solution for it before many more results of experimental investigation are available than we have today. In the meantime the efforts toward derivation of more and more biological laws from physical laws—in the customary formulation: explanation of more and more processes in organisms with the help of physics and chemistry —will be, as it has been, a very fruitful tendency in biological research.

As we have seen before, the fields of psychology and social science are very closely connected with each other. A clear division of the laws of these fields is perhaps still less possible than a division of the terms. If the laws are classified in some way or other, it will be seen that sometimes a psychological law is derivable from those of social science, and sometimes a law of social science from those of psychology. (An example of the first kind is the explanation of the behavior of adults—e.g., in the theories of A. Adler and Freud—by their position within the family or a larger group during childhood; an example of the second kind is the obvious explanation of an increase of the price of a commodity by the reactions of buyers and sellers in the case of a diminished supply.) It is obvious that, at the present time, laws of psychology and social science cannot be derived from those of biology and physics. On the other hand, no scientific reason is known for the assumption that such a derivation should be in principle and forever impossible.

Thus there is at present *no unity of laws*. The construction of one homogeneous system of laws for the whole of science is an aim for the further development of science. This aim cannot be shown to be unattainable. But we do not, of course, know whether it will ever be reached.

On the other hand, there is a *unity of language* in science, viz., a common reduction basis for the terms of all branches of science, this basis consisting of a very narrow and homogeneous class of terms of the physical thing-language. This unity of terms is indeed less far-reaching and effective than the unity of laws would be, but it is a necessary preliminary condition for the unity of laws. We can endeavor to develop science more and more in the direction of a unified system of laws only because we have already at present a unified language. And, in addition, the fact that we have this unity of language is of the greatest practical importance. The practical use of laws consists in making predictions with their help. The important fact is that very often a prediction cannot be based on our knowledge of only one branch of science. For instance, the construction of automobiles will be influenced by a prediction of the presumable number of sales. This number depends upon the satisfaction of the buyers and the economic situation. Hence we have to combine knowledge about the function of the motor, the effect of gases and vibration on the human organism, the ability of persons to learn a certain technique, their willingness to spend so much money for so much service, the development of the general economic situation, etc. This knowledge concerns particular facts and general laws belonging to all the four branches, partly scientific and partly commonsense knowledge. For very many decisions, both in individual and in social life, we need such a prediction based upon a combined knowledge of concrete facts and general laws belonging to different branches of science. If now the terms of different branches had no logical connection between one another, such as is supplied by the homogeneous reduction basis, but were of fundamentally different character, as some philosophers believe, then it would not be possible to connect singular statements and laws of different fields in such a way as to derive predictions from them. Therefore, the unity of the language of science is the basis for the practical application of theoretical knowledge.

9

Physicalism

THOMAS NAGEL

I

It is the purpose of this paper to examine the reasons for believing that physicalism cannot possibly be true.[1] I mean by
physicalism the thesis that a person, with all his psychological
attributes, is nothing over and above his body, with all its
physical attributes. The various theories which make this claim
may be classified according to the identities which they allege
between the mental and the physical.[2] These identities may be
illustrated by the standard example of a quart of water which
is identical with a collection of molecules, each containing
two atoms of hydrogen and one of oxygen.

All states of the water are states of that collection of molecules: for the water to be in a particular bottle is for those
molecules to be in that bottle; for the water to be frozen is for
the molecules to be arranged in a space lattice, with strong
intermolecular attractive force and relatively weak individual
vibratory motion; for the water to be boiling is for the molecules
to have a kinetic energy sufficient to produce a vapor pressure
equal to the atmospheric pressure; and so forth. In addition to
general identities like these, there will be particular ones.[3] One

From *The Philosophical Review*, 74 (1965), 339–356. Reprinted by permission of the author and the editors.

[1] An earlier version of this paper was read at the Pacific Division A.P.A.
meetings in Seattle, September 5, 1964.

[2] I shall not consider behaviorism or reductionism of any kind.

[3] Any identity both of whose terms are universal in form will be called general,

such is the identity between an individual splash of the water and a particular sudden displacement of certain of the molecules —an identity which does not imply that a splash is always identical with that particular type of displacement. In all of these cases we can say something like the following: that the water's splashing is not anything over and above the displacement of those molecules; they are the same occurrence.

It is not clear whether every physicalist theory must assert the identity of each person with his body, nor is the connection between this identity and that of psychological with physical states easy to describe. Still, we can specify a range of possible views in terms of the latter relation alone. (1) An implausibly strong physicalism might assert the existence of a general identity between each psychological condition and a physical counterpart. (2) A weaker view would assert some general identities, particularly on the level of sensation, and particular identities for everything that remains. (3) A still weaker view might not require that a physical condition be found identical even in the particular case with every psychological condition, especially if it were an intensional one. (4) The weakest conceivable view would not even assert any particular identities, but of course it is unclear what other assertion by such a theory about the relation between mental and physical conditions might amount to a contention of physicalism.

I am inclined to believe that some weak physicalist theory of the third type is true, and that any plausible physicalism will include some state and event identities, both particular and general. Even a weak view, therefore, must be defended against objections to the possibility of identifying *any* psychological condition with a physical one. It is with such general objections that we shall be occupied.

I shall contend that they fail as objections to physicalism,

---

even if their specification involves reference to particulars. Thus, "Water is H_2O," "For water to be frozen is for its molecules to be in condition F," and "For *this* water to be frozen is for its molecules to be in condition F" are all general identities. On the other hand, "This water's (now) being frozen is its molecules' being in condition F" is a particular identity.

but I shall also contend that they fail to express properly the real source of unhappiness with that position. This conclusion is drawn largely from my own case. I have always found physicalism extremely repellent. Despite my current belief that the thesis is true, this reaction persists, having survived the refutation of those common objections to physicalism which I once thought expressed it. Its source must therefore lie elsewhere, and I shall make a suggestion about that later.[4] First, however, it will be necessary to show why the standard objections fail, and what kind of identity can hold between mental and physical phenomena.

II

Since Smart refuted them, it has presumably become unnecessary to discuss those objections which rest on the confusion between identity of meaning and identity in fact.[5] We may concentrate rather on two types of objection which seem still to be current.

The first is that physicalism violates Leibniz' law, which requires that if two things are identical they have all their nonintensional and nonmodal properties in common. It is objected that sensory impressions, pains, thoughts, and so forth, have various properties which brain states lack, and vice versa. I shall eventually propose a modification of Leibniz' law, since I do not believe that in its strict form it governs the relation asserted by the identity thesis. At this point, however, the thesis may be defended without resorting to such methods, through a some-

---

[4] In sec. V; of the other sections, II attempts to rebut some standard objections, and III contains a general discussion of identity whose results are applied to physicalism in IV.

[5] J.J.C. Smart, "Sensations and Brain Processes," *The Philosophical Review*, 68 (1959), 141–156, republished in *The Philosophy of Mind*, ed. V.C. Chappell (Englewood Cliffs, N.J.: Prentice-Hall, 1962). See also Smart's book, *Philosophy and Scientific Realism* (London: Routledge and Kegan Paul, 1963), and his article "Materialism," *The Journal of Philosophy*, 60 (1963), 651–662, for further discussion of the identity thesis.

what altered version of a device employed by Smart, and earlier by U.T. Place.[6]

Instead of identifying thoughts, sensations, afterimages, and so forth, with brain processes, I propose to identify a person's having the sensation with his body's being in a physical state or undergoing a physical process. Notice that both terms of this identity are of the same logical type, namely (to put it in neutral terminology) a subject's possessing a certain attribute. The subjects are the person and his body (not his brain), and the attributes are psychological conditions, happenings, and so forth, and physical ones. The psychological term of the identity must be the person's having a pain in his shin rather than the pain itself, because although it is undeniable that pains exist and people have them, it is also clear that this describes a condition of one entity, the person, rather than a relation between two entities, a person and a pain. For pains to exist *is* for people to have them. This seems to me perfectly obvious, despite the innocent suggestions of our language to the contrary.

So we may regard the ascription of properties to a sensation simply as part of the specification of a psychological state's being ascribed to the person. When we assert that a person has a sensation of a certain description B, this is not to be taken as asserting that there exist an x and a y such that x is a person and y is a sensation and $B(y)$, and x *has* y. Rather we are to take it as asserting the existence of only one thing, x, such that x is a person, and moreover $C(x)$, where C is the attribute "has a sensation of description B." The specification of this attribute is accomplished in part by the ascription of properties to the sensation; but this is merely part of the ascription of that psychological state to the person. This position seems to me attractive independently of physicalism, and it can be extended to psychological states and events other than sensations. Any ascription of properties to them is to be taken simply as part of

[6] U.T. Place, "Is Consciousness a Brain Process?," *British Journal of Psychology*, 47 (1956), republished in V.C. Chappell, ed., *The Philosophy of Mind*, pp. 107–109; for Smart, see note 5. My formulation of the physical side of the identity differs from Smart's, and I do not accept his psychological reductionism.

the ascription of other attributes to the person who has them—as *specifying* those attributes.

I deviate from Smart in making the physical side of the identity a condition of the body rather than a condition of the brain,[7] because it seems to me doubtful that anything without a body of some conventional sort could be the subject of psychological states.[8] I do not mean to imply that the presence of a particular sensation need depend on the condition of any part of one's body outside of the brain. Making the physical term of the identity a bodily rather than a brain state merely implies that the brain is *in* a body. To identify the person's having a pain with the brain's being in state X rather than with the body's containing a brain in state X would imply, on the other hand, that if the individual's brain could have been in that state while the rest of his body was destroyed, he would still have been in the corresponding psychological state.

Given that the terms of the identity are as specified, nothing obliges us to identify a sensation or a pain or a thought with anything physical, and this disposes of numerous objections. For although I may have a visual sense impression whose attributes of form and color correspond closely to those which characterize the "Mona Lisa," my *having* the sense impression does not possess those attributes, and it is therefore no cause for worry that nothing in my brain looks like the "Mona Lisa." Given our specification of the psychological side of the identity, the demands on the physical side are considerably lessened. The physical equivalents of auditory impressions may be silent, those of olfactory impressions odorless, and so forth.

Most important, we can be rid of the stubbornest objection of this type, that having to do with location.[9] Brain processes are

---

[7] One might alternatively make it a physical condition of the *person*, so that the two identified attributes would be guaranteed the same subject. I cannot say how such a change would affect the argument.

[8] Cf. Norman Malcolm, "Scientific Materialism and the Identity Theory," *Dialogue*, 3 (1964), 124–125.

[9] Malcolm, "Scientific Materialism," pp. 118–120. See also Jerome Shaffer, "Could Mental States Be Brain Processes?," *The Journal of Philosophy*, 63 (1961). Shaffer thinks the difficulty can be got over, but that this depends on the possibility of a *change* in our concept of mental states, which would make it meaningful to assign them locations.

located in the brain, but a pain may be located in the shin and a thought has no location at all. But if the two sides of the identity are not a sensation and a brain process, but my *having* a certain sensation or thought and my body's *being* in a certain physical state, then they will both be going on in the same place—namely, wherever I (and my body) happen to be. It is important that the physical side of the identity is not a brain process, but rather my *body's* being in that state which may be specified as "having the relevant process going on in its brain." *That* state is not located in the brain; it has been located as precisely as it can be when we have been told the precise location of that of which it is a state—namely, my body. The same is true of my having a sensation: that is going on wherever I happen to be at the time, and its location cannot be specified more precisely than mine can. (That is, even if a pain is located in my right shin, I am *having* that pain in my office at the university.) The location of bodily sensations is a very different thing from the location of warts. It is phenomenal location, and is best regarded as one feature of a psychological attribute possessed by the *whole* person rather than as the spatial location of an event going on in a part of him.

The other type of objection which I shall discuss is that physicalism fails to account for the privacy or subjectivity of mental phenomena. This complaint, while very important, is difficult to state precisely.

There is a trivial sense in which a psychological state is private to its possessor, namely, that since it is his, it cannot be anyone else's. This is just as true of haircuts or, for that matter, of physiological conditions. Its triviality becomes clear when we regard thoughts and sensations as conditions of the person rather than as things to which the person is related. When we see that what is described as though it were a relation between two things is really a condition of one thing, it is not surprising that only one person can stand in the said relation to a given sensation or feeling. In this sense, bodily states are just as private to their possessor as the mental states with which they may be equated.

The private-access objection is sometimes expressed epistemo-logically. The privacy of haircuts is uninteresting because there is lacking in that case a special connection between possession and knowledge which is felt to be present in the case of pains. Consider the following statement of the privacy objection.[10] "When I am in a psychological state—for example, when I have a certain sensation—it is logically impossible that I should fail to know that I am in that state. This, however, is not true of any bodily state. Therefore no psychological state is identical with any bodily state." As it happens, I believe that the first clause of this objection—namely, the incorrigibility thesis—is false, but I do not have to base my attack on that contention, for even if the incorrigibility thesis were true it would not rule out physicalism.

If state x is identical with state y it does not follow by Leibniz' law that if I know I am in state x then I know I am in state y, since the context is intensional. Therefore neither does it follow from "If I am in state x then I know I am in state x" that if I am in state y I know I am in state y. All that follows is that if I am in state y I know I am in state x. Moreover, this connection will not be a necessary one, since only one of the premises—the incorrigibility thesis—is necessary. The other premise—that x is identical with y—is contingent, making the consequence contingent.[11]

[10] See Kurt Baier, "Smart on Sensations," and J.J.C. Smart, "Brain Processes and Incorrigibility," *Australasian Journal of Philosophy*, 40 (1962). This is regarded as a serious difficulty by Smart and other defenders of physicalism. See D.M. Armstrong, "Is Introspective Knowledge Incorrigible?," *The Philosophical Review*, 72 (1963), 418–419. On the other hand, Hilary Putnam has argued that all the problems about privacy and special access which can be raised about persons can be raised about machines as well. See his paper, "Minds and Machines," in *Dimensions of Mind*, ed. Sidney Hook (New York: Collier, 1960).

[11] It is worth noting that if two mental states are necessarily connected, this connection must be mirrored on the level of the physical states with which we identify them. Although the connection between the physical states need not be a logically necessary one, that would be a desirable feature in a physicalistic theory, and it seems in fact to be present in the example of water and molecules: the water's being frozen necessarily includes its being cold, and the specification of the molecular state which *is* its being frozen entails that the molecules will have a low average kinetic energy—which is in fact the same thing as the water's being cold.

There may be more to the special-access objection than this, but I have not yet encountered a version of it which succeeds. We shall later discuss a somewhat different interpretation of the claim that mental states are subjective.

III

Let us now consider the nature of the identity which physicalism asserts. Events, states of affairs, conditions, psychological and otherwise, may be identical in a perfectly straightforward sense which conforms to Leibniz' law as strictly as does the identity between, say, the only horse in Berkeley and the largest mammal in Berkeley. Such identities between events may be due to the identity of two things referred to in their descriptions—for example, my being kicked by the only horse in Berkeley and my being kicked by the largest mammal in Berkeley—or they may not—for example, the sinking of the Titanic and the largest marine disaster ever to occur in peacetime. Whether they hold between things, events, or conditions, I shall refer to them as *strict* identities.

We are interested, however, in identities of a different type— between psychological and physical events, or between the boiling of water and the activity of molecules. I shall call these theoretical identities[12] and shall concentrate for the moment on their application to events and attributes rather than to things, although they hold between things as well. It is a weaker relation than strict identity, and common possession of causal and conditional attributes is crucial for its establishment.[13]

---

[12] Following Hilary Putnam, "Minds and Machines," who says that the "is" in question is that of theoretical identification. The word "identity" by itself is actually too strong for such cases, but I shall adopt it for the sake of convenience.

[13] An attribute, for our purposes, is signified by any sentence-frame containing one free variable (in however many occurrences) where this may be a variable ranging over objects, events, and so forth. (One gets a particular instance of an attribute by plugging in an appropriate particular for the variable and converting to gerundival form.) Thus all three of the following are attributes: ". . . is

Strict identities are likely to be established in other ways, and we can infer the sameness of all causal and conditional attributes. Thus, if being kicked by the only horse in Berkeley gave me a broken leg, then being kicked by the largest mammal in Berkeley had the same effect, given that they are the same creature; and if it is the case that I should not have been kicked by the only horse in Berkeley if I had stayed in my office that afternoon, then it follows that if I had stayed in my office I should not have been kicked by the largest mammal in Berkeley.

But if we lack grounds such as these, we must establish sameness of conditional attributes independently, and this depends on the discovery of general laws from which the particular conditionals follow. Our grounds for believing that a particular quart of water's boiling is the same event as a collection of molecules' behaving in a certain way are whatever grounds we may have for believing that all the causes and effects of one event are also causes and effects of the other, and that all true statements about conditions under which the one event would not have occurred, or about what would have happened if it had not, or about what would happen if it continued, and so forth, are also true of the other.

This is clearly more than mere constant conjunction; it is a fairly strong requirement for identity. Nevertheless it is weaker than the standard version of Leibniz' law in that it does not require possession by each term of *all* the attributes of the other. It does not require that the complex molecular event which we may identify with my being kicked by the only horse in Berkeley be independently characterizable as ridiculous—for example, on the grounds that the latter event was ridiculous and if the former cannot be said to be ridiculous, it lacks an attribute which the latter possesses. There are some attributes from the

---

boiling," ". . . will stop boiling if the kettle is taken off the fire," and ". . . will stop if the kettle is taken off the fire." A particular quart of water has the second of these attributes if and only if that water's boiling has the third, where this can be described as the possession of the third attribute by a particular instance of the first.

common possession of which the identity follows, and others which either do not matter or which we cannot decide whether to ascribe to one of the terms without first deciding whether the identity in question holds.

To make this precise, I shall introduce the notion of independent ascribability. There are certain attributes such as being hot or cold, or boiling or offensive, which cannot significantly be ascribed to a collection of molecules per se. It may be that such attributes *can* be ascribed to a collection of molecules, but such ascription is dependent for its significance on their primary ascription to something of a different kind, like a body of water or a person, with which the molecules are identical. Such attributes, I shall say, are not independently ascribable, to the molecules, though they may be dependently ascribable. Similarly, the property of having eighty-three trillion members is not independently ascribable to a quantity of water, though it may be possessed by a collection of H_2O molecules. Nevertheless, there is in such cases a class of attributes which are independently ascribable to both terms, and the condition for theoretical identity may be stated as follows: that the two terms should possess or lack in common all those attributes which can be independently ascribed to each of them individually—with the qualification that nothing is by this criterion to be identical with two things which are by the same criterion distinct.[14] Actually this will serve as a condition for identity in general; a strict identity will simply be one between terms sufficiently similar in type to allow independent ascription to both of *all* the same attributes, and will include such cases as the sinking of the *Titanic* being the largest marine disaster ever to occur in peacetime, or the morning star being the evening star. The identities I have characterized as theoretical hold across categories of description sufficiently different to prohibit independent ascription to both terms of all the same attributes, although, as I have

---

[14] The qualification takes care of such possibly problematic claims as that I am the square root of 2, for although it may be that we share all attributes which can be independently ascribed to each of us, I also share those attributes with the square root of 3, whose attributes clearly contradict those of the square root of 2.

observed, such ascriptions may be meaningful as *consequences* of the identity.

The question naturally arises, to what extent do particular theoretical identities depend on corresponding general ones? In the examples I have given concerning the case of water, the dependence is obvious. There the particular identities have simply been instances of general ones, which are consequences of the same theory that accounts for the common possession of relevant attributes in the particular cases. Now there is a technical sense in which every particular theoretical identity must be an instance of a general identity, but not all cases need be like that of water. Although it is essential that particular identities must follow from general laws or a general theory, this does not prevent us from distinguishing between cases in which, for example, the molecular counterpart of a macroscopic phenomenon is always the same, and those in which it varies from instance to instance. The common possession of conditional attributes can follow for a particular case from general laws, without its being true that there is a general correlation between macroscopic and microscopic phenomena of that type. For example, it may at the same time follow from general laws that types of microscopic phenomena other than the one present in this case would also share the requisite conditional properties.

The technical sense in which even in such cases the particular identity must be an instance of a general one is that it must be regarded as an instance of the identity between the macroscopic phenomenon and the disjunction of all those microscopic phenomena which are associated with it in the manner described, via general laws. For suppose we have a type of macroscopic phenomenon A and two types of microscopic phenomena B and C so associated with it. Suppose on one occasion particular cases of A and B are occurring at the same place and time, and so forth, and suppose it is asserted that since it follows from general laws that they also have all their conditional attributes in common, A is in this case identical in the specified sense with B. They do not, however, have in common the conditional attribute $F(X)$, defined as follows: "If C and not

B, then X." That is, $F(A)$ but not $F(B)$. Therefore, we must identify the occurrence of A even in this case with the occurrence of the disjunction B or C. This does not prevent us, however, from introducing as a subsidiary sense of identity for particular cases that in which A is B because the disjunction B or C which is properly identical with A is in fact satisfied by B. There is of course a range of cases between the two kinds described, cases in which the disjuncts in the general identity consist of conjunctions which overlap to a greater or lesser degree, and this complicates the matter considerably.[15] Nevertheless we can, despite the technicality, differentiate roughly between particular identities which are in a narrow sense instances of general identities and those which are not—that is, which are instances only of radically disjunctive general identities. Henceforth when I refer to general identities I shall be excluding the latter.

I have concentrated on identities between states, events, and attributes because it is in such terms that physicalism is usually conceived, but if it is also part of physicalism to hold that people are their bodies, it becomes appropriate to inquire into the relation between the theoretical identity of things and the theoretical identity of their attributes. Unfortunately, I do not have a general answer to this question. The case of strict identity presents no problem, for there every attribute of one term is strictly identical with the corresponding attribute of the other; and in our standard example of theoretical identity, each attribute of the water seems to be theoretically identical with some attribute of the molecules, but not vice versa. This may

[15] A fuller treatment would have to include a discussion of the nonsymmetrical relation ". . . consists of . . ." which is distinct from identity. A macroscopic event (the freezing of some water, for example) may be identical with a microscopic event A described in general terms (average kinetic energy, spatial ordering, and the like) while at the same time consisting of a very specific collection B of microscopic events with which it is not identical, since if one of them (the motion of a particular molecule) had been different, that particular complex of microscopic events would not have occurred though both A and the microscopic event would have. (Presumably in such cases the occurrence of B entails the occurrence of A, but more than that needs to be said about the relation between the two.) The same concept applies to the relation between World War II and the immense collection of actions and events of which it consisted, or that between the Eiffel Tower and the girders and rivets which make it up.

be one (asymmetrical) condition for the theoretical identity of things. It is not clear, however, whether the identity of things must always be so closely tied to the identity of their attributes. For example, it might be that everything we could explain in terms of the water and its attributes could be explained in terms of the batch of molecules and their attributes, but that the two systems of explanation were so different in structure that it would be impossible to find a single attribute of the molecules which explained all and only those things explained by a particular attribute of the water.

Whether or not this is true of water, the possibility has obvious relevance to physicalism. One might be able to define a weak criterion of theoretical identity which would be satisfied in such a case, and this might in turn give sense to an identification of persons with their bodies which did not depend on the discovery of a single physical counterpart for every psychological event or condition. I shall, however, forgo an investigation of the subject; this general discussion of identity must remain somewhat programmatic.

IV

It provides us with some things to say, however, about the thesis of physicalism. First, the grounds for accepting it will come from increased knowledge of (a) the explanation of mental events and (b) the physiological explanation of happenings which those mental events in turn explain. Second, in view of the condition of independent ascribability, physicalism need not be threatened by the difficulty that although anger may be, for example, justified, it makes no sense to say this of a physical state with which we may identify it. Third, it does not require general identities at every level: that is, there need not be, either for all persons or even for each individual, a specific physical state of the body which is *in general* identical with intending to vote Republican at the next election, or having a stomachache, in order that physicalism be true. It seems likely that there will be general identities of a rough kind for nonintensional states, such as having particular sensations or sensory impressions, since the physical causes of these are fairly uniform. But one can be practically certain that intensional mental states, even if in each particular case they are identical with some physical state, will not have general physical counterparts, because both the causes and the effects of a given belief or desire or intention are extremely various on different occasions even for the same individual, let alone for different persons. One might as easily hope to find a general equivalent, in molecular terms, of a building's collapsing or a bridge's being unsafe—yet each instance of such an event or circumstance is identical with some microscopic phenomenon.

The relation of intensional mental states to physical states may be even more involved than this. For one thing, if it should be the case that they are dispositional in a classical sense, then physicalism requires only that the events and states to which they are the dispositions be identical with physical events and states. It does not require that they be identical with any additional independent physical state, existing even when the disposition is not being exercised. (In fact, I do not believe that dispositions operate according to the classical Rylean model, and this will affect still further the way in which the identity thesis applies to dispositional mental states; but this is not the place for a discussion of that issue.)

There is still another point: many intensional predicates do not just ascribe a condition to the person himself but have implications about the rest of the world and his relation to it. Physicalism will of course not require that these be identical simply with states of the person's body, narrowly conceived. An obvious case is that of knowledge, which implies not only the truth of what is known but also a special relation between this and the knower. Intentions, thoughts, and desires may also imply a context, a relation with things outside the person. The thesis that all states of a person are states of his body therefore requires a liberal conception of what constitutes a state—one which will admit relational attributes. This is not peculiar to mental states: it is characteristic of intensional attributes wherever they occur. That a sign says that fishing is forbidden does not consist simply in its having a certain geometrically describable distribution of black paint on its surface; yet we are not tempted here to deny that the sign is a piece of wood with paint on it, or to postulate a noncorporeal substance which is the subject of the sign's intensional attributes.

Even with all these qualifications, however, it may be too much to expect a specific physical counterpart for each particular psychological phenomenon. Thus, although it may be the case that what explains and is explained by a particular sensation can also explain and be explained by a particular neurological condition, it may also be that this is not precisely true of an intention, but rather that the various connections which we draw between causes and effects via the intention can be accounted for in terms of many different physical conditions, some of which also account for connections which in psychological discourse we draw via states other than the intention, and no subset of which, proper or improper, accounts for all and only those connections which the intention explains. For this reason a thoroughgoing physicalism might have to fall back on a criterion for identity between things not dependent on the identity of their attributes—a criterion of the sort envisaged at the end of the previous section.

Obviously any physicalistic *theory*, as opposed to the bare philosophical thesis of physicalism, will be exceedingly complex. We are nowhere near a physical theory of how human beings work, and I do not know whether the empirical evidence currently available indicates that we may expect with the advance of neurology to uncover one. My concern here has been only to refute the philosophical position that mental-physical identity is *impossible*, and that *no* amount of further information could constitute evidence for it.

v

Even if what might be called the standard objections have been answered, however, I believe that there remains another source for the philosophical conviction that physicalism is impossible.

It expresses itself crudely as the feeling that there is a fundamental distinction between the subjective and the objective
which cannot be bridged. Objections having to do with privacy
and special access represent attempts to express it, but they fail
to do so, for it remains when they have been defeated. The
feeling is that I (and hence any "I") cannot be a mere physical
object, because I possess my mental states: I am their *subject*,
in a way in which no physical object can possibly be the subject
of its attributes. I have a type of internality which physical
things lack; so in addition to the connection which all my
mental states do admittedly have with my body, they are also
mine—that is, they have a particular *self* as subject, rather than
merely being attributes of an object. Since any mental state
must have a self as subject, it cannot be identical with a mere
attribute of some object like a body, and the self which is its
subject cannot therefore be a body.

Why should it be thought that for *me* to have a certain
sensation—to be in a certain mental state—cannot consist
merely in a physical object's being in some state, having some
attribute? One might put it as follows. States of my body,
physical states, are, admittedly, physical states of me, but this is
not simply because they are states of that body but because in
addition it is my body. And its being my body consists in its
having a certain relation, perhaps a causal one, to the subject of
my mental states. This leads naturally to the conclusion that I,
the subject of my mental states, am something else—perhaps
a mental substance. My physical states are only derivatively
mine, since they are states of a body which is mine in virtue of
being related in the appropriate way to my psychological states.
But this is possible only if those psychological states are mine
in an original, and not merely derivative, sense; therefore *their*
subject cannot be the body which is derivatively mine. The
feeling that physicalism leaves out of account the essential
subjectivity of psychological states is the feeling that nowhere
in the description of the state of a human body could there be
room for a physical equivalent of the fact that *I* (or any self),
and not just that body, am the subject of those states.

This, so far as I can see, is the source of my uneasiness about physicalism. Unfortunately, whatever its merits, it is no more an argument against physicalism than against most other theories of mind, including dualism, and it therefore provides us with no more reason for rejecting the former in favor of the latter than do the standard objections already discussed. It can be shown that if we follow out this type of argument, it will provide us with equally strong reasons for rejecting any view which identifies the subject of psychological states with a substance and construes the states as attributes of that substance. A noncorporeal substance seems safe only because, in retreating from the physical substance as a candidate for the self, we are so much occupied with finding a subject whose states are originally, and not just derivatively, mine—one to which the physical body can be related in a way which explains how *it* can be mine— that we simply postulate such a subject without asking ourselves whether the same objections will not apply to it as well: whether indeed any substance can possibly meet the requirement that its states be *underivatively* mine.

The problem can be shown to be general in the following way: consider everything that can be said about the world without employing any token-reflexive expressions.[16] This will include the description of all its physical contents and their states, activities, and attributes. It will also include a description of all the persons in the world and their histories, memories, thoughts, sensations, perceptions, intentions, and so forth. I can thus describe without token-reflexives the entire world and everything that is happening in it—and this will include a description of Thomas Nagel and what he is thinking and feeling. But there seems to remain one thing which I cannot say in this fashion—namely, which of the various persons in the world I am. Even when everything that can be said in the specified manner has been said, and the world has in a sense been completely described, there seems to remain one fact which has not

[16] I.e., expressions *functioning* as token reflexives. Such words of course lose this function in quotation and in certain cases of *oratio* (or *cogitatio*) *obliqua*: e.g., "John Smith thinks that he is Napoleon."

been expressed, and that is the fact that I am Thomas Nagel. This is not, of course, the fact ordinarily conveyed by those words, when they are used to inform someone else who the *speaker* is—for that could easily be expressed otherwise. It is rather the fact that *I* am the subject of *these* experiences; this body is my body; the subject or center of my world is this person, Thomas Nagel.

Now it follows from this not only that a sensation's being mine cannot consist simply in its being an attribute of a particular body; it follows also that it cannot consist in the sensation's being an attribute of a particular soul which is joined to that body; for nothing in the specification of that soul will determine that *it* is mine, that I am *that* person. So long as we construe psychological states as attributes of a substance, no matter what substance we pick, it can be thrown, along with the body, into the "objective" world; its states and its relation to a particular body can be described completely without touching upon the fact that I am that person.[17] It turns out therefore that, given the requirements which led us to reject physicalism, the quest for the self, for a substance which *is* me and whose possession of a psychological attribute will *be* its being mine, is a quest for something which could not exist. The only possible conclusion is that the self is not a substance, and that the special kind of possession which characterizes the relation between me and my psychological states cannot be represented as the possession of certain attributes by a subject, no matter what that subject may be. The subjectivity of the true psychological subject is of a different kind from that of the mere subject of attributes. And if I am to extend this to cases other than my own, I must conclude that for no person is it the case that his having a particular sensation consists in some occupant of the world having a particular attribute or being in a certain state.

I shall not discuss the reasons for rejecting this position. My attitude toward it is precisely the reverse of my attitude toward physicalism, which repels me, although I am persuaded of its truth. The two are of course related, since what bothers me

[17] Cf. Wittgenstein, *Tractatus*, 5.64.

about physicalism is the thought that I cannot be a mere physical object, cannot in fact be anything *in* the world at all, and that my sensations and so forth cannot be simply the attributes of some substance.

But if we reject this view (as it seems likely that we must) and accept the alternative that a person is something in the world and that his mental states are states of that thing, then there is no a priori reason why it should not turn out to be a physical body and those states physical states. We are thus freed to investigate the possibility, and to seek the kind of understanding of psychological states which will enable us to formulate specific physicalistic theories as neurology progresses.

POSTSCRIPT, 1968

I now believe that theoretical identity is not distinct from strict identity, and that the device by which I formerly defined theoretical identity can be used to explain how Leibniz' law is satisfied by identities whose terms are of disparate types. Suppose boiling is independently ascribable to a quart of water but not to the molecules which compose it. Nevertheless, we can say that the molecules are boiling if they bear a certain relation to the water and the water is boiling. The relation in question, call it R, is simply that which I formerly described as theoretical identity. It holds between a and b if (i) they possess or lack in common all those attributes which can be independently ascribed to each of them individually (call this relation S), and (ii) neither a nor b bears relation S to any third term which does not bear relation S to the other.[1] Let F range over nonintensional and nonmodal attributes, and let us symbolize the modal statement "F is independently ascribable (truly or falsely) to a" as $I(F,a)$. Then

(1) $S(a,b) = df (F) (I(F,a) \cdot I(F,b) \cdot \supset \cdot F(a) = F(b))$

---

[1] Condition (ii) is added for the reason cited in footnote 14 of my 1965 paper.

(2) $R(a,b) \equiv df$ (i) $S(a,b)$ & (ii) a true statement results when-
      ever a name or definite description is substituted for "x"
      in the schema $S(a,x) \equiv S(b,x)$

I claim that a true statement results whenever names or definite descriptions are substituted for 'x' and 'y' in the following schema:

(3) $(F)\ (I(F,x) \cdot F(x) \cdot R(x,y) \cdot \supset F(y)\ )$

If this is correct, then when a and b are related by R they will share all the attributes independently ascribed to either of them. By Leibniz' law, therefore,

(4) $R(a,b) \equiv a = b$

10

Dispositions

GILBERT RYLE

(1) FOREWORD

I have already had occasion to argue that a number of the words which we commonly use to describe and explain people's behavior signify dispositions and not episodes. To say that a person knows something, or aspires to be something, is not to say that he is at a particular moment in process of doing or undergoing anything, but that he is able to do certain things, when the need arises, or that he is prone to do and feel certain things in situations of certain sorts.

This is, in itself, hardly more than a dull fact (almost) of ordinary grammar. The verbs 'know', 'possess', and 'aspire' do not behave like the verbs 'run', 'wake up', or 'tingle'; we cannot say 'he knew so and so for two minutes, then stopped and started again after a breather', 'he gradually aspired to be a bishop', or 'he is now engaged in possessing a bicycle'. Nor is it a peculiarity of people that we describe them in dispositional terms. We use such terms just as much for describing animals, insects, crystals, and atoms. We are constantly wanting to talk about what can be relied on to happen as well as to talk about what is actually happening; we are constantly wanting to give explanations of incidents as well as to report them; and we are constantly wanting to tell how things can be managed as well as to tell what is now going on in them. Moreover, merely to

From *The Concept of Mind* (London: Hutchinson and Co., 1949), ch. V, secs. 1–3. Reprinted by permission of the author and the publishers.

classify a word as signifying a disposition is not yet to say much more about it than to say that it is not used for an episode. There are lots of different kinds of dispositional words. Hobbies are not the same sort of thing as habits, and both are different from skills, from mannerisms, from fashions, from phobias, and from trades. Nest-building is a different sort of property from being feathered, and being a conductor of electricity is a different sort of property from being elastic.

There is, however, a special point in drawing attention to the fact that many of the cardinal concepts in terms of which we describe specifically human behavior are dispositional concepts, since the vogue of the paramechanical legend has led many people to ignore the ways in which these concepts actually behave and to construe them instead as items in the descriptions of occult causes and effects. Sentences embodying these dispositional words have been interpreted as being categorical reports of particular but unwitnessable matters of fact instead of being testable, open hypothetical, and what I shall call 'semihypothetical' statements. The old error of treating the term 'force' as denoting an occult force-exerting agency has been given up in the physical sciences, but its relatives survive in many theories of mind and are perhaps only moribund in biology.

The scope of this point must not be exaggerated. The vocabulary we use for describing specifically human behavior does not consist only of dispositional words. The judge, the teacher, the novelist, the psychologist, and the man in the street are bound also to employ a large battery of episodic words when talking about how people do, or should, act and react. These episodic words, no less than dispositional words, belong to a variety of types, and we shall find that obliviousness to some of these differences of type has both fostered, and been fostered by, the identification of the mental with the ghostly. Later in this chapter I shall discuss two main types of mental episodic-words. I do not suggest that there are no others.

(2) THE LOGIC OF DISPOSITIONAL STATEMENTS

When a cow is said to be a ruminant, or a man is said to be a cigarette-smoker, it is not being said that the cow is ruminating now or that the man is smoking a cigarette now. To be a ruminant is to tend to ruminate from time to time, and to be a cigarette-smoker is to be in the habit of smoking cigarettes.

The tendency to ruminate and the habit of cigarette-smoking could not exist, unless there were such processes or episodes as ruminating and smoking cigarettes. 'He is smoking a cigarette now' does not say the same sort of thing as 'he is a cigarette-smoker', but unless statements like the first were sometimes true, statements like the second could not be true. The phrase 'smoke a cigarette' has both episodic uses and, derivative from them, tendency-stating uses. But this does not always occur. There are many tendency-stating and capacity-stating expressions which cannot also be employed in reports of episodes. We can say that something is elastic, but when required to say in what actual events this potentiality is realized, we have to change our vocabulary and say that the object is contracting after being stretched, is just going to expand after being compressed, or recently bounced on sudden impact. There is no active verb corresponding to 'elastic', in the way in which 'is ruminating' corresponds to 'is a ruminant'. Nor is the reason for this nonparallelism far to seek. There are several different reactions which we expect of an elastic object, while there is, roughly, only one sort of behavior that we expect of a creature that is described to us as a ruminant. Similarly there is a wide range of different actions and reactions predictable from the description of someone as 'greedy', while there is, roughly, only one sort of action predictable from the description of someone as 'a cigarette-smoker'. In short, some dispositional words are highly generic or determinable, while others are highly specific or determinate; the verbs with which we report the different exercises of generic tendencies, capacities, and liabilities are apt to differ from the verbs with which we name the dispositions, while the episodic verbs corresponding to the highly specific dispositional verbs are apt to be the same. A baker can be baking now, but a grocer is not described as 'grocing' now, but only as selling sugar now, or weighing tea now, or wrapping up butter now. There are halfway houses. With qualms we will speak of a doctor as engaged now in doctoring someone, though not of a solicitor as now solicitoring, but only as now drafting a will, or now defending a client.

Dispositional words like 'know', 'believe', 'aspire', 'clever', and 'humorous' are determinable dispositional words. They signify abilities, tendencies, or pronenesses to do, not things of one unique kind, but things of lots of different kinds. Theorists who recognize that 'know' and 'believe' are commonly used as dispositional verbs are apt not to notice this point, but to assume that there must be corresponding acts of knowing or apprehending and states of believing; and the fact that one person can never find another person executing such wrongly postulated acts, or being in such states, is apt to be accounted for by locating these acts and states inside the agent's secret grotto.

A similar assumption would lead to the conclusion that since being a solicitor is a profession, there must occur professional activities of solicitoring, and, as a solicitor is never found doing any such unique thing, but only lots of different things like drafting wills, defending clients, and witnessing signatures, his unique professional activity of solicitoring must be one which he performs behind locked doors. The temptation to construe dispositional words as episodic words and this other temptation to postulate that any verb that has a dispositional use must also have a corresponding episodic use are two sources of one and the same myth. But they are not its only sources.

It is now necessary to discuss briefly a general objection that is sometimes made to the whole program of talking about capacities, tendencies, liabilities, and pronenesses. Potentialities, it is truistically said, are nothing actual. The world does not contain, over and above what exists and happens, some other things which are mere would-be things and could-be happenings. To say of a sleeping man that he can read French, or of a piece of dry sugar that it is soluble in water, seems to be pretending at once to accord an attribute and to put that attribute into cold storage. But an attribute either does, or does not, characterize something. It cannot be merely on deposit account. Or, to put it in another way, a significant affirmative indicative sentence must be either true or false. If it is true, it asserts that something has, or some things have, a certain character; if it is false, then its subject lacks that character. But there is no halfway house between a statement's being true and its being false, so there is no way in which the subject described by a statement can shirk the disjunction by being merely able or likely to have or lack the character. A clock can strike the hour that it is, or strike an hour that it is not; but it cannot strike an hour that might be the correct one but is neither the correct nor an incorrect one.

This is a valid objection to one kind of account of such statements as that the sugar is soluble, or the sleeper can read French, namely an account which construes such statements as asserting extra matters of fact. This was indeed the mistake of the old faculty theories which construed dispositional words as denoting occult agencies or causes, i.e., things existing, or processes taking place, in a sort of limbo world. But the truth that sentences containing words like 'might', 'could', and 'would . . . if' do not report limbo facts does not entail that such sentences have not got proper jobs of their own to perform. The job of reporting matters of fact is only one of a wide range of sentence-jobs.

It needs no argument to show that interrogative, imperative, and optative sentences are used for other ends than that of notifying their recipients of the existence of occurrence of things. It does, unfortunately, need some argument to show that there are lots of significant (affirmative and negative) indicative sentences which have functions other than that of reporting facts. There still survives the preposterous assumption that every true or false statement either asserts or denies that a mentioned object or set of objects possesses a specified attribute. In fact, some statements do this and most do not. Books of arithmetic, algebra, geometry, jurisprudence, philosophy, formal logic, and economic theory contain few, if any, factual statements. That is why we call such subjects 'abstract'. Books on physics, meteorology, bacteriology, and comparative philology contain very few such statements, though they may tell us where they are to be found. Technical manuals, works of criticism, sermons, political speeches, and even railway guides may be more or less instructive, and instructive in a variety of ways, but they teach us few singular, categorical, attributive, or relational truths.

Leaving on one side most of the sorts of sentences which have other than fact-reporting jobs, let us come straight to laws. For though assertions that mentioned individuals have capacities, liabilities, tendencies, and the rest are not themselves statements of laws, they have features which can best be brought out after some peculiarities of law sentences have been discussed.

Laws are often stated in grammatically uncomplex indicative sentences, but they can also be stated in, among other constructions, hypothetical sentences of such patterns as 'Whatever is so and so, is such and such' or 'If a body is left unsupported, it falls at such and such a rate of acceleration'. We do not call a hypothetical sentence a 'law', unless it is a 'variable' or 'open' hypothetical statement, i.e., one of which the protasis can embody at least one expression like 'any' or 'whenever'. It is in virtue of this feature that a law applies to instances, though its statement does not mention them. If I know that any pendulum that is longer by any amount than any other pendulum will swing slower than the shorter pendulum by an amount proportional to its excess length, then on finding a particular pendulum three inches longer than another particular pendulum, I can infer how much slower it will swing. Knowing the law does not involve already having found these two pendulums; the statement of the law does not embody a report of their existence. On the other hand, knowing or even understanding the law does involve knowing that there could be particular matters of fact satisfying the protasis and therefore also satisfying the apodosis of the law. We have to learn to use statements of particular matters of fact, before we can learn to use the law-statements which do or might apply to them. Law-statements belong to a different and more sophisticated level of discourse from that, or those, to which belong the statements of the facts that satisfy them. Algebraical statements are in a similar way on a different level of discourse from the arithmetical statements which satisfy them.

Law-statements are true or false but they do not state truths or falsehoods of the same type as those asserted by the statements of fact to which they apply or are supposed to apply. They have different jobs. The crucial difference can be brought out in this way. At least part of the point of trying to establish laws is to find out how to infer from particular matters of fact to other particular matters of fact, how to explain particular matters of fact by reference to other matters of fact, and how to bring about or prevent particular states of affairs. A law is used as, so to speak, an inference-ticket (a season ticket) which licenses its possessors to move from asserting factual statements to asserting other factual statements. It also licenses them to provide explanations of given facts and to bring about desired states of affairs by manipulating what is found existing or happening. Indeed we should not admit that a student has learned a law, if all he were prepared to do were to recite it. Just as a student, to qualify as knowing rules of grammar, multiplication, chess, or etiquette, must be able and ready to apply these rules in concrete operations, so, to qualify as knowing a law, he must be able and ready to apply it in making concrete inferences from and to particular matters of fact, in explaining them and, perhaps also, in bringing them about, or preventing them. Teaching a law is, at least *inter alia*, teaching how to do new things, theoretical and practical, with particular matters of fact.

It is sometimes urged that if we discover a law, which enables us to infer from diseases of certain sorts to the existence of bacteria of certain sorts, then we have discovered a new existence, namely a causal connection between such bacteria and such diseases; and that consequently we now know, what we did not know before, that there exist not only diseased persons and bacteria, but also an invisible and intangible bond between them. As trains cannot travel, unless there exist rails for them to travel on, so, it is alleged, bacteriologists cannot move from the clinical observation of patients to the prediction of microscopic observations of bacteria, unless there exists, though it can never be observed, an actual tie between the objects of these observations.

Now there is no objection to employing the familiar idiom 'causal connection'. Bacteriologists do discover causal connections between bacteria and diseases, since this is only another way of saying that they do establish laws and so provide themselves with inference-tickets which enable them to infer from diseases to bacteria, explain diseases by assertions about bacteria, prevent and cure diseases by eliminating bacteria, and so forth. But to speak as if the discovery of a law were the finding of a third, unobservable existence is simply to fall back into the old habit of construing open hypothetical statements as singular categorical statements. It is like saying that a rule of grammar is a sort of extra but unspoken noun or verb, or that a rule of chess is a sort of extra but invisible chessman. It is to fall back into the old habit of assuming that all sorts of sentences do the same sort of job, the job, namely, of ascribing a predicate to a mentioned object.

The favorite metaphor 'the rails of inference' is misleading in just this way. Railway lines exist in just the same sense that trains exist, and we discover that rails exist in just the way that we discover that trains exist. The assertion that trains run from one place to another does imply that a set of observable rails exists between the two places. So to speak of the 'rails of inference' suggests that inferring from diseases to bacteria is really not inferring at all, but describing a third entity; not *arguing* 'because so and so, therefore such and such', but *reporting* 'there exists an unobserved bond between this observed so and so and that observed such and such'. But if we then ask 'What is this third, unobserved entity postulated for?' the only answer given is 'to warrant us in arguing from diseases to bacteria'. The legitimacy of the inference is assumed all the time. What is gratuitously desiderated is a story that shall seem to reduce 'therefore' sentences and 'if any . . .' sentences to sentences of the pattern 'Here is a . . .'; i.e., of obliterating the functional differences between arguments and narratives. But much as railway tickets cannot be 'reduced' to queer counterparts of the railway journeys that they make possible, and much as railway journeys cannot be 'reduced' to queer counterparts of the railway stations at which they start and finish, so law-statements cannot be 'reduced' to counterparts of the inferences and explanations that they license, and inferences and explanations cannot be 'reduced' to counterparts of the factual statements that constitute their termini. The sentence-job of stating facts is different from the job of stating an argument from factual statement to factual statement, and both are different from the job of giving warrants for such arguments. We have to learn to use sentences for the first job before we can learn to use them for the second, and we have to learn to use them for the first and the second jobs before we can learn to use them for the third. There are, of course, plenty of other sentence-jobs, which it is not our present business to consider. For example, the sentences which occupy these pages have not got any of the jobs which they have been describing.

We can now come back to consider dispositional statements, namely statements to the effect that a mentioned thing, beast, or person, has a certain capacity, tendency, or propensity, or is subject to a certain liability. It is clear that such statements are not laws, for they mention particular things or persons. On the other hand they resemble laws in being partly 'variable' or 'open'. To say that this lump of sugar is soluble is to say that it would dissolve, if submerged anywhere, at any time and in any parcel of water. To say that this sleeper knows French, is to say that if, for example, he is ever addressed in French, or shown any French newspaper, he responds pertinently in French, acts appropriately, or translates it correctly into his own tongue. This is, of course, too precise. We should not withdraw our statement that he knows French on finding that he did not respond pertinently when asleep, absent-minded, drunk, or in a panic; or on finding that he did not correctly translate highly technical treatises. We expect no more than that he will ordinarily cope pretty well with the majority of ordinary French-using and French-following tasks. 'Knows French' is a vague expression and, for most purposes, none the less useful for being vague.

The suggestion has been made that dispositional statements about mentioned individuals, while not themselves laws, are deductions from laws, so that we have to learn some perhaps crude and vague laws before we can make such dispositional statements. But in general the learning process goes the other way. We learn to make a number of dispositional statements about individuals before we learn laws stating general correlations between such statements. We find that some individuals are both oviparous and feathered, before we learn that any individual that is feathered is oviparous.

Dispositional statements about particular things and persons are also like law-statements in the fact that we use them in a partly similar way. They apply to, or they are satisfied by, the actions, reactions, and states of the object; they are inference-tickets, which license us to predict, retrodict, explain, and modify these actions, reactions, and states.

Naturally, the addicts of the superstition that all true indicative sentences either describe existents or report occurrences will demand that sentences such as 'this wire conducts electricity', or 'John Doe knows French', shall be construed as conveying factual information of the same type as that conveyed by 'this wire is conducting electricity' and 'John Doe is speaking French'. How could the statements be true unless there were something now going on, even though going on, unfortunately, behind the scenes? Yet they have to agree that we do often know that a wire conducts electricity and that individuals know French, without having first discovered any undiscoverable goings on. They have to concede, too, that the theoretical utility of discovering these hidden goings on would consist only in its entitling us to do just that predicting, explaining, and modifying which we already do and often know that we are entitled to do. They would have to admit, finally, that these postulated processes are themselves, at the best, things the existence of which they themselves infer from the fact that we can predict, explain, and modify the observable actions and reactions of individuals. But if they demand actual 'rails' where ordinary inferences are made, they will have to provide some further actual 'rails' to justify their own peculiar inference from the legitimacy of ordinary inferences to the 'rails' which they postulate to carry them. The postulation of such an endless hierarchy of 'rails' could hardly be attractive even to those who are attracted by its first step.

Dispositional statements are neither reports of observed or observable states of affairs, nor yet reports of unobserved or unobservable states of affairs. They narrate no incidents. But their jobs are intimately connected with narratives of incidents, for, if they are true, they are satisfied by narrated incidents. 'John Doe has just been telephoning in French' satisfies what is asserted by 'John Doe knows French', and a person who has found out that John Doe knows French perfectly needs no further ticket to enable him to argue from his having read a telegram in French to his having made sense of it. Knowing that John Doe knows French is being in possession of that ticket, and expecting him to understand this telegram is traveling with it.

It should be noticed that there is no incompatibility in saying that dispositional statements narrate no incidents and allowing the patent fact that dispositional statements can have tenses. 'He was a cigarette-smoker for a year' and 'the rubber began to lose its elasticity last summer' are perfectly legitimate dispositional statements; and if it were never true that an individual might be going to know something, there could exist no teaching profession. There can be short-term, long-term, or termless inference-tickets. A rule of cricket might be in force only for an experimental period, and even the climate of a continent might change from epoch to epoch.

(3) MENTAL CAPACITIES AND TENDENCIES

There is at our disposal an indefinitely wide range of dispositional terms for talking about things, living creatures and human beings. Some of these can be applied indifferently to all sorts of things; for example, some pieces of metal, some fishes, and some human beings weigh 140 lbs., are elastic and combustible, and all of them, if left unsupported, fall at the same rate of acceleration. Other dispositional terms can be applied only to certain kinds of things; 'hibernates', for example, can be applied with truth or falsity only to living creatures, and 'Tory' can be applied with truth or falsity only to nonidiotic, non-infantile, nonbarbarous human beings. Our concern is with a restricted class of dispositional terms, namely those appropriate only to the characterization of human beings. Indeed, the class we are concerned with is narrower than that, since we are concerned only with those which are appropriate to the characterization of such stretches of human behavior as exhibit qualities of intellect and character. We are not, for example, concerned with any mere reflexes which may happen to be peculiar to men, or with any pieces of physiological equipment which happen to be peculiar to human anatomy.

Of course, the edges of this restriction are blurred. Dogs as well as infants are drilled to respond to words of command, to pointing, and to the ringing of dinner bells; apes learn to use and even construct instruments; kittens are playful and parrots are imitative. If we like to say that the behavior of animals is instinctive while part of the behavior of human beings is rational, though we are drawing attention to an important difference or family of differences, it is a difference the edges of which are, in their turn, blurred. Exactly when does the instinctive imitativeness of the infant develop into rational histrionics? By which birthday has the child ceased ever to respond to the dinner bell like a dog and begun always to respond to it like an angel? Exactly where is the boundary line between the suburb and the country?

Since this book as a whole is a discussion of the logical behavior of some of the cardinal terms, dispositional and occurrent, in which we talk about minds, all that is necessary in this section is to indicate some general differences between the uses of some of our selected dispositional terms. No attempt is made to discuss all these terms, or even all of the types of these terms.

Many dispositional statements may be, though they need not be, and ordinarily are not, expressed with the help of the words 'can', 'could', and 'able'. 'He is a swimmer', when it does not signify that he is an expert, means merely that he can swim. But the words 'can' and 'able' are used in lots of different ways, as can be illustrated by the following examples. 'Stones can float (for pumice-stone floats)'; 'that fish can swim (for it is not disabled, although it is now inert in the mud)'; 'John Doe can swim (for he has learned and not forgotten)'; 'Richard Roe can swim (if he is willing to learn)'; 'you can swim (when you try hard)'; 'she can swim (for the doctor has withdrawn his veto)', and so on. The first example states that there is no license to infer that because this is a stone, it will not float; the second denies the existence of a physical impediment; the last asserts the cessation of a disciplinary impediment. The third, fourth, and fifth statements are informative about personal qualities, and they give different sorts of information.

To bring out the different forces of some of these different uses of 'can' and 'able', it is convenient to make a brief disquisition on the logic of what are sometimes called the 'modal words', such as 'can', 'must', 'may', 'is necessarily', 'is not necessarily', and 'is not necessarily not'. A statement to the effect that something must be, or is necessarily, the case functions as what I have called an 'inference-ticket'; it licenses the inference to the thing's being the case from something else which may or may not be specified in the statement. When the statement is to the effect that something is necessarily not, or cannot be, the case, it functions as a license to infer to its not being the case. Now sometimes it is required to refuse such a license to infer that something is not the case, and we commonly word this refusal by saying that it can be the case, or that it is possibly the case. To say that something can be the case does not entail that it is the case, or that it is not the case, or, of course, that it is in suspense between being and not being the case, but only that there is no license to infer from something else, specified or unspecified, to its not being the case.

This general account also covers most 'if-then' sentences. An 'if-then' sentence can nearly always be paraphrased by a sentence containing a modal expression, and *vice versa*. Modal and hypothetical sentences have the same force. Take any ordinary 'if-then' sentence, such as 'if I walk under that ladder, I shall meet trouble during the day', and consider how we should colloquially express its contradictory. It will not do to attach a 'not' to the protasis verb, to the apodosis verb, or to both at once, for the results of all three operations would be equally superstitious statements. It would do, but it would not be convenient or colloquial to say 'No, it is not the case that if I walk under a ladder I shall have trouble'. We should ordinarily reject the superstition by saying 'No, I might walk under the ladder and not have trouble' or 'I could walk under it without having trouble' or, to generalize the rejection, 'trouble does not necessarily come to people who walk under ladders'. Conversely the original superstitious statement could have been worded 'I could not walk under a ladder without experiencing trouble during the day'. There is only a stylistic difference between the 'if-then' idiom and the modal idioms.

It must, however, not be forgotten that there are other uses of 'if', 'must', and 'can' where this equivalence does not hold. 'If' sometimes means 'even though'. It is also often used in giving conditional undertakings, threats, and wagers. 'Can' and 'must' are sometimes used as vehicles of nontheoretical permissions, orders, and vetoes. True, there are similarities between giving or refusing licenses to infer and giving or refusing licenses to do other things, but there are big differences as well. We do not, for instance, naturally describe as true or false the doctor's ruling 'the patient must stay in bed, can dictate letters, but must not smoke'; whereas it is quite natural to describe as true or false such sentences as 'a syllogism can have two universal premisses', 'whales cannot live without surfacing from time to time', 'a freely falling body must be accelerating', and 'people who walk under ladders need not come to disaster during the day'. The ethical uses of 'must', 'may', and 'may not' have affinities with both. We are ready to discuss the truth of ethical statements embodying such words, but the point of making such statements is to regulate parts of people's conduct, other than their inferences. In having both these features they resemble the treatment recommendations given to doctors by their medical textbooks, rather than the regimen-instructions given by doctors to their patients. Ethical statements, as distinct from particular *ad hominem* behests and reproaches, should be regarded as warrants addressed to any potential givers of behests and reproaches, and not to the actual addresses of such behests and reproaches, i.e., not as personal action-tickets but as impersonal injunction-tickets; not imperatives but 'laws' that only such things as imperatives and punishments can satisfy. Like statute laws they are to be construed not as orders, but as licenses to give and enforce orders.

We may now return from this general discussion of the sorts of jobs performed by modal sentences to consider certain specific differences between a few selected 'can' sentences, used for describing personal qualities.

To say that John Doe can swim differs from saying of a puppy that it can swim. For to say that the puppy can swim is compatible with saying that it has never been taught to swim, or had practice in swimming, whereas to say that a person can swim implies that he has learned to swim and has not forgotten. The capacity to acquire capacities by being taught is not indeed a human peculiarity. The puppy can be taught or drilled to beg, much as infants are taught to walk and use spoons. But some kinds of learning, including the way in which most people learn to swim, involve the understanding and application either of spoken instructions or at least of staged demonstrations; and a creature that can learn things in these ways is unhesitatingly conceded to have a mind, where the teachability of the dog and infant leaves us hesitant whether or not to say that they yet qualify for this certificate.

To say that Richard Roe can swim (for he can learn to swim) is to say that he is competent to follow and apply such instructions and demonstrations, though he may not yet have begun to do so. It would be wrong to predict about him, what it would be right to predict about an idiot, that since he now flounders helplessly in the water, he will still flounder helplessly after he has been given tuition.

To say that you can swim (if you try) is to use an interesting intermediate sort of 'can'. Whereas John Doe does not now have to try to swim, and Richard Roe cannot yet swim, however hard he tries, you know what to do but only do it when you apply your whole mind to the task. You have understood the instructions and demonstrations, but still have to give yourself practice in the application of them. This learning to apply instructions by deliberate and perhaps difficult and alarming practice is something else which we regard as peculiar to creatures with minds. It exhibits qualities of character, though qualities of a different order from those exhibited by the puppy that shows tenacity and courage even in its play, since the novice is making himself do something difficult and alarming with the intention to develop his capacities. To say that he can swim if he tries is, therefore, to say both that he can understand instructions and also that he can intentionally drill himself in applying them.

It is not difficult to think of many other uses of 'can' and 'able'. In 'John Doe has been able to swim since he was a boy, but now he can invent new strokes', we have one such use. 'Can invent' does not mean 'has learned and not forgotten how to invent'. Nor is it at all like the 'can' in 'can sneeze'. Again the 'can' in 'can defeat all but champion swimmers' does not have the same force as either that in 'can swim' or that in 'can invent'. It is a 'can' which applies to race horses.

There is one further feature of 'can' which is of special pertinence to our central theme. We often say of a person, or of a performing animal, that he can do something, in the sense that he can do it correctly or well. To say that a child can spell a word is to say that he can give, not merely some collection or other of letters, but the right collection in the right order. To say that he can tie a reef-knot is to say not merely that when he plays with bits of string, sometimes reef-knots and sometimes granny-knots are produced, but that reef-knots are produced whenever, or nearly whenever, reef-knots are required, or at least that they are nearly always produced when required and when the child is trying. When we use, as we often do use, the phrase 'can tell' as a paraphrase of 'know', we mean by 'tell', 'tell correctly'. We do not say that a child can tell the time, when all that he does is deliver random time-of-day statements, but only when he regularly reports the time of day in conformity with the position of the hands of the clock, or with the position of the sun, whatever these positions may be.

Many of the performance-verbs with which we describe people and, sometimes with qualms, animals, signify the occurrence not just of actions but of suitable or correct actions. They signify achievements. Verbs like 'spell', 'catch', 'solve', 'find', 'win', 'cure', 'score', 'deceive', 'persuade', 'arrive', and countless others signify not merely that some performance has been gone through, but also that something has been brought off by the agent going through it. They are verbs of success. Now successes are sometimes due to luck; a cricketer may score a boundary by making a careless stroke. But when we say of a person that he can bring off things of a certain sort, such as solve anagrams or cure sciatica, we mean that he can be relied on to succeed reasonably often even without the aid of luck. He knows how to bring it off in normal situations.

We also use corresponding verbs of failure, like 'miss', 'misspell', 'drop', 'lose', 'foozle', and 'miscalculate'. It is an important fact that if a person can spell or calculate, it must also be possible for him to misspell and miscalculate; but the sense of 'can' in 'can spell' and 'can calculate' is quite different from its sense in 'can misspell' and 'can miscalculate'. The one is a competence, the other is not another competence but a liability. For certain purposes it is also necessary to notice the further difference between both these senses of 'can' and the sense in which it is true to say that a person cannot solve an anagram incorrectly, win a race unsuccessfully, find a treasure unavailingly, or prove a theorem invalidly. For this 'cannot' is a logical 'cannot'. It says nothing about people's competences or limitations, but only that, for instance, 'solve incorrectly' is a self-contradictory expression. We shall set later that the epistemologist's hankering for some incorrigible sort of observation derives partly from his failure to notice that in one of its senses 'observe' is a verb of success, so that in this sense, 'mistaker observation' is as self-contradictory an expression as 'invalid proof', or 'unsuccessful cure'. But just as 'invalid argument' and 'unsuccessful treatment' are logically permissible expressions, so 'inefficient' or 'unavailing observation' is a permissible expression, when 'observe' is used not as a 'find' verb but as a 'hunt' verb.

Enough has been said to show that there is a wide variety of types of 'can' words, and that within this class there is a wide variety of types of capacity-expressions and liability-expressions. Only some of these capacity-expressions and liability-expressions are peculiar to the description of human beings, but even of these there are various types.

Tendencies are different from capacities and liabilities. 'Would if . . .' differs from 'could'; and 'regularly does . . . when . . .' differs from 'can'. Roughly, to say 'can' is to say that it is not a certainty that something will not be the case, while, to say 'tends', 'keeps on', or 'is prone', is to say that it is a good bet that it will be, or was, the case. So 'tends to' implies 'can', but is not implied by it. 'Fido tends to howl when the moon shines' says more than 'it is not true that if the moon shines, Fido is silent'. It licenses the hearer not only not to rely on his silence, but positively to expect barking.

But there are lots of types of tendency. Fido's tendency to get mange in the summer (unless specially dieted) is not the same sort of thing as his tendency to bark when the moon shines (unless his master is gruff with him). A person's blinking at fairly regular intervals is a different sort of tendency from his way of flickering his eyelids when embarrassed. We might call the latter, what we should not call the former, a 'mannerism'.

We distinguish between some behavior tendencies and some others by calling some of them 'pure habits', others of them 'tastes', 'interests', 'bents', and 'hobbies', and yet others of them 'jobs' and 'occupations'. It might be a pure habit to draw on the right sock before the left sock, a hobby to go fishing when work and weather permit, and a job to drive lorries. It is, of course, easy to think of borderline cases of regular behavior which we might hesitate to classify; some people's jobs are their hobbies and some people's jobs and hobbies are nearly pure habits. But we are fairly clear about the distinctions between the concepts themselves. An action done from pure habit is one that is not done on purpose and is one that the agent need not be able to report having done even immediately after having done it; his mind may have been on something else. Actions performed as parts of a person's job may be done by pure habit; still, he does not perform them when not on the job. The soldier does not march, when home on leave, but only when he knows that he has got to march, or ought to march. He resumes and drops the habit when he puts on and takes off his uniform.

Exercises of hobbies, interests, and tastes are performed, as we say, 'for pleasure'. But this phrase can be misleading, since it suggests that these exercises are performed as a sort of investment from which a dividend is anticipated. The truth is the reverse, namely that we do these things because we like doing them, or want to do them, and not because we like or want something accessory to them. We invest our capital reluctantly in the hope of getting dividends which will make the outlay worthwhile, and if we were offered the chance of getting the dividends without investing the capital, we should gladly abstain from making the outlay. But the angler would not accept or understand an offer of the pleasures without the activities of angling. It is angling that he enjoys, not something that angling engenders.

To say that someone is now enjoying or disliking something

entails that he is paying heed to it. There would be a contradiction in saying that the music pleased him though he was
paying no attention to what he heard. There would, of course,
be no contradiction in saying that he was listening to the music
but neither enjoying nor disliking it. Accordingly, to say that
someone is fond of or keen on angling entails not merely that
he tends to wield his rod by the river when he is not forced or
obliged not to do so, but that he tends to do so with his mind
on it, that he tends to be wrapped up in daydreams and memories of angling, and to be absorbed in conversations and books
on the subject. But this is not the whole story. A conscientious
reporter tends to listen intently to the words of public speakers,
even though he would not do this, if he were not obliged to do
it. He does not do it when off duty. In these hours he is, perhaps,
wont to devote himself to angling. He does not have to try to
concentrate on fishing as he has to try to concentrate on
speeches. He concentrates without trying. This is a large part of
what 'keen on' means.

Besides pure habits, jobs, and interests there are many other
types of higher level tendencies. Some behavior regularities are
adherences to resolutions or policies imposed by the agent on
himself; some are adherences to codes or religions inculcated
into him by others. Addictions, ambitions, missions, loyalties,
devotions, and chronic negligences are all behavior tendencies,
but they are tendencies of very different kinds.

Two illustrations may serve to bring out some of the differences between capacities and tendencies, or between competences and pronenesses. (*a*) Both skills and inclinations can
be simulated, but we use abusive names like 'charlatan' and
'quack' for the frauds who pretend to be able to bring things off,
while we use the abusive word 'hypocrite' for the frauds who
affect motives and habits. (*b*) Epistemologists are apt to perplex
themselves and their readers over the distinction between knowledge and belief. Some of them suggest that these differ only in
degree of something or other, and some that they differ in the
presence of some introspectible ingredient in knowing which is
absent from believing, or *vice versa*. Part of this embarrassment

is due to their supposing that 'know' and 'believe' signify occurrences, but even when it is seen that both are dispositional
verbs, it has still to be seen that they are dispositional verbs of
quite disparate types. 'Know' is a capacity verb, and a capacity
verb of that special sort that is used for signifying that the
person described can bring things off, or get things right. 'Believe', on the other hand, is a tendency verb and one which does
not connote that anything is brought off or got right. 'Belief' can
be qualified by such adjectives as 'obstinate', 'wavering', 'un-
swerving', 'unconquerable', 'stupid', 'fanatical', 'whole-hearted',
'intermittent', 'passionate', and 'childlike', adjectives some or all
of which are also appropriate to such nouns as 'trust', 'loyalty',
'bent', 'aversion', 'hope', 'habit', 'zeal', and 'addiction'. Beliefs,
like habits, can be inveterate, slipped into and given up; like
partisanships, devotions, and hopes they can be blind and
obsessing; like aversions and phobias they can be unacknowl-
edged; like fashions and tastes they can be contagious; like
loyalties and animosities they can be induced by tricks. A
person can be urged or entreated not to believe things, and he
may try, with or without success, to cease to do so. Sometimes a
person says truly 'I cannot help believing so and so'. But none
of these dictions, or their negatives, are applicable to knowing,
since to know is to be equipped to get something right and not
to tend to act or react in certain manners.

Roughly, 'believe' is of the same family as motive words,
where 'know' is of the same family as skill words; so we ask how
a person knows this, but only why a person believes that, as
we ask how a person ties a clove-hitch, but why he wants to tie
a clove-hitch, or why he always ties granny-knots. Skills have
methods, where habits and inclinations have sources. Similarly,
we ask what makes people believe or dread things but not what
makes them know or achieve things.

Of course, belief and knowledge (when it is knowledge *that*)
operate, to put it crudely, in the same field. The sorts of things
that can be described as known or unknown can also be described as believed or disbelieved, somewhat as the sort of
things that can be manufactured are also the sorts of things that

can be exported. A man who believes that the ice is dangerously thin gives warnings, skates warily, and replies to pertinent questions in the same ways as the man who knows that it is dangerously thin; and if asked whether he knows it for a fact, he may unhesitatingly claim to do so, until embarrassed by the question how he found it out.

Belief might be said to be like knowledge and unlike trust in persons, zeal for causes, or addiction to smoking, in that it is 'propositional'; but this, though not far wrong, is too narrow. Certainly to believe that the ice is dangerously thin is to be unhesitant in telling oneself and others that it is thin, in acquiescing in other people's assertions to that effect, in objecting to statements to the contrary, in drawing consequences from the original proposition, and so forth. But it is also to be prone to skate warily, to shudder, to dwell in imagination on possible disasters and to warn other skaters. It is a propensity not only to make certain theoretical moves but also to make certain executive and imaginative moves, as well as to have certain feelings. But these things hang together on a common propositional hook. The phrase 'thin ice' would occur in the descriptions alike of the shudders, the warnings, the wary skating, the declarations, the inferences, the acquiescences, and the objections.

A person who knows that the ice is thin, and also cares whether it is thin or thick, will, of course, be apt to act and react in these ways too. But to say that he keeps to the edge, because he knows that the ice is thin, is to employ quite a different sense of 'because', or to give quite a different sort of 'explanation', from that conveyed by saying that he keeps to the edge because he believes that the ice is thin.

PART III

Intentionality, Thought, and Language

11

Empiricism and the Philosophy of Mind

WILFRID SELLARS

I. THOUGHTS: THE CLASSICAL VIEW

1. Recent empiricism has been of two minds about the status of *thoughts*. On the one hand, it has resonated to the idea that insofar as there are *episodes* which are thoughts, they are *verbal* or *linguistic* episodes. Clearly, however, even if candid overt verbal behaviors by people who had learned a language *were* thoughts, there are not nearly enough of them to account for all the cases in which it would be argued that a person was thinking. Nor can we plausibly suppose that the remainder is accounted for by those inner episodes which are often very clumsily lumped together under the heading "verbal imagery."

On the other hand, they have been tempted to suppose that the *episodes* which are referred to by verbs pertaining to thinking include all forms of "intelligent behavior," verbal as well as nonverbal, and that the "thought episodes" which are supposed to be manifested by these behaviors are not really episodes at all, but rather hypothetical and mongrel hypothetical-categorical facts about these and still other behaviors. This, however, runs into the difficulty that whenever we try to explain what we mean by calling a piece of *nonhabitual* behavior intelligent, we seem

From *Minnesota Studies in the Philosophy of Science*, vol. I, eds. Herbert Feigl and Michael Scriven, secs. 46–59 (pp. 307–321). Copyright, 1956, by the University of Minnesota. Reprinted by permission of the author and the University of Minnesota Press. Sections and parts have been renumbered from 1 to 14 and from I to V respectively.

to find it necessary to do so in terms of *thinking*. The uncomfortable feeling will not be downed that the dispositional account of thoughts in terms of intelligent behavior is covertly circular.

2. Now the classical tradition claimed that there is a family of episodes, neither overt verbal behavior nor verbal imagery, which are *thoughts*, and that both overt verbal behavior and verbal imagery owe their meaningfulness to the fact that they stand to these *thoughts* in the unique relation of "expressing" them. These episodes are introspectable. Indeed, it was usually believed that they could not occur without being known to occur. But this can be traced to a number of confusions, perhaps the most important of which was the idea that *thoughts* belong in the same general category as sensations, images, tickles, itches, etc. This misassimilation of thoughts to sensations and feelings was equally, as we saw in Sections 26 ff. above,[1] a misassimilation of sensations and feelings to thoughts, and a falsification of both. The assumption that if there are thought episodes, they must be immediate experiences is common both to those who propounded the classical view and to those who reject it, saying that they "find no such experiences." If we purge the classical tradition of these confusions, it becomes the idea that to each of us belongs a stream of episodes, not themselves immediate experiences, to which we have privileged, but by no means either invariable or infallible, access. These episodes can occur without being "expressed" by overt verbal behavior, though verbal behavior is—in an important sense—their natural fruition. Again, we can "hear ourselves think," but the verbal imagery which enables us to do this is no more the thinking itself than is the overt verbal behavior by which it is expressed and communicated to others. It is a mistake to suppose that we must be having verbal imagery— indeed, any imagery—when we "know what we are thinking"— in short, to suppose that "privileged access" must be construed on a perceptual or quasi-perceptual model.

---

[1] This and following sectional references in the text are to the unedited version of this essay in *Minnesota Studies in the Philosophy of Science*, vol. I.

Now, it is my purpose to defend such a revised classical analysis of our commonsense conception of thoughts, and in the course of doing so I shall develop distinctions which will later contribute to a resolution, in principle, of the puzzle of *immediate experience*. But before I continue, let me hasten to add that it will turn out that the view I am about to expound could, with equal appropriateness, be represented as a modified form of the view that thoughts are *linguistic* episodes.

II. OUR RYLEAN ANCESTORS

3. But, the reader may well ask, in what sense can these episodes be "inner" if they are not immediate experiences? and in what sense can they be "linguistic" if they are neither overt linguistic performances, nor verbal imagery *"in foro interno"*? I am going to answer these and the other questions I have been raising by making a myth of my own, or, to give it an air of up-to-date respectability, by writing a piece of science fiction—anthropological science fiction. Imagine a stage in prehistory in which humans are limited to what I shall call a Rylean language, a language of which the fundamental descriptive vocabulary speaks of public properties of public objects located in space and enduring through time. Let me hasten to add that it is also Rylean in that although its basic resources are limited (how limited I shall be discussing in a moment), its total expressive power is very great. For it makes subtle use not only of the elementary logical operations of conjunction, disjunction, negation, and quantification, but especially of the subjunctive conditional. Furthermore, I shall suppose it to be characterized by the presence of the looser logical relations typical of ordinary discourse which are referred to by philosophers under the headings "vagueness" and "open texture."

I am beginning my myth *in medias res* with humans who have already mastered a Rylean language, because the philosophical situation it is designed to clarify is one in which we are not puzzled by how people acquire a language for referring to public properties of public objects, but are very puzzled indeed about how we learn to speak of inner episodes and immediate experiences.

There are, I suppose, still some philosophers who are inclined to think that by allowing these mythical ancestors of ours the use *ad libitum* of subjunctive conditionals, we have, in effect, enabled them to say anything that we can say when we speak of *thoughts, experiences* (seeing, hearing, etc.), and *immediate experiences*. I doubt that there are many. In any case, the story I am telling is designed to show exactly *how* the idea that an intersubjective language *must* be Rylean rests on too simple a picture of the relation of intersubjective discourse to public objects.

4. The questions I am, in effect, raising are "What resources would have to be added to the Rylean language of these talking animals in order that they might come to recognize each other and themselves as animals that *think, observe,* and have *feelings* and *sensations,* as we use these terms?" and "How could the addition of these resources be construed as reasonable?" In the first place, the language would have to be enriched with the fundamental resources of semantical discourse—that is to say, the resources necessary for making such characteristically semantical statements as " 'Rot' means red," and " 'Der Mond ist rund' is true if and only if the moon is round." It is sometimes said, e.g., by Carnap,[2] that these resources can be constructed out of the vocabulary of formal logic, and that they would therefore already be contained, in principle, in our Rylean language. I have criticized this idea in another place[3] and shall not discuss it here. In any event, a decision on this point is not essential to the argument.

Let it be granted, then, that these mythical ancestors of ours are able to characterize each other's verbal behavior in semantical terms; that, in other words, they not only can talk about each other's predictions as causes and effects, and as

[2] Rudolf Carnap, *Introduction to Semantics* (Chicago: Univ. of Chicago Press, 1942).

[3] Wilfrid Sellars, "Empiricism and Abstract Entities," in Paul A. Schilpp, ed., *The Philosophy of Rudolf Carnap* (La Salle, Ill.: Open Court, 1963).

indicators (with greater or less reliability) of other verbal and nonverbal states of affairs, but can also say of these verbal productions that they *mean* thus and so, that they say *that* such and such, that they are true, false, etc. And let me emphasize that to make a semantical statement about a verbal event is not a shorthand way of talking about its causes and effects, although there is a sense of "imply" in which semantical statements about verbal productions do *imply* information about the causes and effects of these productions. Thus, when I say " '*Es regnet*' means it is raining," my statement "implies" that the causes and effects of utterances of "*Es regnet*" beyond the Rhine parallel the causes and effects of utterances of "It is raining" by myself and other members of the English-speaking community. And if it didn't imply this, it couldn't perform its role. But this is not to say that semantical statements are definitional shorthand for statements about the causes and effects of verbal performances.

5. With the resources of semantical discourse, the language of our fictional ancestors has acquired a dimension which gives considerably more plausibility to the claim that they are in a position to talk about *thoughts* just as we are. For characteristic of thoughts is their *intentionality, reference,* or *aboutness,* and it is clear that semantical talk about the meaning or reference of verbal expressions has the same structure as mentalistic discourse concerning what thoughts are about. It is therefore all the more tempting to suppose that the intentionality of *thoughts* can be traced to the application of semantical categories to overt verbal performances, and to suggest a modified Rylean account according to which talk about so-called "thoughts" is shorthand for hypothetical and mongrel categorical-hypothetical statements about overt verbal and nonverbal behavior, and that talk about the *intentionality* of these "episodes" is correspondingly reducible to semantical talk about the verbal components.

What is the alternative? Classically it has been the idea that not only are there overt verbal episodes which can be characterized in semantical terms, but, *over* and *above* these, there are certain inner episodes which are properly characterized by the traditional vocabulary of *intentionality*. And, of course, the classical scheme includes the idea that semantical discourse about overt verbal performances is to be analyzed in terms of talk about the intentionality of the mental episodes which are "expressed" by these overt performances. My immediate problem is to see if I can reconcile the classical idea of thoughts as inner episodes which are neither overt behavior nor verbal imagery and which are properly referred to in terms of the vocabulary of intentionality, with the idea that the categories of intentionality are, at bottom, semantical categories pertaining to overt verbal performances.[4]

III. THEORIES AND MODELS

6. But what might these episodes be? And, in terms of our science fiction, how might our ancestors have come to recognize their existence? The answer to these questions is surprisingly straightforward, once the logical space of our discussion is enlarged to include a distinction, central to the philosophy of science, between the language of *theory* and the language of *observation*. Although this distinction is a familiar one, I shall take a few paragraphs to highlight those aspects of the distinction which are of greatest relevance to our problem.

Informally, to construct a theory is, in its most developed or sophisticated form, to postulate a domain of entities which behave in certain ways set down by the fundamental principles of the theory, and to correlate—perhaps, in a certain sense to identify—complexes of these theoretical entities with certain nontheoretical objects or situations; that is to say, with objects or situations which are either matters of observable fact or, in principle at least, describable in observational terms. This "correlation" or "identification" of theoretical with observational states of affairs is a tentative one "until further notice,"

---

[4] An earlier attempt along these lines is to be found in Wilfrid Sellars, "Mind, Meaning and Behavior," *Philosophical Studies*, 3 (1952), 83–94, and "A Semantical Solution of the Mind-Body Problem," *Methodos*, 5(1953), 45–84.

and amounts, so to speak, to erecting temporary bridges which permit the passage from sentences in observational discourse to sentences in the theory, and vice versa. Thus, for example, in the kinetic theory of gases, empirical statements of the form "Gas g at such and such a place and time has such and such a volume, pressure, and temperature" are correlated with theoretical statements specifying certain statistical measures of populations of molecules. These temporary bridges are so set up that inductively established laws pertaining to gases, formulated in the language of observable fact, are correlated with derived propositions or theorems in the language of the theory, and that no proposition in the theory is correlated with a falsified empirical generalization. Thus, a good theory (at least of the type we are considering) "explains" established empirical laws by deriving theoretical counterparts of these laws from a small set of postulates relating to unobserved entities.

These remarks, of course, barely scratch the surface of the problem of the status of theories in scientific discourse. And no sooner have I made them, than I must hasten to qualify them —almost beyond recognition. For while this by now classical account of the nature of theories (one of the earlier formulations of which is due to Norman Campbell, and which is to be bound more recently in the writings of Carnap, Reichenbach, Hempel, and Braithwaite)[5] does throw light on the logical status of theories, it emphasizes certain features at the expense of others. By speaking of the construction of a theory as the elaboration of a postulate system which is tentatively correlated with observational discourse, it gives a highly artificial and unrealistic picture of what scientists have actually done in the process of

[5] Norman Campbell, *Physics: The Elements* (Cambridge: Cambridge Univ. Press, 1953); Rudolf Carnap, "The Interpretation of Physics," in H. Feigl and M. Brodbeck, eds., *Readings in the Philosophy of Science* (New York: Appleton-Century-Crofts, 1953), pp. 309–318 (this selection consists of pp. 59–69 of his *Foundations of Logic and Mathematics* [Chicago: Univ. of Chicago Press, 1939]); H. Reichenbach, *Philosophie der Raum-Zeit-Lehre* (Berlin: de Gruyter, 1928), and *Experience and Prediction* (Chicago: Univ. of Chicago Press, 1938); C.G. Hempel, *Fundamentals of Concept Formation in Empirical Science* (Chicago: Univ. of Chicago Press, 1938); R.B. Braithwaite, *Scientific Explanation* (Cambridge: Cambridge Univ. Press, 1953).

constructing theories. I don't wish to deny that logically sophisticated scientists today *might* and perhaps, on occasion, *do*
proceed in true logistical style. I do, however, wish to emphasize two points:

(1) The first is that the fundamental assumptions of a theory
are usually developed not by constructing uninterpreted calculi
which might correlate in the desired manner with observational
discourse, but rather by attempting to find a *model*, i.e., to
describe a domain of familiar objects behaving in familiar ways
such that we can see how the phenomena to be explained would
arise if they consisted of this sort of thing. The essential thing
about a model is that it is accompanied, so to speak, by a
commentary which *qualifies* or *limits*—but not precisely nor in
all respects—the analogy between the familiar objects and the
entities which are being introduced by the theory. It is the descriptions of the fundamental ways in which the objects in the
model domain, thus qualified, behave, which, transferred to the
theoretical entities, correspond to the postulates of the logistical
picture of theory construction.

(2) But even more important for our purposes is the fact
that the logistical picture of theory construction obscures the
most important thing of all, namely that the process of devising
"theoretical" explanations of observable phenomena did not
spring full-blown from the head of modern science. In particular,
it obscures the fact that not all commonsense inductive inferences are of the form

All observed A's have been B, *therefore* (*probably*) all A's are B,

or its statistical counterparts, and leads one mistakenly to suppose that so-called "hypothetic-deductive" explanation is limited
to the sophisticated stages of science. The truth of the matter, as
I shall shortly be illustrating, is that science is continuous with
commonsense, and the ways in which the scientist seeks to
explain empirical phenomena are refinements of the ways in
which plain men, however crudely and schematically, have attempted to understand their environment and their fellow men
since the dawn of intelligence. It is this point which I wish to

stress at the present time, for I am going to argue that the distinction between theoretical and observational discourse is involved in the logic of concepts pertaining to inner episodes. I say "involved in" for it would be paradoxical and, indeed, incorrect, to say that these concepts are theoretical concepts.

7. Now I think it fair to say that some light has already been thrown on the expression "inner episodes"; for while it would indeed be a category mistake to suppose that the inflammability of a piece of wood is, so to speak, a hidden burning which becomes overt or manifest when the wood is placed on the fire, not all the unobservable episodes we suppose to go on in the world are the offspring of category mistakes. Clearly it is by no means an illegitimate use of "in"—though it is a use which has its own logical grammar—to say, for example, that "in" the air around us there are innumerable molecules which, in spite of the observable stodginess of the air, are participating in a veritable turmoil of episodes. Clearly, the sense in which these episodes are "in" the air is to be explicated in terms of the sense in which the air "is" a population of molecules, and this, in turn, in terms of the logic of the relation between theoretical and observational discourse.

I shall have more to say on this topic in a moment. In the meantime, let us return to our mythical ancestors. It will not surprise my readers to learn that the second stage in the enrichment of their Rylean language is the addition of theoretical discourse. Thus we may suppose these language-using animals to elaborate, without methodological sophistication, crude, sketchy, and vague theories to explain why things which are similar in their observable properties differ in their causal properties, and things which are similar in their causal properties differ in their observable properties.

IV. METHODOLOGICAL *versus* PHILOSOPHICAL BEHAVIORISM

8. But we are approaching the time for the central episode in our myth. I want you to suppose that in this neo-Rylean culture

there now appears a genius—let us call him Jones—who is an unsung forerunner of the movement in psychology, once revolutionary, now commonplace, known as Behaviorism. Let me emphasize that what I have in mind is Behaviorism as a methodological thesis, which I shall be concerned to formulate. For the central and guiding theme in the historical complex known by this term has been a certain conception, or family of conceptions, of how to go about building a science of psychology.

Philosophers have sometimes supposed that Behaviorists are, as such, committed to the idea that our ordinary mentalistic concepts are *analyzable* in terms of overt behavior. But although behaviorism has often been characterized by a certain metaphysical bias, it is not a thesis about the *analysis* of *existing* psychological concepts, but one which concerns the construction of new concepts. As a methodological thesis, it involves no commitment whatever concerning the logical analysis of commonsense mentalistic discourse, nor does it involve a denial that each of us has a privileged access to our state of mind, nor that these states of mind can properly be described in terms of such commonsense concepts as believing, wondering, doubting, intending, wishing, inferring, etc. If we permit ourselves to speak of this privileged access to our states of mind as "introspection," avoiding the implication that there is a "means" whereby we "see" what is going on "inside," as we see external circumstances by the eye, then we can say that Behaviorism, as I shall use the term, does not deny that there is such a thing as introspection, nor that it is, on some topics, at least, quite reliable. The essential point about 'introspection' from the standpoint of Behaviorism is that *we introspect in terms of commonsense mentalistic concepts*. And while the Behaviorist admits, as anyone must, that much knowledge is embodied in commonsense mentalistic discourse, and that still more can be gained in the future by formulating and testing hypotheses in terms of them, and while he admits that it is perfectly legitimate to call such a psychology "scientific," he proposes, for his own part, to make no more than a heuristic use of mentalistic discourse, and to construct his concepts "from scratch" in the course of developing his own scientific account of the observable behavior of human organisms.

9. But while it is quite clear that scientific Behaviorism is *not* the thesis that commonsense psychological concepts are *analyzable* into concepts pertaining to overt behavior—a thesis which has been maintained by some philosophers and which may be called 'analytical' or 'philosophical' Behaviorism—it is often thought that Behaviorism is committed to the idea that the concepts of a Behavioristic psychology must be so analyzable, or, to put things right side up, that properly introduced behavioristic concepts must be built by explicit definition—in the broadest sense—from a basic vocabulary pertaining to overt behavior. The Behaviorist would thus be saying "Whether or not the mentalistic concepts of everyday life are definable in terms of overt behavior, I shall ensure that this is true of the concepts that I shall employ." And it must be confessed that many behavioristically oriented psychologists have believed themselves committed to this austere program of concept formation.

Now I think it reasonable to say that, *thus conceived,* the behavioristic program would be unduly restrictive. Certainly, nothing in the nature of sound scientific procedure requires this self-denial. Physics, the methodological sophistication of which has so impressed—indeed, overly impressed—the other sciences, does not lay down a corresponding restriction on its concepts, nor has chemistry been built in terms of concepts explicitly definable in terms of the observable properties and behavior of chemical substances. The point I am making should now be clear. The behavioristic requirement that all concepts should be *introduced* in terms of a basic vocabulary pertaining to overt behavior is compatible with the idea that some behavioristic concepts are to be introduced as *theoretical* concepts.

10. It is essential to note that the theoretical terms of a behavioristic psychology are not only *not* defined in terms of overt behavior, they are also *not* defined in terms of nerves, synapses, neural impulses, etc. A behavioristic theory of behavior is not, as such, a physiological explanation of behavior.

The ability of a framework of theoretical concepts and propositions successfully to explain behavioral phenomena is logically independent of the identification of these theoretical concepts with concepts of neurophysiology. What *is* true—and this is a logical point—is that each special science dealing with some aspect of the human organism operates within the frame of a certain regulative ideal, the ideal of a coherent system in which the achievements of each have an intelligible place. Thus, it is part of the Behaviorist's business to keep an eye on the total picture of the human organism which is beginning to emerge. And if the tendency to premature identification is held in check, there may be considerable heuristic value in speculative attempts at integration; though, until recently, at least, neurophysiological speculations in behavior theory have not been particularly fruitful. And while it is, I suppose, noncontroversial that when the total scientific picture of man and his behavior is in, it will involve *some* identification of concepts in behavior theory with concepts pertaining to the functioning of anatomical structures, it should not be assumed that behavior theory is committed *ab initio* to a physiological identification of *all* its concepts,—that its concepts are, so to speak, physiological from the start.

We have, in effect, been distinguishing between two dimensions of the logic (or 'methodologic') of theoretical terms: (a) their role in explaining the selected phenomena of which the theory is the theory; (b) their role as candidates for integration in what we have called the "total picture." These roles are equally part of the logic, and hence the "meaning," of theoretical terms. Thus, at any one time the terms in a theory will carry with them as part of their logical force that which it is reasonable to envisage—whether schematically or determinately—as the manner of their integration. However, for the purposes of my argument, it will be useful to refer to these two roles as though it were a matter of a distinction between what I shall call *pure theoretical concepts*, and hypotheses concerning the relation of these concepts to concepts in other specialties. What we *can* say is that the less a scientist is in a position to conjecture about the way in which a certain theory can be expected to integrate with other specialities, the more the concepts of his theory approximate to the status of pure theoretical concepts. To illustrate: We can imagine that chemistry developed a sophisticated and successful theory to explain chemical phenomena before either electrical or magnetic phenomena were noticed; and that chemists developed as pure theoretical concepts certain concepts which it later became reasonable to identify with concepts belonging to the framework of electromagnetic theory.

V. THE LOGIC OF PRIVATE EPISODES: THOUGHTS

11. With these all too sketchy remarks on Methodological Behaviorism under our belts, let us return once again to our fictional ancestors. We are now in a position to characterize the original Rylean language in which they described themselves and their fellows as not only a *behavioristic* language, but a behavioristic language which is restricted to the *nontheoretical* vocabulary of a behavioristic psychology. Suppose, now, that in the attempt to account for the fact that his fellow men behave intelligently not only when their conduct is threaded on a string of overt verbal episodes—that is to say, as we would put it, when they "think out loud"—but also when no detectable verbal output is present, Jones develops a *theory* according to which overt utterances are but the culmination of a process which begins with certain inner episodes. *And let us suppose that his model for these episodes* which initiate the events which culminate in overt verbal behavior *is that of overt verbal behavior itself. In other words, using the language of the model, the theory is to the effect that overt verbal behavior is the culmination of a process which begins with "inner speech."*

It is essential to bear in mind that what Jones means by "inner speech" is not to be confused with *verbal imagery*. As a matter of fact, Jones, like his fellows, does not as yet even have the concept of an image.

It is easy to see the general lines a Jonesean theory will take. According to it the true cause of intelligent nonhabitual behavior is "inner speech." Thus, even when a hungry person overtly says "Here is an edible object" and proceeds to eat it, the true—theoretical—cause of his eating, given his hunger, is not the overt utterance, but the "inner utterance of this sentence."

12. The first thing to note about the Jonesean theory is that, as built on the model of speech episodes, *it carries over to these inner episodes the applicability of semantical categories.* Thus, just as Jones has, like his fellows, been speaking of overt utterances as *meaning* this or that, or being *about* this or that, so he now speaks of these inner episodes as *meaning* this or that, or being *about* this or that.

The second point to remember is that although Jones' theory involves a *model*, it is not identical with it. Like all theories formulated in terms of a model, it also includes a *commentary* on the model; a commentary which places more or less sharply drawn restrictions on the analogy between the theoretical entities and the entities of the model. Thus, while his theory talks of "inner speech," the commentary hastens to add that, of course, the episodes in question are not the wagging of a hidden tongue, nor are any sounds produced by this "inner speech."

13. The general drift of my story should now be clear. I shall therefore proceed to make the essential points quite briefly:

(1) What we must suppose Jones to have developed is the germ of a theory which permits many different developments. We must not pin it down to any of the more sophisticated forms it takes in the hands of classical philosophers. Thus, the theory need not be given a Socratic or Cartesian form, according to which this "inner speech" is a function of a separate substance; though primitive peoples may have had good reason to suppose that humans consist of two separate things.

(2) Let us suppose Jones to have called these discursive entities *thoughts*. We can admit at once that the framework of thoughts he has introduced is a framework of "unobserved," "nonempirical," "inner" episodes. For we can point out im-

mediately that in these respects they are no worse off than the particles and episodes of physical theory. For these episodes are "in" language-using animals as molecular impacts are "in" gases, not as "ghosts" are in "machines." They are "nonempirical" in the simple sense that they are *theoretical*—not definable in observational terms. Nor does the fact that they are, *as introduced*, unobserved entities imply that Jones could not have good reason for supposing them to exist. Their "purity" is not a *metaphysical* purity, but, so to speak, a *methodological* purity. As we have seen, the fact that they are not introduced as physiological entities does not preclude the possibility that at a later methodological stage, they may, so to speak, "turn out" to be such. Thus, there are many who would say that it is already reasonable to suppose that these *thoughts* are to be "identified" with complex events in the cerebral cortex functioning along the lines of a calculating machine. Jones, of course, has no such idea.

(3) Although the theory postulates that overt discourse is the culmination of a process which begins with "inner discourse," this should not be taken to mean that overt discourse stands to "inner discourse" *as voluntary movements stand to intentions and motives*. True, overt linguistic events *can* be produced as means to ends. But serious errors creep into the interpretation of both language and thought if one interprets the idea that overt linguistic episodes *express* thoughts, on the model of the use of an instrument. Thus, it should be noted that Jones' theory, as I have sketched it, is perfectly compatible with the idea that the ability to have thoughts is acquired in the process of acquiring overt speech and that only after overt speech is well established, can "inner speech" occur without its overt culmination.

(4) Although the occurrence of overt speech episodes which are characterizable in semantical terms is explained by the theory in terms of *thoughts* which are *also* characterized in semantical terms, this does not mean that the idea that overt speech "has meaning" is being *analyzed* in terms of the intentionality of thoughts. It must not be forgotten that *the*

semantical characterization of overt verbal episodes is the pri-
mary use of semantical terms, and that overt linguistic events as
semantically characterized are the model for the inner episodes
introduced by the theory.

(5) One final point before we come to the *dénouement* of
the first episode in the saga of Jones. It cannot be emphasized
too much that although these theoretical discursive episodes or
thoughts are introduced as *inner* episodes—which is merely to
repeat that they are introduced as *theoretical* episodes—they
are *not* introduced as *immediate experiences*. Let me remind the
reader that Jones, like his neo-Rylean contemporaries, does not
as yet have this concept. And even when he, and they, acquire
it, by a process which will be the second episode in my myth, it
will only be the philosophers among them who will suppose
that the inner episodes introduced for one theoretical purpose—
thoughts—must be a subset of immediate experiences, inner
episodes introduced for another theoretical purpose.

14. Here, then, is the *dénouement*. I have suggested a number of times that although it would be most misleading to say
that concepts pertaining to thinking are theoretical concepts, yet
their status might be illuminated by means of the contrast between theoretical and nontheoretical discourse. We are now in a
position to see exactly why this is so. For once our fictitious ancestor, Jones, has developed the theory that overt verbal behavior
is the expression of thoughts, and taught his compatriots to
make use of the theory in interpreting each other's behavior, it
is but a short step to the use of this language in self-description.
Thus, when Tom, watching Dick, has behavioral evidence
which warrants the use of the sentence (in the language of the
theory) "Dick is thinking 'p' " (or "Dick is thinking that
p"), Dick, using the same behavioral evidence, can say, in the
language of the theory, "I am thinking 'p' " (or "I am thinking
that p"). And it now turns out—need it have?—that Dick can
be trained to give reasonably reliable self-descriptions, using
the language of the theory, without having to observe his overt
behavior. Jones brings this about, roughly, by applauding
utterances by Dick of "I am thinking that p" when the be-

havioral evidence strongly supports the theoretical statement "Dick is thinking that p"; and by frowning on utterances of "I am thinking that p," when the evidence does not support this theoretical statement. Our ancestors begin to speak of the privileged access each of us has to his own thoughts. *What began as a language with a purely theoretical use has gained a reporting role.*

As I see it, this story helps us understand that concepts pertaining to such inner episodes as thoughts are primarily and essentially *intersubjective*, as intersubjective as the concept of a positron, and that the reporting role of these concepts—the fact that each of us has a privileged access to his thoughts—constitutes a dimension of the use of these concepts which *is built on* and *presupposes* this intersubjective status. My myth has shown that the fact that language is essentially an *intersubjective* achievement, and is learned in intersubjective contexts—a fact rightly stressed in modern psychologies of language, thus by B.F. Skinner, and by certain philosophers, e.g., Carnap, Wittgenstein[6]—is compatible with the "privacy" of "inner episodes." It also makes clear that this privacy is not an "absolute privacy." For if it recognizes that these concepts have a reporting use in which one is not drawing inferences from behavioral evidence, it nevertheless insists that the fact that overt behavior is evidence for these episodes *is built into the very logic of these concepts,* just as the fact that the observable behavior of gases is evidence for molecular episodes is built into the very logic of molecule talk.

[6] B.F. Skinner, "The Operational Analysis of Psychological Terms," *Psychological Review,* 52 (1945), 270–277 (reprinted in *Readings in the Philosophy of Science,* pp. 585–594); Rudolf Carnap, "Psychologie in Physikalischer Sprache," *Erkenntnis,* 3 (1933), 107–142 (English trans.: "Psychology in Physical Language," in A.J. Ayer, ed., *Logical Positivism* [New York: Free Press, 1959], pp. 165–198).

12

The Chisholm-Sellars Correspondence on Intentionality

RODERICK M. CHISHOLM AND
WILFRID SELLARS

July 30, 1956

Professor Wilfrid Sellars
Department of Philosophy
University of Minnesota
Minneapolis 14, Minnesota

Dear Sellars:

Thanks very much indeed for sending me the page proofs of your "Empiricism and the Philosophy of Mind." I have not yet had time to read all of it with care or to follow the entire thread of the argument, but I have read a good part of it and have looked through it all. I have seen enough to know that it is a challenging manuscript with which I should spend some time and that you and I are interested in pretty much the same problems. I agree with much of what you have to say about the "myth of the given" and with what you say about the "rightness" of statements or assertions. I would talk, though, about the rightness of believing rather than of saying, and this difference, I think, reflects the general difference between us in our attitude toward intentionality.

From *Minnesota Studies in the Philosophy of Science*, vol. II, eds. H. Feigl, M. Scriven, and G. Maxwell, pp. 529–539. Copyright 1958 by the University of Minnesota. Reprinted by permission of the authors and the University of Minnesota Press.

With respect to intentionality, I think you locate the issue between us pretty well on page 311. Among the central questions, as you intimate at the bottom of 319, are these: (1) Can we explicate the intentional character of believing and of other psychological attitudes by reference to certain features of language; or (2) must we explicate the intentional characteristics of language by reference to believing and to other psychological attitudes? In my Aristotelian Society paper[1] I answer the first of these questions in the negative and the second in the affirmative; see also my paper on Carnap in the recent *Philosophical Studies*.[2] I would gather that the fundamental difference between us concerned our attitude toward these two questions, and I don't feel clear about what your belief is in respect to the analysis of "semantical statements." . . .

With best wishes,
Roderick M. Chisholm

August 3, 1956

Professor Roderick Chisholm
Philosophy Department
Brown University
Providence, Rhode Island

Dear Chisholm:

Your friendly remarks encourage me to call attention to some features of my treatment of intentionality which may not have stood out on a first reading. Their importance, as I see it, is that they define a point of view which is, in a certain sense, intermediate between the alternatives envisaged in your letter. I

---

[1] Roderick Chisholm, "Sentences about Believing" (included in this volume). The page references above are to "Empiricism and the Philosophy of Mind" (referred to, hereafter, as EPM), in H. Feigl and M. Scriven, eds., *Minnesota Studies in the Philosophy of Science*, vol. I (Minneapolis: Univ. of Minnesota Press, 1956). Part of EPM is reprinted in this volume, and where appropriate page or section references to this volume will be included. Page 311 corresponds in this volume to p. 201; page 319 to p. 211.

[2] Roderick Chisholm, "A Note on Carnap's Meaning Analysis," *Philosophical Studies*, 6 (1955), 87–89.

shall not attempt to summarize in a few words the complex argument by which, in the concluding third of my essay, I sought to recommend this analysis. It may be worthwhile, however, to state its essentials in a way which emphasizes the extent to which I would go along with many of the things you want to say.[3]

A-1. Unlike Ryle, I believe that meaningful *statements* are the expression of inner episodes, namely *thoughts*, which are not to be construed as mongrel categorical-hypothetical facts pertaining to overt behavior.

A-2. I speak of *thoughts* instead of *beliefs* because I construe *believing that p* as the disposition to have *thoughts that p*; though actually, of course, the story is more complicated. The *thought that p* is an episode which might also be referred to as the *"mental assertion" that p*. In other words, "having the thought that p" is not equivalent to "thinking that p" (as in "Jones thinks that p") for the latter is a cousin of "believing that p" and like the latter has a dispositional force.

A-3. Thought episodes are essentially characterized by the categories of intentionality.

A-4. Thought episodes, to repeat, are not speech episodes. They are *expressed* by speech episodes.

A-5. *In one sense of "because"*, statements are meaningful utterances *because* they express thoughts. In one sense of "because", Jones' statement, s, means *that p, because s* expresses Jones' thought *that p*.

So far I have been highlighting those aspects of my view which, I take it, are most congenial to your own ways of thinking. It is in the following points, if anywhere, that our differences are to be found.

A-6. Although statements mean states of affairs *because* they express thoughts which are about states of affairs, this *because* is not the *because* of analysis. Notice as

---

[3] The following paragraphs have been renumbered A-1, A-2, etc., in order to avoid confusion and permit ready reference. A similar procedure has been followed in the case of subsequent groups of numbered paragraphs or sentences.

something to keep in mind that physical objects move because the subatomic particles which make them up move; yet obviously the idea that physical objects move is not to be *analyzed* in terms of the idea that subatomic particles move.

A-7. Thoughts, of course, are not theoretical entities. We have direct (noninferential) knowledge, on occasion, of what we are thinking, just as we have direct (noninferential) knowledge of such nontheoretical states of affairs as the bouncings of tennis balls.

A-8. Yet if thoughts are not theoretical entities, it is because they are more than *merely* theoretical entities.

A-9. In my essay I picture the framework of thoughts as one which was developed "once upon a time" as a *theory*[4] to make intelligible the fact that silent behavior could be as effective as behavior which was (as we should say) *thought through out loud* step by step.

A-10. But though we initially used the framework merely as a theory, we came to be able[5] to describe ourselves as having such and such thoughts without having to *infer* that we had them from the evidence of our overt, publicly accessible behavior.

A-11. The *model* for the theory is overt speech. Thoughts

---

[4] The nature and role of theories and models in behavioristic psychology is discussed in §§ 51–55 of EPM. See this volume, §§ 6–10, pp. 202–209.

[5] ". . . once our fictitious ancestor, Jones, has developed the theory that overt verbal behavior is the expression of thoughts, and taught his compatriots to make use of the theory in interpreting each other's behavior, it is but a short step to the use of this language in self-description. Thus, when Tom, watching Dick, has behavioral evidence which warrants the use of the sentence (in the language of the theory) 'Dick is thinking 'p' ' (or 'Dick is thinking that p'), Dick, using the same behavioral evidence, can say, in the language of the theory, 'I am thinking 'p' ' (or 'I am thinking that p'). And it now turns out—need it have?—that Dick can be trained to give reasonably reliable self-descriptions, using the language of the theory, without having to observe his overt behavior. Jones brings this about, roughly, by applauding utterances by Dick of 'I am thinking that p' when the behavioral evidence strongly supports the theoretical statement 'Dick is thinking that p'; and by frowning on utterances of 'I am thinking that p,' when the evidence does not support this theoretical statement. Our ancestors begin to speak of the privileged access each of us has to his own thoughts. What began as a language with a purely theoretical use has gained a reporting role" (EPM, p. 320; this volume, pp. 212–213).

are construed as "inner speech"—i.e., as episodes which are (roughly) as like overt speech as something which is *not* overt speech can be.[6]

A-12. The argument presumes that the meta*linguistic* vocabulary in which we talk about linguistic episodes can be analyzed in terms which do not presuppose the framework of mental acts; in particular, that

". . ." means p

is not to be analyzed as

". . ." expresses t and t is about p

where t is a thought.

A-13. For my claim is that the categories of intentionality are nothing more nor less than the metalinguistic categories in terms of which we talk epistemically about overt speech as they appear in the framework of thoughts construed on the model of overt speech.

A-14. Thus I have tried to show in a number of papers— most successfully, I believe, in these lectures—that the role of

". . ." means - - -

can be accounted for without analyzing this form in terms of mental acts.[7] . . .

Sincerely,
Wilfrid Sellars

[6] See the distinction between the model around which a theory is built, and the 'commentary' on the model in EPM, §§ 51 and 57 (this volume, §§ 6 and 12).

[7] See EPM, Part VI, "The Logic of Means"; also § 80 of my essay "Counterfactuals, Dispositions and the Causal Modalities," in H. Feigl *et al.*, eds., *Minnesota Studies in the Philosophy of Science,* vol. II (Minneapolis: Univ. of Minnesota Press, 1958), pp. 225–308; see also below, pp. 224 ff. and pp. 233 ff. in this volume.

August 12, 1956

Dear Sellars:

Thanks very much indeed for your letter of August 3, and for the clear statement of the essentials of your position. I think I can locate fairly well now the point at which we diverge.

I certainly have no quarrel with your first five points, and I can accept A-7 and A-8. As for A-6 and points that follow A-8, acceptability of these seems to depend upon A-12. If you could persuade me of A-12, perhaps you could persuade me of the rest.

I know that you have written at length about the thesis of paragraph A-12, but I do not think you have satisfied what my demands would be. In order to show that

". . ." means p

is not to be analyzed as

". . ." expresses t and t is about p

your metalinguistic vocabulary must contain only locutions (1) which, according to the criteria of my Aristotelian Society paper,[8] are not intentional and (2) which can be defined in physicalistic terms. This is the point we would have to argue at length, for I believe you must introduce some term which, if it means anything at all, will refer to what you call thoughts. This term could be disguised by calling it "a primitive term of semantics" or "a primitive term of pragmatics," or something of that sort, but this would be only to concede that, when we analyze the kind of meaning that is involved in natural language, we need some concept we do not need in physics or in "behavioristics." Until you can succeed in doing what Carnap tried to do in the *Philosophical Studies* paper[9] I criticized last year[10] (namely to analyze the semantics or pragmatics of natural language in the physicalistic vocabulary of a behavioristic psychology, with no undefined semantical terms and no reference to

---

[8] See note 1 above.

[9] Rudolf Carnap, "Meaning and Synonymy in Natural Languages," *Philosophical Studies*, 6 (1955), 33–47.

[10] See note 2 above.

thoughts), I think I will remain unconverted to your views about intentionality.

Perhaps you will agree that if one rejects your paragraph A-12, then it is not unreasonable for him also to reject what you say on the bottom of page 319 of the lectures: "It must not be forgotten that the semantical characterization of overt verbal episodes is the primary use of semantical terms, and that overt linguistic events as semantically characterized are the model for the inner episodes introduced by the theory." . . .

Cordially yours,
Roderick M. Chisholm

August 15, 1956

Dear Chisholm:

. . . we have communicated so well so far, that I cannot forbear to make one more try for complete understanding (if not agreement).

You write, ". . . I believe you must introduce some term which, if it means anything at all, will refer to what you call thoughts. This term could be disguised by calling it 'a primitive term of semantics', or 'a primitive term of pragmatics', or something of the sort, but this would only be to concede that, when we analyze the kind of meaning that is involved in natural language, we need some concept we do not need in physics or in 'behavioristics.' " Now I certainly agree that semantical statements about statements in natural languages, i.e., statements in actual use, cannot be constructed out of the resources of behavioristics. I have insisted on this in a number of papers (e.g., "A Semantical Solution . . ."),[11] and, most recently in my essay[12] for the Carnap volume and my London lectures. I quite agree that one

---

[11] Wilfrid Sellars, "A Semantical Solution of the Mind-Body Problem," *Methodos*, 5 (1953), 45–84; see also Wilfrid Sellars, "Mind, Meaning and Behavior," *Philosophical Studies*, 3 (1953), 83–95.

[12] Wilfrid Sellars, "Empiricism and Abstract Entities," in *The Philosophy of Rudolf Carnap*, ed. P.A. Schilpp (La Salle, Ill.: Open Court, 1963).

additional expression must be taken as primitive, specifically "means" or "designates" with its context

". . ." means - - -.

What I have emphasized, however, is that although "means" is *in a grammatical sense* a "relation word," it is no more to be assimilated to words for descriptive *relations* than is "ought", and that though it is a "descriptive" predicate if one means by "descriptive" that it is not a *logical* term nor constructible out of such, it is not *in any more interesting* (*or usual*) *sense* a descriptive term.

I think that I have made these points most effectively in the Lectures, pp. 291–293, and 310. I mention them only because, since they are in a section which occurs much earlier than the section on *thoughts* which has been the subject of our correspondence to date, you may not have noticed them.

I am also sending under separate cover a mimeograph of my Carnap paper,[13] which discusses at length, pp. 30 ff, the point about the irreducibility of "designates" to the concepts of formal logic.

Cordially yours,
Wilfrid Sellars

August 24, 1956

Dear Sellars:

The philosophic question which separates us is, in your terms, the question whether

(1) ". . ." means p

is to be analyzed as

(2) ". . ." expresses t and t is about p.

I would urge that the first is to be analyzed in terms of the second, but you would urge the converse. But we are in agreement, I take it, that we need a semantical (or intentional) term which is not needed in physics.

---

[13] See note 12 above.

How are we to decide whether (1) is to be analyzed in terms of (2), or conversely? If the question were merely of constructing a language, the answer would depend merely upon which would give us the neatest language. But if we take the first course, analyzing the meaning of noises and marks by reference to the thoughts that living things have, the "intentionalist" will say: "Living things have a funny kind of characteristic that ordinary physical things don't have." If we take the second course, there could be a "linguisticist" who could say with equal justification, "Marks and noises have a funny kind of characteristic that living things and other physical things don't have."

Where does the funny characteristic belong? (Surely, it doesn't make one whit of difference to urge that it doesn't stand for a "descriptive relation." Brentano said substantially the same thing, incidentally, about the ostensible relation of "thinking about," etc.) Should we say there is a funny characteristic (i.e., a characteristic which would not be labeled by any physicalistic adjective) which belongs to living things— or that there is one which belongs to certain noises and marks?

When the question is put this way, I should think, the plausible answer is that it's the living things that are peculiar, not the noises and marks. I believe it was your colleague Hospers who proposed this useful figure: that whereas both thoughts and words have meaning, just as both the sun and the moon send light to us, the meaning of the words is related to the meaning of the thoughts just as the light of the moon is related to that of the sun. Extinguish the living things and the noises and marks wouldn't shine any more. But if you extinguish the noises and marks, people can still think about things (but not so well, of course). Surely it would be unfounded psychological dogma to say that infants, mutes, and animals cannot have beliefs and desires until they are able to use language.

In saying "There is a characteristic . . ." in paragraph 2 above, I don't mean to say, of course, that there are abstract entities.

I don't expect you to agree with all the above. But do you

agree that the issue described in paragraph one is an important one and that there is no easy way to settle it? . . .

Cordially yours,
Roderick M. Chisholm

August 31, 1956

Dear Chisholm:

Your latest letter, like the preceding ones, raises exactly the right questions to carry the discussion forward. (The points made in a fruitful philosophical discussion must, so to speak, be "bite size.") Let me take them up in order.

1. The contrast you draw between the "intentionalist" and the "linguisticist" is, in an essential respect, misleading. You write: ". . . if we take the first course, analyzing the meaning of noises and marks by reference to the thoughts that living things have, the 'intentionalist' will say: 'Living things have a funny kind of characteristic that ordinary physical things don't have.' If we take the second course, there could be a 'linguisticist' who could say with equal justification: 'Marks and noises have a funny kind of characteristic that living things and other physical things don't have.'

"Where does the funny characteristic belong? . . ." This is misleading because (although I am sure you did not mean to do so) it evokes a picture of the "linguisticist" as tracing the aboutness of thoughts to characteristics which marks and noises can have as *marks* and *noises* (e.g., serial order, composition out of more elementary marks and noises belonging to certain mutually exclusive classes, etc.). But while these "sign design" characteristics of marks and noises make it possible for them to function as expressions in a language, they do not, of course, constitute this functioning. Marks and noises are, in a *primary* sense, linguistic expressions only as "nonparrotingly" produced by a language-using animal.

2. Thus the "linguisticist" no less than the "intentionalist" will say that (certain) living things are able to produce marks

and noises of which it can correctly be said that they *refer* to such and such and say such and such, whereas "ordinary physical things" are not. The problem, therefore, is not (as you put it), "Should we say that there is a funny characteristic (i.e., a characteristic which would not be labeled by any physicalistic adjective) which belongs to living things—or that there is one which belongs to certain noises and marks?" but rather, *granted* that the primary mode of existence of a language is in meaningful verbal performances by animals that can *think, desire, intend, wish, infer,* etc., and *granted* (a) that these verbal performances *mean* such and such, and (b) that the *mental acts* which they express are about such and such; how are these concepts—*all of which pertain to certain living things rather than "ordinary physical things"*—to be explicated.

3. What persuades you that the "means" of

"..." means - - -

must stand for a *characteristic*, even if a "funny" one? If, like Brentano, you conclude (rightly) that it is only ostensibly a relation, must it therefore be a characteristic of some other kind, perhaps a kind all its own? (Perhaps you would prefer to say that it is *means-p* rather than *means* simpliciter which, in the case of propositions, is the characteristic.) I am not, of course, denying that the term "characteristic" can, with a certain initial plausibility, be extended to cover it. I do, however, claim that to use "characteristic" in such an extended way is to blur essential distinctions; but more of this in a moment.

I would be the last person to say that "the meaning of a term is its use," for there is no sense of "use" which *analyzes* the relevant sense of "means". Yet this W—nian maxim embodies an important insight, and, used with caution, is a valuable tool. Suppose there were an expression which, though it clearly didn't designate an item belonging to one of your other ontological categories, you were reluctant to speak of as standing for a characteristic, though you granted that it played a systematic or ruleful role in discourse. What about "yes" as in "yes, it is raining." Suppose "yes" always occurred in the context "yes, p." Could we not even here make a meaningful use of the rubric " '. . .' means - - -," thus

"Ja, p" means yes, p

and

"Ja" means yes.

(I take it that you would not be tempted to say that yes is a characteristic of propositions.) But if "yes" doesn't stand for a characteristic, the fact that

"x bedeutet y" means x *means* y

is not a *conclusive* reason for supposing that *means* is a characteristic. To be sure, from

"Bedeutet" means *means*

we can infer that

There is something (i.e., *means*) which "bedeutet" means. But that *means* is, in this broadest of senses, a 'something' tells us precious little indeed, for in this sense yes is a something too!

Where E is any expression in L, whatever its role, and E' is the translation of E in English, then we can properly say both

E (in L) means - - -

and

There is something, namely - - -, which E means,

where what goes in the place held by '- - -' is the English expression named by 'E', thus

There is something, namely *I shall,* which "Ich werde" (in German) means

and

There is something, namely *this* which "dieser (diese, dieses)" (in German) means.

(In this last example, of course, the context brackets *'this'* so that it is not playing its "pointing" role, though the semantical rubric mobilizes its pointing role in its own way.)

4. You write: "the meaning of the words is related to the meaning of the thoughts just as the light of the moon is related to that of the sun. Extinguish the living things and the noises and marks wouldn't shine any more. But if you extinguish the noises and marks, people can still think about things (but not so well, of course)." Now I agree, *of course*, that marks in books and noises made by phonographs 'have meaning' only by virtue of their relation to 'living' verbal episodes in which language is the direct expression of thought (e.g., conversation, writing on a blackboard.) And, as I have emphasized in an earlier letter (3 August), I agree that 'living' verbal episodes are meaningful *because* they express thoughts. Our difference concerns the *analysis* of this 'because'.

Let me have another try at making the essential points. I have argued that it is *in principle* possible to conceive of the characteristic forms of semantical discourse being used by a people who have not yet arrived at the idea that there are such things as *thoughts*. They *think*, but they *don't know that they think*. Their use of language is meaningful *because* it is the expression of thoughts, but they *don't know* that it is the expression of thoughts; that is to say, they don't know that overt speech is the culmination of inner episodes of a kind which we conceive of as thoughts.

(Compare them, for a moment, with a people who have, as yet, no *theoretical* concepts in terms of which to give *theoretical* explanations of the observable behavior of physical objects. They are, nevertheless, able to explain particular events by means of that general knowledge which is embodied in dispositional concepts pertaining to thing kinds.)

Now, in order to communicate, such a people would, of course, have to appreciate both the norms which, by specifying what may not be said without withdrawing what, delimit the syntactical or 'intralinguistic' structure of the language, as well as such facts as that, *ceteris paribus*, a person who says "This is green" is in the presence of a green object, and a person who says "I shall do A" proceeds to do it. This understanding springs from the routines by which the language is learned and passed on from generation to generation by "social inheritance." It constitutes their mastery of the language.

If you grant that they *could* get this far without having arrived at the *concept of a thought* (though, of course, not without *thinking*), the crucial question arises, Could they come to make use of semantical discourse while remaining untouched by the idea that overt verbal behavior is the culmination of inner episodes, let alone that it is the expression of thoughts? To this question *my* answer is 'Yes.'

In your first letter you expressed agreement "with much of what [I] have to say about the 'myth of the given.'" Well, of a piece with my rejection of this myth is my contention that before these people could come to know *noninferentially* (by 'introspection') that they have thoughts, they must *first* construct the concept of what it is to be a thought.[14] Thus, while I agree with you that the rubric

". . ." means - - -

is not constructible in Rylean terms ('Behaviorese,' I have called it), I also insist that it is not to be analyzed in terms of

". . ." expresses t, and t is about - - -.

My solution is that "'. . .' means - - -" is the core of a unique mode of discourse which is as distinct from the *description* and *explanation* of empirical fact, as is the language of *prescription* and *justification*.

I have probably lost you somewhere along the line. But, if not, from this point on the argument is clear. To put it in a way

[14] ". . . once we give up the idea that we begin our sojourn in this world with any—even a vague, fragmentary, and undiscriminating—awareness of the logical space of particulars, kinds, facts, and resemblances, and recognize that even such 'simple' concepts as those of colors are the fruit of a long process of publicly reinforced responses to public objects (including verbal performances) in public situations . . . we . . . *recognize that instead of coming to have a concept of something because we have already noticed that sort of thing, to have the ability to notice a sort of thing is already to have the concept of that sort of thing, and cannot account for it*" (EPM, p. 306).

which artificially separates into stages a single line of conceptual development, thoughts began by being conceived as theoretical episodes on the analogy of overt verbal behavior ("inner speech"). Men came, however, in a manner which I pictured in my Jonesean myth, to be able to say what they are thinking without having to draw theoretical inferences from their own publicly observable behavior.[15] They now not only think, but know that they think; and can not only *infer* the thoughts of others, but have direct (noninferential) knowledge of what is going on in their own minds.

5. You write: "Surely it would be unfounded psychological dogma to say that infants, mutes, and animals cannot have beliefs, and desires until they are able to use language." Here I shall limit myself to a few brief points:

(a) Since I do not *define* thoughts in terms of overt verbal behavior, and grant that thought *episodes* occur without overt linguistic expression, there is, on my view, no *contradiction* in the idea of a being which *thinks*, yet has no language to serve as the overt expression of his thoughts.

(b) Not only do the subtle adjustments which animals make to their environment tempt us to say that they do what they do because they *believe* this, *desire* that, *expect* such and such, etc.; we are able to *explain* their behavior by ascribing to them these beliefs, desires, expectations, etc. *But,* and this is a key point, we invariably find ourselves *qualifying* these explanations in terms which would amount, in the case of a human subject, to the admission that he wasn't *really* thinking, believing, desiring, etc. For in the explanation of animal behavior the mentalistic framework is used as a *model* or *analogy* which is modified and restricted to fit the phenomena to be explained. It is as though we started out to explain the behavior of macroscopic objects, in particular, gases, by saying that they are made up of minute bouncing billiard balls, and found ourselves forced to add, "but, of course. . . ."

(c) The use of the mentalistic framework as the point of de-

---

[15] See EPM, p. 212, this volume; quoted above in note 5.

parture for the explanation of animal behavior can be characterized as the approach "from the top down." Recent experimental psychology has been attempting an approach "from the bottom up," and while it is still barely under way, it has been made amply clear that *discrimination* is a far more elementary phenomenon than *classification,* and that chains of stimulus-response connections can be extremely complex and, in principle, account for such 'sophisticated' forms of adjustment as, for example, the learning of a maze, where these same adjustments, approached "from the top down," would be explained in terms of a qualified use of the framework of beliefs, desires, expectations, etc. To make this point in an extreme form, what would once have been said about an earthworm which comes to take the right hand turn in a T-shaped tube, as a result of getting a mild shock on turning to the left?

(d) I think you will agree with me that the ability to have thoughts entails the ability to do *some* classifying, see *some* implications, draw *some* inferences. I think you will also agree that it is a bit strong to conclude that a white rat must be *classifying* objects because it reacts in similar ways to objects which are similar in certain respects, and in dissimilar ways to objects which are dissimilar in certain respects; or that an infant must be *inferring* that his dinner is coming because he waves his spoon when his mother puts on his bib. While I do not wish to cut off, as with a knife, inner episodes which are below the level of thoughts, from inner episodes which are thoughts, I remain convinced that we approach the ability to have thoughts in the course of approaching the ability to use a language in interpersonal discourse, and that the ability to have thoughts without expressing them is a subsequent achievement.

(e) This brings me to my concluding point which, though last, is by no means least. It has been taken for granted in the above paragraphs that the languages the learning of which is the acquiring the ability to have thoughts, are languages in the sense of the highly conventionalized, socially sanctioned, and inherited, systems of symbols, such as French, German, English, etc. (and the closely related sign languages' for handicapped persons). But it would be naive to suppose that the only forms of overt behavior which can play the role of symbols in the classifying, inferring, intending behavior of a human organism, are languages in this narrow sense. It is, therefore, even on my view, by no means impossible that there be a mature deaf-mute who has "beliefs and desires" and can "think about things (but not so well, of course)" though he has learned no *language*. I would merely urge (with W—n and others) that the overt behavior which has come to play the role of more conventional symbol-behavior must have been selectively *sanctioned* ("reinforced") by his fellows; and that a certain mode of behavior, B, can correctly be said to express the thought *that* p, only if B is the *translation* in his 'language' of the sentence in our *language*, call it S, represented by 'p'. This implies that B plays a role in his behavioral economy which parallels that of S in ours.

(f) I must add two footnotes to this "concluding point" which will relate it to what has gone before. The first concerns a topic of fundamental importance in current controversies about *meaning*. The mutual translatability of two expressions in actual usage is, with certain exceptions, an "ideal." By this I mean that whenever we make use of the rubric

E (in L) means - - -

we would, in most cases, have to admit that "strictly speaking" E and E′ (see above, end of paragraph 3) "do not have quite the same meaning." Yet our use of this rubric is sensible for two reasons, of which the second is the more important. (a) The difference between the roles of E and E′ may be irrelevant to the context in which this rubric is used. (b) This rubric, which treats E and E′ as mutually translatable is the *base* or *point of departure* from which, by the addition of *qualifications*, we explain the use of expressions in L to someone who does not know how to use them.

(g) This second footnote is to point out that there is no substitute for the subtly articulated languages which have been developed over the millenia of human existence. To pick up a point made in (b) above, if we said of our deaf-mute that he had the thought *that* p, we would, do you not agree, find it necessary to add, *if pressed*, "but of course . . ."?

6. You write: "I don't expect you to agree with all the above." If I write "ditto," I am putting it mildly. On the other hand, though I don't agree with some of the things you say, I have attempted to make it clear that I do agree with a great deal of what you say, more than you might think possible, while remaining on my side of the fence.

Cordially,
Wilfrid Sellars

P.S. This really started out to be a letter!

September 12, 1956

Dear Sellars:

Thanks for your latest letter which I have enjoyed reading. I think the differences between us now seem to boil down to two points, both rather difficult to argue about.

First, we are apparently in some disagreement about what is meant by such terms as "analysis" and "explication." Consider the following sentences:

(B-1) The meaning of thoughts is to be analyzed in terms of the meaning of language, and not conversely.

(B-2) Language is meaningful *because* it is the expression of thoughts—of thoughts which are *about* something.

(B-3) The people in your fable "come to make use of semantical discourse while remaining untouched by the idea that overt verbal behavior is the culmination of inner episodes, let alone that it is the expression of thoughts."

When I first wrote you I took (B-1) to be inconsistent with (B-2), but you affirm both (B-1) and (B-2). And apparently you take (B-3) to imply (B-1), but I accept (B-3) and deny (B-1).

Perhaps we could do better if we resolved not to use such technical terms as "analysis" and "explication." What would (B-1) come to then?

The second point of disagreement between us lies in the fact that I am more skeptical than you are about the content of such "solutions" as the one you propose on page 4. "My solution is that ' ". . ." means - - -' is the core of the unique mode of discourse which is as distinct from the *description* and *explanation* of empirical fact as is the language of *prescription* and *justification.*" I am inclined to feel that the technical philosophical term "descriptive" is one which is very much over used, and I am not sure I can attach much meaning to it. Indeed I would be inclined to say that if the locution "Such and such a sentence is *not* descriptive" means anything at all, it means that the sentence in question (like "Do not cross the street" and "Would that the roses were blooming") is neither true nor false. But the sentence " 'Hund' means *dog* in German" is a sentence which is *true*. And anyone who denied it would be making a *mistake*—in the same sense, it seems to me, that he would be making a mistake if he said "Berlin is part of Warsaw." Hence it does not illuminate any of my problems to say that the sentence is not descriptive or that it embodies a unique mode of discourse.

But I hope we haven't reached an impasse quite yet.

Cordially yours,
R.M.   Chisholm

September 19, 1956

Dear Chisholm:

Many thanks for your letter of September 12 which, as usual, brings things back to a sharp focus. I shall take up the two points to which, as you see it, our differences "boil down" in the order in which you state them.

You ask what your sentence (B-1) would come to "if we resolved not to use such technical terms as 'analysis' and 'explication'." Good. I quite agree that these terms are dangerous unless carefully watched. "Analysis" now covers everything from definition to explanations of the various dimensions of the use of a term which are anything but definitions. Let me therefore begin by pointing out that I have been careful *not* to say that "the meaning of thoughts is to be analyzed in terms of the meaning of language, and not conversely." This formulation stems from your letter of August 24,[16] and I should have taken exception to it in my last letter. I have, of course, denied that the meaning of language is to be analyzed in terms of the meaning of thoughts—see, for example, paragraph A-12 in my letter of August 3—but the closest I have come to affirming the converse is when I wrote—paragraph A-13 in that letter— "the categories of intentionality are nothing more nor less than the metalinguistic categories in terms of which we talk epistemically about overt speech *as they appear* in the framework of thoughts *construed on the model of overt speech*." (No italics in the original.) Indeed, I have explicitly denied (point 5(b) in my letter of August 31)[17] that thoughts (and consequently their aboutness) are to be *defined* in terms of language. I have, however, argued that the aboutness of thoughts is to be *explained* or *understood* by reference to the categories of semantical discourse about language. It is only if "analysis" is stretched to (and, I think, beyond) the limits of its usefulness, that I would be prepared to accept your (B-1).

Let me interrupt the above train of thought to pull together some of the implications of the fable according to which the framework of thoughts was cooked up by Jones with semantical discourse about overt speech as his model. The fable has it, for example, that

x is a token of S and S (in L) means p

(where x ranges over overt linguistic episodes) was his model for

x is a case of T and T is the thought that p

(where x ranges over the inner episodes introduced by the 'theory'). Now if I am correct in my interpretation of semantical statements, sentences of the form

---

[16] P. 221 above.
[17] P. 228 above.

S (in L) means p

as used by one of Jones' contemporaries *imply* but do not *assert* certain Rylean facts about the place of S in the behavioral economy of the users of L.

> (They imply these facts in that "S (in L) means p" said by x would not be *true* unless the sentence named by 'S' plays the same Rylean role in the behavior of those who use L, as the sentence abbreviated by 'p' plays in the behavior of the speaker. They do not *assert* Rylean facts, for "S means p" is not (re-)constructible out of Rylean resources.)

The Rylean facts which, in this sense, 'underlie' semantical statements about the expressions of a language—and which must be appreciated not only by anyone who is going to be in a position to make semantical statements about the language, but also by anyone who is going to use the language to communicate with another user of the language—are (roughly) correlations between (a) environmental situation and verbal behavior, (b) verbal behavior and other verbal behavior, and (c) verbal behavior and nonverbal behavior. And when semantical discourse about overt speech is taken as the model for the inner episodes which Jones postulates to account for the fact that the behavior of his fellows can be just as intelligent when they are silent as when (as we would put it) they think it through out loud, it is these correlations—*more accurately, these correlations as they would be if all such behavior were "thought through out loud"*—which are the effective model for the roles played by these inner episodes. It is these roles—though not, of course, the framework of intentionality which conveys them— which today we (reasonably) expect to interpret in terms of neurophysiological connections, as we have succeeded in interpreting the 'atoms,' 'molecules,' etc., of early chemical theory in terms of contemporary physical theory.[18]

---

[18] In other words, one must distinguish two dimensions in the role played by semantical statements about overt linguistic performances as models for the concept of thoughts as episodes having aboutness or reference: (a) the dimension involving the semantical form itself, "S means p" being the model for "T is about p"; (b) the dimension in which the verbal-behavioral facts *implied* by

I am beginning, however, to touch on topics which presuppose agreement on more fundamental issues (though the above may be useful as giving a more definite picture of the direction of my thought.) So back to your questions! Before this interruption I was making the point that I could not accept sentence (B-1) without qualifying it so radically that the term "analysis" would have to be stretched to the breaking point. In effect, then, we *both* deny (B-1). I, however, am prepared to accept a "first cousin" of (B-1). You, I take it, are not. Or are you? (See below.)

Again, though we both accept (B-2) *as a sentence,* I accept it only if the "because" it contains is (roughly) the "because" of theoretical explanation, whereas you have interpreted it as the "because" of analysis. How, then, will you interpret your acceptance of (B-2) now that we have "resolved not to use such technical terms as 'analysis' and 'explication' "? The sense in which I accept (B-2), on the other hand, is part and parcel of the sense in which I accept (B-1).

This brings me to the heart of the first part of this letter. In your letter of August 12[19] you wrote, "I certainly have no quarrel with your first five points, and I can accept A-7 and A-8. As for A-6 and points that follow A-8, acceptability of these seems to depend upon A-12. If you could persuade me of A-12, perhaps you could persuade me of the rest." The question I wish to raise is this. You now write that you accept (B-3). You wrote

semantical statements about overt linguistic performances are the model for the factual or descriptive character of mental episodes, their relationship *in the causal* order to one another and to overt behavior.

It is the descriptive structure of mental episodes which, as was written above, "we (reasonably) expect to interpret in terms of neurophysiological connections, as we have succeeded in interpreting the 'atoms,' 'molecules,' etc. of early chemical theory in terms of contemporary physical theory." For a discussion of the logic of this 'interpretation' or 'fusion,' as it is sometimes called, see EPM, §§ 55, 58 [this volume, §§ 10, 13] and 40–41; also §§ 47–50 of my essay "Counterfactuals, Dispositions and the Causal Modalities," particularly § 49. For a reprise of the above analysis of the sense in which thoughts are *really* neurophysiological states of affairs, an analysis which defends the substance of the naturalistic-materialistic tradition while avoiding the mixing of categories characteristic of earlier formulations, see below, p. 246 f.

[19] P. 219 above.

on August 12[20] that if I could persuade you of A-12 perhaps I could persuade you of the rest. *But doesn't (B-3) entail A-12?*

I turn now to "the second point of disagreement between us." Let me say right at the beginning that I share your mistrust of "solutions" of philosophical puzzles which simply find a new category for the expressions which raise them. It is because I believe that what I propose amounts to something more than this that I have ventured to call it a solution.

Perhaps the most important thing that needs to be said is that I not only *admit,* I have never *questioned* that

'Hund' means *dog* in German

is *true* in what, for our purposes, is exactly the same sense as

Berlin is part of Warsaw

would be if the facts of geography were somewhat different.

" 'Hund' means dog in German" is true ≡ 'Hund' means dog in German

just as

"Berlin is part of Warsaw" is true ≡ Berlin is part of Warsaw.

There is just no issue between us on this point. When I have said that semantical statements *convey* descriptive information but do not *assert* it, I have not meant to imply that semantical statements *only* convey and do not assert. They make semantical assertions. Nor is "convey," as I have used it, a synonym for "evince" or "express" as emotivists have used this term. I have certainly not wished to assimilate semantical statements to ejaculations or symptoms.

It might be worth noting at this point that, as I see it, it is just as proper to say of statements of the form "Jones ought to do A" that they are *true,* as it is to say this of mathematical, geographical, or semantical statements. This, of course, does not preclude me from calling attention to important differences in the 'logics' of these statements.

[20] *Ibid.*

I quite agree, then, that it is no more a solution of our problem simply to say that semantical statements are "unique," than it would be a solution of the corresponding problem in ethics simply to say that prescriptive statements are "unique." What is needed is a painstaking exploration of statements belonging to various (prima facie) families, with a view to discovering *specific* similarities and differences in the ways in which they behave. Only *after* this has been done can the claim that a certain family of statements is, in a certain respect, unique, be anything more than a promissory note. But while I would be the last to say that the account I have given of semantical and mentalistic statements is more than a beginning, I do think that it *is* a beginning, and that I have paid at least the first installment on the note.

I also agree that the term "descriptive" is of little help. Once the "journeyman" task (to use Ayer's expression) is well under way, it may be possible to give a precise meaning to this technical term. (Presumably this technical use would show some measure of continuity with our ordinary use of "describe.") I made an attempt along this line in my Carnap paper, though I am not very proud of it. On the other hand, as philosophers use the term today, it means little that is definite apart from the logician's contrast of "descriptive expression" with "logical expression" (on this use "ought" would be a descriptive term!) and the moral philosopher's contrast of "descriptive" with "prescriptive." According to both these uses, "S means p" would be a descriptive statement.

It is, then, the *ordinary* force of "describe," or something very like it, on which I have wished to draw when I have said that " 'Hund' means *dog* in German" is not a *descriptive* assertion. I have wished to say that there is an important sense in which this statement does not describe the role of "Hund" in the German language, though it *implies* such a description.

(Remote parallel: When I *express the intention* of doing A, I am not *predicting* that I will do A, yet there is a sense in which the expression of the intention *implies* the corresponding prediction.)

Well, then, what *is* the business of such statements as "'Hund' means *dog* in German"? I wish I could add to my previous attempts with which you are already familiar, but I can't, unless it be the following negative point. It is tempting (though it clearly won't do) to suppose that

S_1: "Hund" means *dog* in German

really makes the same statement as

S_2: "Hund" plays in German the same role as "dog" plays in English.

If so, it would describe the role of "Hund" as

Tom resembles (in relevant respects) Dick

describes Tom. But a simple use of Church's translation test makes it clear that S_1 and S_2 are not equivalent. We who *use* "dog" (as an English word) use S_1 to explain to another *user* of "dog" the role of "Hund" in German, by holding out to him, so to speak, as an exhibit the word which plays the corresponding role in English, our language. And while a person could not correctly be said to have understood S_1 unless, *given that he uses the word "dog,"* he knows that S_1 is true *if and only if* S_2 is true, nevertheless he can clearly know that S_2 is true without having that piece of knowledge the proper expression of which is "'Hund' means *dog* in German." For he may know that S_2 is true without "having learned the word 'dog'."

(Parallel: A person may know that Tom resembles (in relevant respects) Dick, without knowing what either Tom or Dick is like.)

Well, once again what turned out to be a short, clear-cut letter has gotten out of hand. Writing it has been helpful to me in clearing up some of my own ideas. I hope you find it of some use.

Cordially yours,
Wilfrid Sellars

October 3, 1956

Dear Sellars:

Excuse the delay in replying to your letter of September 19th.

I have the feeling that perhaps we should start again, for in formulating a reply to your letter I find that we are in danger of an impasse. You conclude the first part of your letter by asking "But doesn't B-3 entail A-12?" Since one of these sentences contains the technical term "analysis" which we decided to avoid, this question needs reformulation. I would not have said that B-3 entails A-12; hence we are using "analysis" in different ways. And, with respect to the second part of your letter, my natural temptation would be to say that the business of the sentence " 'Hund' means dog in German" where "German" is used to designate the language spoken by German people, is to tell us that German-speaking people use the word "Hund" to express their thoughts about dogs. But saying this, of course, would not get us very far!

I think that everything I want to say is expressible in the following seven sentences:

(C-1) Thoughts (i.e., beliefs, desires, etc.) are intentional— they are about something.

(C-2) Linguistic entities (sentences, etc.) are also intentional.

(C-3) Nothing else is intentional.

(C-4) Thoughts would be intentional even if there were no linguistic entities. (This is a sentence about psychology. I concede that if we had no language, our thoughts would be considerably more crude than they are.)

(C-5) But if there were no thoughts, linguistic entities would not be intentional. (If there were no people, then the mark or noise "Hund"—if somehow occasionally it got produced—would not mean dog.)

(C-6) Hence thoughts are a "source of intentionality"—i.e., nothing would be intentional were it not for the fact that thoughts are intentional. (When I used "because" in an earlier letter I meant it this way; I did not intend it, as you assumed, to be the "because" of analysis.)

(C-7) Hence—and this would be Brentano's thesis—thoughts are peculiar in that they have an important characteristic which nothing else in the world has—namely, the characteristic described in C-6.

Hospers' sun-moon analogy holds in all of this. For if we forget about stars and meteors, the above sentences will hold if "thoughts" is replaced by "sun", "linguistic entities" is replaced by "moon", and "are intentional" is replaced by "is a source of light".

Conceivably a man who was very well informed about the moon and knew very little about the sun could be helped in understanding the sun by learning of its resemblance to the moon. Such a man would be like the man of your fable. But the fact that there could be such a man, it seems to me, has no bearing upon the important astronomical truth expressed by C-7. And hence your fable, as I interpret it, does not lead me to question the truth of C-7.

Is there any hope now of our even seeing eye to eye?

With best wishes,
R.M. Chisholm

October 19, 1956

Dear Chisholm:

I admit that in the very letter in which I was agreeing that we should try to say what we want to say without relying on the technical term "analysis", I relied on this term to settle a point when I asked "But doesn't B-3 entail A-12?" I think you will agree, however, that if the point *could* have been settled in the framework of our earlier letters, we would be that much further along. We would now be trying to determine how matters stood when the point was restated without the use of this technical term. Although things haven't worked out that way, a word about the background of my argument may be useful.

I think that in my first letter I was so using the term "analysis" that to say that X is to be analyzed in terms of Y entails that it would be incorrect to say of anyone that he had the concept of X but lacked the concept of Y. (The converse entailment does not seem to hold.) Thus, when, in A-12, I denied that

". . ." means p

is to be *analyzed* as

". . ." expresses t, and t is about p

I intended, in effect, to deny that the fact that a person lacked the concept of a statement's expressing the thought that p would be a conclusive reason against supposing him to have the concept of a statement's meaning that p. And if that part of A-12, the negative part, which begins "in particular . . ." and which says all that I wanted to say in advancing A-12, is interpreted in this manner, surely it is entailed by B-3. For what, in effect, does B-3 say? Let me try the following paraphrase:

> It is conceivable that people might have made semantical statements about one another's overt verbal behavior before they had arrived at the idea that there are such things as *thoughts* of which overt verbal behavior is the expression.

Thus, as I saw it, the fact that you formulated and accepted B-3, taken together with your earlier statement that if I could persuade you of A-12 *perhaps* I could persuade you of the rest, gave good grounds for hoping that we might be approaching a substantial measure of agreement.

But even if we leave A-12 aside and stick with B-3, I think I can make my point. For once B-3 is granted, what alternatives are left for an account of the relation of the framework of thoughts and their aboutness to the framework of semantical statements about linguistic episodes? (Clearly it rules out the classical account according to which to say of a statement that it means that p is simply a concise way of saying that the statement expresses the thought that p.) One alternative, the alternative to which you seem to be committed, can be put—rather bluntly, to be sure—as follows:

> To say of a verbal performance that it means that p is to attribute to the performance a certain property, namely the property of meaning that p, and hence the generic property of meaning *something*. It is conceivable that a people might have come to recognize this property of verbal performances, their own and those of others, without realizing that there are such things as thoughts (not that they haven't been thinking all along), just as they might have come to recognize the moon's property of being luminous without having discovered the sun. Subsequently they (introspectively) notice thoughts and, on examining them, discover that they have in common the property of being about something. Comparing this property with that of meaning something, they discover that they are, if not the same property, at least properties of the same sort, in that being about something (in the case of thoughts) and meaning something (in the case of linguistic expressions) are alike ways of being intentional. They then establish that verbal expressions have the property of meaning something, and hence of being intentional, only if they stand in a certain relation to thoughts, whereas thoughts can have the property of being about something, and hence of being intentional, regardless of any relation they have to verbal expressions (though they also establish that thoughts above a certain level of crudity do not occur unless the thinker has learned a language capable of expressing them.) They conclude that thoughts are the source of intentionality.

I have put this alternative as bluntly as I have because I want to hammer away on the theme that the traditional puzzles about intentionality arise, in large measure, from the presupposition that because statements of the form "S means p" are often *true*, they must be capable of being gripped by such philosophical wrenches as "property", "relation", "attribute", "describe", etc.

For given this presupposition one will either say (assuming that one doesn't fall into the trap of philosophical behaviorism) that in the sense in which to be a bachelor is to be an unmarried man, or to be an *uncle of* is to be the *brother of a parent of*,

> for a verbal performance to mean something is for it to be the expression of a thought which is about that something,

or one will say, as in effect you do, not in your seven sentences, to be sure, but in that "eighth sentence" which is your commentary on the other seven, that

> the being about something of thoughts and the meaning something of verbal performances are similar properties (both ways of being intentional) not, however, by virtue of anything like the above, but rather by virtue of being, if not the same property, then two properties which stand to one another much as the luminosity of the sun to the luminosity of the moon.

You see, it isn't so much that I disagree with your seven sentences, for I can use each of them separately, with varying degrees of discomfort, to say something which needs to be said. (I think we would both have reservations about C-3, but none that is relevant to our problem.) It is rather that I am unhappy about the force they acquire in the over-all framework in which you put them.

What is the second alternative left open by B-3? It is, needless to say, the one I have been trying to recommend. It enables me to say so many of the things you want to say, that I can't help feeling at times that there must be some happy formulation which, if only I could hit upon it, would convince you. At other times I realize that our failure to agree may spring from a more radical difference in our general philosophical outlooks than *appears* to exist. If so, I doubt very much that the trouble lies in the area of "synonymy", "analysis", and such fashionable perplexities. Unless I am very much mistaken, it lies in the area of "fact", "property", "describe", and their kindred. A discussion of "ought" would provide a test case, but I hesitate to start that here. I will, however, send you (without obligation, as merchants say) a copy of a forthcoming paper of mine on the subject.[21] It would be interesting to see how you react to it.

As for your remark that "[your] natural temptation would be to say that the business of the sentence ' "Hund" means dog in German' . . . is to tell us that German-speaking people use this word 'Hund' to express their thoughts about dogs," I quite agree (a) that the sentence does tell us this, and (b) that to say that *the business of* the sentence is to tell us this is, as you imply, to espouse a certain philosophical interpretation of the sentence, and hence prejudge the question at issue between us. What catches my eye, however, is the fact that the philosophical interpretation which seems to be implicit in your "natural temptation" is none other than what I have called the "classical account," i.e., that for a verbal performance to mean something is for it to be the expression of a thought which is about that something. But this interpretation is incompatible with B-3, which you accept.

Now I grant, indeed insist, that there is a very intimate relation between " 'Hund' means dog in German" and "Statements involving the word 'Hund' made by German-speaking people express thoughts about dogs." Although my piece of historical—prehistorical—fiction expresses my conviction that semantical sentences could play their characteristic role even if those who used them lacked the framework of thoughts, I would not for one moment wish to deny that as we use these sentences there is a legitimate sense in which "x makes meaningful assertions" logically implies "x has thoughts." An example from another area may illuminate this point. As people once used the word "water", "x is a piece of water" clearly did not imply "x consists of molecules of H_2O"; but as chemist Jones (1937—) uses the word "water", even in everyday life, does or does not "x is a piece of water" imply—in a sense of "imply" which it is quite legitimate to call "logical"—"x consists of molecules of H_2O"?

It might be thought that by saying that the framework of thoughts developed as a theory develops, with semantical dis-

<hr />

[21] "Imperatives, Intentions and the Logic of 'Ought'," *Methodos*, 8 (1956). See also §§ 77–78 of my essay "Counterfactuals, Dispositions and the Causal Modalities."

course about overt verbal episodes as its model, I am making a purely *historical* point. One might be tempted to say that regardless of how there *came to be* such a thing as discourse about thoughts, discourse about thoughts now refer to thoughts as discourse about linguistic episodes refers to linguistic episodes and discourse about physical objects to physical objects. But the point I am making is both a historical *and* a logical one, nor can the two be separated as with a knife. Clearly the mere fact that

> "Thought" refers to thoughts
> "Sneeze" refers to sneezes
> "Ought" refers to obligations

are all of them true, has not the slightest tendency to show that the 'logic' of "thought" is like that of "sneeze" or "ought". To say that the term "molecule" was introduced by means of 'postulates' and 'coordinating definitions' (so called) is to make, however clumsily, a logical point about the way in which this expression *is* used, and this use is quite other than that of expressions which refer to macro-observables, even though

> "Molecule" refers to molecules

is every whit as true as the statements listed above.

Although I compare the framework of thoughts to that of theoretical entities I qualify this comparison in (at least) the following two respects:

(1) I grant, as anyone must, that each of us has direct (and privileged) access to his thoughts in the sense that we can (on occasion) know what we think without inferring this from our overt behavior as we infer what the molecules in a gas are doing from the behavior of the gas as a macro-observable object. Broad epistemological considerations, however, expounded in "Empiricism and the Philosophy of Mind," lead me to conclude that this "direct access" is to be interpreted as an *additional* role which the language of thoughts (and the thoughts about thoughts which they express) has come to play, rather than as a matter of a *replacement* of an original 'theoretical' framework of thoughts by a framework in which they are on a par with public observables.

(2) In saying that the framework of thoughts developed with semantical discourse about overt linguistic episodes as its model (thoughts being 'inner speech'), I have distinguished between two roles played by the model:

(a) The semantical concepts of the model appear in the framework of thoughts as the basic categories of intentionality. The fact that the model is *semantical* rather than Rylean discourse about overt linguistic episodes accounts for the fact that intentionality is a *necessary* feature of thoughts, it being absurd to say of anything that it is a *thought* but lacks intentionality.

(b) The Rylean facts about linguistic expressions which, as I have put it, are implied though not asserted by semantical statements about them are the model for the behavior of thoughts as episodes 'in the order of causes.'

This distinction between the two roles of the model is elaborated in my letter of September 19, and can be summed up by the formula that just as semantical statements about linguistic episodes do not describe, but imply a description, of these episodes, so statements about the 'content' or 'intentional object' of thoughts do not describe thoughts, though they imply a description of them.[22] It is the fact that these implied descriptions do not do more than draw an analogy between the way in which thoughts are connected with one another and with the world (in observation and conduct), and the way in which overt linguistic episodes are so connected—an analogy, however, which is qualified *inter alia* by the idea that overt linguistic episodes are the culmination of causal chains initiated

---

[22] I now (March, 1957) find it somewhat misleading (though not, as I am using these terms, incorrect) to say that statements as to what a person is thinking about *do not describe*, but, rather, *imply a description of* the person. This, however, is not because I reject any aspect of the above analysis, but because I am more conscious of the extent to which my use of the term "describe" is a technical use which departs, in certain respects, from ordinary usage. For the sort of thing I have in mind, see the discussion in §§ 78–79 of my essay "Counterfactuals, Dispositions and the Causal Modalities" of the sense in which the world can 'in principle' be described without the use of either prescriptive or modal expressions. It is in a parallel sense that I would wish to maintain that the world can 'in principle' be described without mentioning either the *meaning* of expressions or the *aboutness* of thoughts.

by thoughts—it is this fact which makes it sensible to envisage the identification of thoughts *in their descriptive character* with neurophysiological episodes in the central nervous system, in that sense of "identify" which we have in mind when we speak of the identification of chemical episodes with certain complex episodes involving nuclear particles.

Well, I don't know that we are any further along toward agreement than we were last August. I hope, however, that I have succeeded in clearing up some points about my interpretation of intentionality. And let me be the first to say that my present account, inadequate though it may be, is a substantial improvement over the gropings of my earlier papers. I can only plead that if they are looked at from the present vantage point, they seem to reach out for formulations which elude them.

Cordially,
Wilfrid Sellars

November 19, 1956

Dear Sellars:

I find very little to disagree with in your last letter (dated, I fear, October 19th). Most of what I wanted to say was in the points made in my previous letter, which you assent to. I think there is only one matter left.

I do concede your statement B-3, especially in the paraphrase you gave it in your last letter; that is, I concede that it is conceivable that people might make semantical statements about one another's verbal behavior before arriving at the conception that there are such things as thoughts; and I also concede that, given your account of analysis ("to say that X is to be analyzed in terms of Y entails that it would be incorrect to say of anyone that he had the concept of X but lacked the concept of Y") your statement A-12 follows. There is no point, so far as our present questions are concerned, in debating about the proper use of the technical term "analysis"; but I would note that, given this definition of "analysis", possibly we cannot say that Russell's definition of "cardinal number" is an analysis.

The only point I wish to make is this. The "paradox of analysis" reminds us that it is conceivable that people might have referred to certain things as "cubes" before they had arrived at the idea that there is anything having six sides. And, to borrow the example which Hempel borrowed from Neurath (see Feigl-Sellars, page 380),[23] it is conceivable that people might have referred to the fact that watches run well before they had arrived at the idea that there is a sun which bears certain relations of motion to the earth, etc., and I am inclined to feel that the sense in which I have conceded your statement B-3 is this "paradox-of-analysis" sense. If the people of your myth were to give just a little bit of thought to the semantical statements they make, wouldn't they then see that these semantical statements entail statements about the thoughts of the people whose language is being discussed?

With best wishes,
R.M. Chisholm

[23] Herbert Feigl and Wilfrid Sellars, eds., *Readings in Philosophical Analysis* (New York: Appleton-Century-Crofts, 1949).

13

Thinking

1. THE DISTINGUISHING FEATURES OF CONCEPTUAL THINKING

In opposition to the traditional assumption that there is a basic concept of thinking, contemporary philosophers have been inclined to argue that there is actually a very wide variety of activities properly called "thinking" but no special feature that is either common or peculiar to all of them.[1] I shall not begin by disputing this view, although I do hope to destroy its plausibility in the course of my argument. Initially, I shall concentrate on what is at least a distinguishable form of thinking, which philosophers might term "conceptual." Such remarks as "It just occurred to me that if I am to catch the evening train, I must leave at once" may be regarded as overt verbal expressions of this kind of thinking.

An obvious feature of the mental processes involved in conceptual thinking is that they are typically unobservable. When a man says that it occurred to him, on a certain occasion, that it would soon rain, the presumption is that something did occur to him, and that what occurred to him need not have been expressed in audible terms. Since the basis for general philosophical doubts about the presence of such inner occurrencies was undermined in Chapter V, there is no need to

From *Knowledge, Mind, and Nature*, by Bruce Aune, pp. 177–211. Copyright 1967 by Random House, Inc. Reprinted by permission of the author and the publisher.

[1] For a survey of this variety, see Gilbert Ryle, *The Concept of Mind* (London: Hutchinson, 1949), pp. 280 ff.

air such doubts here. We can rather safely assume that we do generally regard thinking as, at least sometimes, an unobservable activity or process, and turn our attention to the peculiar feature that such processes are thought to have.

One of these peculiar features—indeed the feature that, according to Franz Brentano, is really distinctive of the mental —is that thinking is necessarily "intentional," in the sense of referring to something.[2] Thus, in the example initially given, the object of the thought was (roughly) the thinker's catching the evening train. That thinking always has an object is perhaps disputable, if everything that might be called "thinking" is considered, such as letting your mind wander, or ransacking your memory for a name you cannot recall. But there certainly is a recognizable form of thinking, the conceptual form that concerns me here, which is such that if you claim to be thinking at all, you must be able to specify what you are thinking about.

Another feature of this kind of thinking is that it is "judgmental" in character, involving the forms of thought discussed by Kant. Excepting cases of reverie and the like, to think of something is at least to entertain the idea of something's being so, being done, and so forth. I say "at least entertain the idea" because not all thinking involves belief: one may consider possibilities, rehearse intentions, and imagine fanciful situations. What is common to these cases is that the distinguishable thoughts involved are best expressed by complete sentences. If you ask a man what he is thinking (not what he is thinking *of*), a full answer will have the form, "I was thinking *that* such and such." If the man merely responded with a single word, such as "Mary," it would be entirely reasonable, though not perhaps polite, to ask: "Well, what about her? I asked 'What were you thinking?' not 'What were you thinking about?' "

According to Kant, the forms of thought are as various as the

---

[2] Franz Brentano, *Psychologie vom Empirischen Standpunkt* (Hamburg: Meiner, 1955), vol. I, bk. 2, ch. 1. An English translation of this section (by D.B. Terrell) is included in R.M. Chisholm, ed., *Realism and the Background of Phenomenology* (New York: The Free Press, 1960). For a recent defense of "Brentano's thesis" concerning the mental, see R.M. Chisholm, *Perceiving* (Ithaca, N.Y.: Cornell Univ. Press, 1957), ch. 11.

possible forms of sentences. Like sentences, thoughts may be categorical (I thought that all A's are B's), hypothetical (I thought that if p then q), disjunctive (I thought that p or q), negative, and so on. Since Kant restricted the possible forms of thought to those abstract forms recognized by the logicians of his day as standard sentential forms, he recognized far less variety than one would today. Just as there is (to take a single example) an important difference between a "material" and a subjunctive conditional, so there is an important difference between the forms of thought involving these two kinds of conditionality: "I thought that p only if q" describes a very different thought from "I thought that if p were so, q would be so as well." In general, whatever can be said can also be thought; and for every logically distinct form of sentence, there is a logically distinct form of thought.

A final feature of conceptual thinking concerns what Collingwood called its "bipolarity."[3] Anyone who engages in this kind of thinking will do it well or carelessly, logically or illogically, intelligently or stupidly. In trying to solve a problem, the thinking he does will also be successful or unsuccessful. He will normally succeed or fail in reaching the desired conclusion to a theoretical problem, and he will typically succeed or fail in making up his mind what to do in some practical situation. This bipolarity will also extend to the "contents" of the man's thoughts. What he affirms or conjectures may be true or false; what he explicitly chooses to do may or may not be realized; and what he hopes may or may not come to pass.

Taking all of these distinctive features into account, we may characterize the kind of thinking termed "conceptual" by reference to four chief marks: (1) it may be silent or "inner," (2) it has reference or intentionality, (3) it involves some kind of form or conceptual articulation, and (4) it is essentially bipolar. Although these four marks are easily understandable in a general way, they are unfortunately difficult to understand in detail; in fact, most traditional accounts of conceptual thinking

[3] R.G. Collingwood, *The Principles of Art* (Oxford: Clarendon Press, 1938), p. 157.

have failed to do justice to them all.[4] In recent years, however, a theory of thinking popular in the fourteenth century has been revived, reworked, and polished to the extent that it seems tailormade to account for these peculiar marks. This is another analogy theory, and its recent development has again been independently initiated by Wilfrid Sellars and by Peter Geach.[5] In its new guise the theory is extremely complicated, with far-reaching, highly controversial implications. Because of this complexity, I shall begin by outlining the theory's chief contentions in a rather loose, informal way, and only later, by considering a number of penetrating objections, go on to probe its fine-grained structure. As in my treatment of the analogy theory of sense impressions, I shall concentrate on Sellars' version of the theory, since in my judgment he has worked it out in more satisfactory detail.

2. GENERAL FEATURES OF THE ANALOGY THEORY OF THINKING

The root idea of the analogy theory of thinking is that intelligent speech may itself be a form of thinking, a species of "thinking out loud." This form of thinking is also held to be far less problematic, philosophically, than anything silent or unobservable. This is not just because overt speech is audible to the public ear; it is also because the form of the speech and even its reference is normally aboveboard and checkable. It may be granted that a man's audible discourse is not always a pastiche of clichés and that it is frequently not possible to determine the reference and functional form of his words merely

[4] For a discussion of this point, see Bruce Aune, "Thinking," in P. Edwards, ed., *The Encyclopedia of Philosophy*, vol. 8 (New York: Crowell-Collier and Macmillan, 1967), pp. 100–104; also see ch. 3 of my book, *Knowledge, Mind and Nature*.

[5] See Wilfrid Sellars, "Empiricism and the Philosophy of Mind," in H. Feigl and M. Scriven, eds., *Minnesota Studies in the Philosophy of Science*, vol. I (Minneapolis: Univ. of Minnesota Press, 1956), pp. 253–329 (pp. 307–321 are included in this volume); also Sellars and Roderick M. Chisholm, "Chisholm-Sellars Correpondence on Intentionality" (included in this volume). For P.T. Geach's views, see his *Mental Acts* (London: Routledge and Kegan Paul, 1957).

by consulting grammars and dictionaries. Yet as we saw in Chapter IV, in order actually to *mean* something by a form of words a man must utter them in a certain frame of mind, which will involve a readiness to recognize other utterances as relevant or irrelevant to the content of what he said. Hence, if a man means something unusual by a familiar word, we can demand that he clarify his meaning in words we can understand. If he cannot do this, and if a persistent dialectical probing fails to disclose a consistent sense to his words, the presumption will be that he did not really mean anything at all, but perhaps just uttered his words in a state of confusion.[6]

An extremely important feature of intelligent speech is that it conforms to certain patterns, particular elements of which are often intelligible only by reference to others. If a man makes a certain remark, the point of what he is saying is often comprehensible only in relation to other things he has said or will say. These "other things" will often reflect his reasons for saying what he did, and they will be related to the latter by some principle of inference or relevance to which his speech normally conforms. As I have argued in Chapter IV especially, in mastering a language a man learns not just to identify various objects in his environment but also to conform his extended remarks to the general principles of inference, relevance, and the like, that his group accepts. If a man does not conform to these principles—if there is no connection between his remarks that can be followed by a sophisticated speaker of his language—then the presumption is that he is either linguistically incompetent or else, as would be more likely if his utterances were long and complicated, out of his mind. In either case his words would not

---

[6] The line between nonsense and bizarre sense is actually not as easy to survey as this might suggest. But it is at any rate only because of a kind of dialectical probing (in this case, of poems) that we can be assured that the following lines are only bizarre and not truly nonsensical: "I am aware of the damp souls of housemaids/ Sprouting despondently at area gates" and "what if a much of a which of a wind/ gives the truth to summer's lie . . . ?" (These lines are reprinted with the permission of Harcourt, Brace and World, Inc.; they are taken respectively, from T.S. Eliot's "Morning at the Window," in *The Complete Poems and Plays, 1909–1950*, and from E.E. Cummings, "what if a much of a which of a wind," in *Poems, 1923–1954*.)

be taken as an example of intelligent, thoughtful speech unless some special explanation were provided.

I do not propose to dwell on these principles here. I merely wish to point out at this stage that some utterances are often inexplicable except with reference to others. Some of the most striking cases of this are found in abstract studies such as logic and mathematics, but they are in fact ubiquitous in common life. If a man announces the solution to a difficult logical or mathematical problem, his announcement, let alone his confidence that he is right, is generally baffling until he indicates the line of reasoning that led him to it. When he outlines the steps he took in reaching his conclusion, we feel that his announcement is explained: we know how his solution was worked out. The idea that it might not have been worked out at all, but rather grasped intuitively as if presented by the gods, is always highly suspect and would not, in any case, explain the man's confidence that his solution is sound. So far as common life is concerned, it is axiomatic that solutions to difficult problems are reached only by a complex process of reasoning, involving sometimes a very large number of subordinate steps.

Similar considerations apply to the explanations advanced for any distinctive human action. Such actions are not just peculiar movements explainable in stimulus-response terms; they are rather goal-directed activities, conforming to the general principles of practical reasoning. If, to take an extreme case, a man kills another in one of the farfetched ways described in detective fiction, then his action is understandable, in all its particularity, only when the line of reasoning that led him to it is laid bare. The suggestion that the man's action did not spring from a process of reasoning is just as peculiar as the idea that one can discover the solution to a difficult mathematical problem without any calculation at all. In everyday life, it would never be taken seriously.

It can scarcely be denied that most of the reasoning we do, whether in calculation or deliberation, is not audibly expressed. Yet the fundamental contention of the analogy theorist is that the principles involved in this reasoning are fully exemplified in overt discourse and that, even if we had no conception of an inner mental episode, we could fully understand these principles and use them in explaining human behavior. Thus, he would contend that if a certain part of our community did all their thinking in audible terms, the same general principles of intelligent action would be exemplified in their behavior and would enable us to explain why they say and do the things that they do. The fact that these principles could be fully exemplified in overt behavior indicates, he would add, that they may be understood as entirely public principles, of a kind that could reasonably be expected to arise in a public context.

The claim of the analogy theorist is, accordingly, that these principles of reasoning are conceptually more basic than the notion of silent thinking, that the latter is, in fact, usefully regarded as an analogical development of the concept of saying, of making utterances in accordance with these principles. The utility of interpreting the concept of silent thought in this way is that it can then be understood as a thoroughly intersubjective one, even though it applies to publicly unobservable episodes. Thus, if we assume that at some time in the distant past we had absolutely no conception of an inner mental occurrence but were entirely at home with the notion of intelligent speech, it is easy to imagine how the concept of silent thought could be forced upon us by thoroughly intersubjective considerations. Men occasionally say or do things of such surprising complexity that it is hard to envisage their being done without calculation or deliberation. Rather than accepting it as a brute fact that such actions really are done on impulse, it would be natural to assume that the relevant thinking actually was done, though no one happened to hear it. Some men speak more softly than others, and the supposition would not be absurd that some deliberation or calculation can be done so softly that it is not audible at all. Pressing this line of thought would lead one to a conception of something like a Platonic dialogue in the soul or of a Biblical speech in the heart.

A conception of silent thought as literally a dialogue in the soul is of course highly imaginative and no more acceptable to a tough-minded philosopher than the idea that a man's soul is a subtle inner agent that duplicates, though at a deeper level, the intellectual features he himself has. Yet the conception of such an inner dialogue is capable of development and qualification, and its imaginative trappings can quickly be stripped away. The first bit of stripping to be done would involve the natural assumption that silent thinking is like overt speech in the way that the verbal imagery we typically have in reading a poem to ourselves is like the words we utter in reading the poem out loud. This assumption must be done away with because a good share of the hard thinking a man does is often not accompanied by any imagery at all, verbal or otherwise.[7] As a consequence of this, we can only hold that while thinking may sometimes involve imagery of a certain sort, this imagery is not essential to our thinking what we do think.

The immediate result of treating imagery, or other sensuous *qualia,* as logically inessential to silent thought is that we are left with the possibility of only a very tenuous and abstract similarity between overt and silent contemplation or deliberation. Yet if there is sense in saying that we might conceivably have been driven to speak of silent thought in order to explain what is generally explainable by reference to overt activities of reasoning, a mere *formal analogy* between thinking and speaking would be entirely adequate. To conceive of silent thinking in this way would be to regard it as involving a sequence of covert episodes, elements of which have the definitive property of being related to one another in the formal manner that elements of overt discourse are interrelated. By saying that the elements of silent thought are *formally* analogous to the elements of a corresponding line of intelligent speech, I mean that while the pairs may differ materially or empirically even more radically than utterances of "It is raining" differ from utterances of *"Es regnet,"* they nevertheless play analogous roles in, for instance, taking one from a given premise to a given conclusion. In this regard they would be, like arguments formulated in German or French, subject to criticism by the same formal or logical principles as apply to overt English speech.

It is important to observe that the kind of formal analogy

---

[7] On this see H.H. Price, *Thinking and Experience* (London: Hutchinson, 1953), pp. 235 ff.

discussed in Chapter VI was rather different from the kind involved here: there it was concerned with material character- istics while here it is concerned with what might be called functional characteristics. Thus, unlike the formal analogy be- tween a red triangle and a certain sort of visual image, the analogy between uses of the English "It is raining" and the German *"Es regnet"* does not concern their material (their phonemic or orthographic) features but the jobs they do or the roles they play in discourse. To the extent that these em- pirically different uses do play analogous roles in English and German, there is a sense in which the roles they play are ab- stractly the same. This "abstract sameness" is what is involved when one says that the English "I," the French *"je,"* and the German *"Ich"* all play the role of a first-person singular pro- noun.[8] I shall attempt to make the notion of such roles more precise as my discussion proceeds. For the moment, I shall merely say that while the concept of a linguistic role is most easily grasped by reference to the functional similarities of linguistic activities publicly carried on in different languages, it may also be applied to those inner episodes that, issuing in intelligent remarks and actions, *do the work* of audible or visible calculation and deliberation.

Assuming that the formal analogy in point is at least in- tuitively clear, it may nevertheless be objected that if thoughts are to be conceived as episodes that play certain formal or functional roles, we must surely conceive them as particular role-players, as determinate episodes that involve these roles. But this objection is easily met, since, aside from their purely formal features, these episodes are distinguished from one another by coordination with the distinctive verbal forms that are said to "express" them, and the frames of mind appropriate to standard uses of these verbal forms. Thus, just as an utter- ance is a particular linguistic move only when it is produced in a certain frame of mind—that is, in a certain state of readiness to make additional remarks, movements, and the like of a certain sort—so an inner episode is a particular thought only to the extent that it occurs when the thinker is in a similar frame

[8] See *Knowledge, Mind, and Nature,* p. 88, footnote.

of mind. In general, the frame of mind appropriate to the thought that *p* is essentially the same, formally, as the frame of mind appropriate to the assertion (or whatever) that *p*.[9] If this were not so, the notion of a silent thought could not serve, as it does, to explain the peculiarities of intelligent, non-habitual remarks and actions.

Notice that the preceding account allows both an intrinsic and an extrinsic characterization of silent thought, just as the account of impressions in the last chapter allowed both intrinsic and extrinsic modes of characterization. Intrinsically, thoughts are characterized by the formal role they exemplify; extrinsically, they are characterized by reference to their connection with specific verbal and other dispositions ("frames of mind") in ways that parallel the connection between their overt verbal expression and the frames of mind that are distinctive of them. The thought that *p* is, moreover, identifiable or describable by reference to the words we use for asserting, conjecturing, and the like, that *p*. This is done by two familiar verbal constructions: *oratio obliqua*, as in "The thought *that p*," and *oratio recta*, as in "He said in his heart something tantamount to '*p*'," where Geach's expression "said in his heart something tantamount to '*p*'" may be understood as equivalent to "having made a silent move that is a formal analogue of the assertion, '*p*'."

3. FUNDAMENTAL CLAIMS OF THE ANALOGY THEORY OF THINKING

When I first introduced the analogy theory of thinking, I remarked that it seems tailormade to account for the four distinctive features of conceptual thinking. It should be clear by

---

[9] I say "essentially the same" because the frame of mind appropriate to the assertion that *p* normally involves the disposition to repeat or rephrase the assertion if one is asked what one said or meant. But the frame of mind appropriate to the thought that *p* would be rather different in this connection: if asked what he said, a man who had silently thought that *p* but did not say that *p* would tend to reply, if English were his language, "I didn't say anything at all."

now what the criteria for the presence of these features can be. Consider, to begin with, the unobservable character of most conceptual thinking. According to the analogy theory, our basis for claiming that such thinking occurs in persons generally (not just in ourselves)[10] is that while nonhabitual intelligent action presupposes a certain amount of reasoning, many instances of such action are not accompanied by sufficient overt speech to account for this reasoning. Since the requisite reasoning was not done overtly, it must have been done covertly, *in foro interno*.[11] Thus, the fact that Jones was able to solve the problem without overt calculation suggests that he must have done this calculation in some covert manner. The problem was too difficult to be solved without calculation, and when Jones was asked to justify his answer, he outlined the steps of a complicated proof with no hesitation whatever. The wholly public behavior of announcing a solution or of outlining a proof is therefore acceptable as a normic indicator of silent thought because our conception of such thought, like our conception of pain, involves an explanatory hypothesis—this time one concerning the possible occasions of *nonhabitual, intelligent* human behavior.

As one might expect, a report of having thought that *p* may also be regarded as a normic indicator of having thought that *p*. The basic reason for this is that a man's admission or avowal that he thought that *p*, is a bit of intelligent behavior itself. As such, it also is to be explained by reference to various principles of intelligent action. Since the man's admission or honest avowal cannot be explained by reference to his wish to deceive or amuse, the relevant explanation will no doubt allude to his memory of what he thought. The general reliability of his memory is what justifies his confidence, and therefore our opinion, that his thought was this rather than that.

[10] Here we have to remember that the concept of silent thinking is an intersubjective one, whose general applicability is not established by reference to each man's subjective certainty that he engages in silent thought.

[11] This consideration seems to weigh heavily with contemporary psychologists who discuss silent thought. See esp. B.F. Skinner, *Verbal Behavior* (New York: Appleton-Century-Crofts, 1957), pp. 434 ff., and John Dollard and Neal E. Miller, *Personality and Psychotherapy* (New York: McGraw-Hill, 1950), pp. 110–115.

Next, consider the matter of a thought's form. Since a silent thought is conceived as an unobservable counterpart to an overt speech act, involving the same linguistic role, the form of the thought is the same as the form of the words used to express it. If, for instance, the overt verbal expression has the logical form of a conditional, then the form of the thought is also a conditional. The role of the thought in the thinking game is essentially the same as the role of the words in the asserting game. To appreciate this point, we must, of course, not confuse logical form with the sensible form—the visible or audible patterns—of linguistic tokens. The claim that a silent thought may have the form of a conditional does not in any sense imply that a man who has such a thought contains in his head a faint symbol of conditionality, such as a dagger or horseshoe.

The question of the sense in which thoughts are bipolar is closely related to the question of their form, since some of these forms are truth-functional. If a thought has the form of a so-called material conditional, it must play the same role in thought as a corresponding statement of that form plays in audible discourse. And whether something plays this role is logically independent of its specific empirical features. What is essential to such a role-player is that certain inferences may be drawn from it, whether these inferences are carried out in thought or in speech. Thus, if T is a thought of this conditional sort, then, assuming that it is true, there are other thoughts, P and Q, such that if P is known to be true, one is entitled to infer that Q is true, whether one actually does so or not. So far as the analogy theory of thinking is concerned, the truth-values of T, P, and Q will depend on the truth-values of their analogues in overt discourse.

It is important to note in this connection that the term "thought" is ambiguous in just the way that the term "statement" is ambiguous.[12] Both of these terms can refer either to an intelligent act (of thinking or of saying) or to the so-called

---

[12] The term "thought" is also ambiguous in that it sometimes means "belief." The concept of belief is analyzed on pp. 213–218 of *Knowledge, Mind, and Nature*.

intentional object of the act (to what is thought or to what is said). Strictly, the values true and false are not attributable to the act but to the object of the act. This does not imply, however, that these values are properly attributable only to abstract entities. On the contrary, my thoughts and my statements (in the sense in which others cannot think my thoughts or make my statements) are like the smiles discussed in Chapter VI: they are merely nominal objects, talk about which is reducible to talk about me. Roughly, "My statement or thought is true" is equivalent to "I speak or think truly" in just the way that "My smile is wan" is equivalent to "I smile wanly." Of course, there are other senses of "thought" and "statement" that are not susceptible of this kind of treatment. But the ambiguity of the terms relevant to the present discussion can be clarified without plunging into the questions of what propositions are and of how they are related to human acts of thinking and statement-making.[13]

The final distinctive feature of conceptual thought, namely its reference, is also readily understandable according to the analogy theorist's approach. Since the concept of thought is regarded as an analogical development of the concept of an assertion or statement, the reference of a thought will be conceived by analogy with the reference of an assertion or statement.[14] In the standard case, where the reference of an assertion is determined by general conventions, the words used will include a term or phrase that plays a certain referring role; and this role may be involved in thought as well. Thus, my thought, that p, may be a thought of the largest golden mountain because it involves a formal element that plays the same referring role as the expression "the largest golden mountain" plays in,

[13] On this see Wilfrid Sellars' papers, "Abstract Entities," *Review of Metaphysics*, 16 (1963), 627–671, and "Notes on Intentionality" (included in this volume).

[14] One must recall at this point that analogies are not identities or out-and-out similarities; they hold only in certain respects, and not in others. One respect in which a thought differs from an assertion is that thoughts cannot strictly be brought about voluntarily. The importance of this disanalogy between thoughts and assertions is indicated on p. 214, footnote, of *Knowledge, Mind, and Nature*.

say, a standard use of "The largest golden mountain exists in Peru."[15]

4. INTROSPECTION

One special virtue of the analogy theory of thinking is that it allows us to do justice to the notion of introspection. Ever since Wittgenstein's critique of private objects became influential, philosophers have been extremely suspicious of this idea on the ground that if there were such things as inner, episodic thoughts, there could be no intersubjective means of deciding whether or not they correspond appropriately to their objects.[16] Thus, if P represents a covert thought, there could be no way (it is argued) of ascertaining whether P does or does not consistently occur in connection with, say, pains rather than feelings of joy. Indeed, it would be hard to see how the assumed reference of these inner processes could ever be understood.

While it is admittedly impossible to determine the character of silent thoughts by any direct intersubjective means, the argument against behaviorism made it plain that a direct checkup on covert occurrences cannot reasonably be demanded. Given this, if we proceed to consider the familiar fact that children must typically *learn* to keep their thoughts to themselves, to avoid blurting out everything that comes to mind, we can see the point of regarding thoughts as often barely suppressed responses that take the place, for purposes of calculation or deliberation, of overt speech. And just as overt speech is the product of a well-entrenched set of habits of right response, so, we can say, is the inner episode that replaces audible utterance the product of a set of habits—in fact the same set of habits.

[15] On the notion of a referring role, see *Knowledge, Mind, and Nature*, ch. III, pp. 79 ff.; also p. 279 below.
[16] See Ludwig Wittgenstein, *Philosophical Investigations*, trans. G.E.M. Anscombe (Oxford: Blackwell, 1953), pt. I, secs. 256–273; and Norman Malcolm, "Knowledge of Other Minds," *The Journal of Philosophy*, 15 (1958), 976, reprinted in Malcolm, *Knowledge and Certainty* (Englewood Cliffs, N.J.: Prentice-Hall, 1963), p. 139.

Because these habits of response can plainly be checked for reliability, Wittgenstein's objection to inner identifications will not apply to the thoughts springing from them. On the present view there is no conceptual gap between thinking and speaking, and our readiness to say certain things on certain occasions will provide a normic criterion for the thoughts we happen to have. Any question, accordingly, about whether a man's silent thoughts may be only wildly related to his present feelings is easily answered on this approach, since he can always express them aloud and thereby allow others to determine, if necessary, whether they do or do not correctly describe his state.

Any theory able to account satisfactorily for thought's publicly unobservable character, for its intentionality, its logical form, and bipolarity, and do justice to the age-old notion of introspection as well, plainly has a good deal in its favor. Although I think that the preceding remarks go a long way toward showing that the analogy theory provides a fully acceptable theory of conceptual thought, I cannot claim that it goes all the way. I must therefore pursue its analysis on a much finer grained level. I shall attempt this by considering a number of objections, most of which are familiar in the literature. The first objection is a very simple one, but it will serve to prepare the way for the others, which have very deep roots indeed.

5. IS THE ANALOGY THEORY OF THINKING EXCESSIVELY CONTRIVED?

This objection arises from the consideration that the idea of an occurrence whose essential features are mainly formal is far too abstract and sophisticated to belong to the commonsense picture of man and the world. Surely, the commonsense notion of thinking has more substance than this. The tendency of philosophers to advance imagistic theories must obviously have some roots in common conceptions, and these conceptions are assuredly reflected in the lady's reply to the professor who announced his opposition to all such theories: "But you really do think, don't you?"

Although this objection no doubt does justice to the initial claims of the popular consciousness, it nevertheless goes astray in a crucial respect. For we do ordinarily—that is, all of us, in our nonprofessional moments—speak of silent thought, and we characterize it in *verbal* terms, by giving its gist in indirect discourse. The idea of thinking as something like overt speech is not only common to Plato, the Bible, and so-called stream-of-consciousness novels like Joyce's *Ulysses*, but it is implicit in our ordinary manner of conveying the content of our thought, which is propositional rather than pictorial. Besides, the imaginative conception of thought as a series of mental pictures is never really pressed in ordinary conversation. No one would actually insist that there are images such that if a man does not have them he cannot be thinking of Siberia or infinity, and a man's failure to detect images would never be taken as a proof that he was not thinking at all. Hence, while it is undoubtedly difficult for an ordinary person to regard thinking as occasionally wholly transparent, it is not unreasonable to say that the analogy theorist's claim is enshrined in the total corpus of our well-considered admissions regarding thought—that it may occasionally be entirely free of imagery, that its gist is best, most accurately and completely captured by sentences rather than pictures, and so on. Since the theory does have the advantage of laying bare the thoroughly intersubjective principles according to which the concept of silent thinking may have developed in a social context, it is surely not unreasonable to take it as advancing a plausible analysis of a common notion, an analysis that is perhaps surprising only because it is purified or clarified, its imaginative trappings cleanly cut away.

6. AN ALLEGED CIRCULARITY IN THE ANALOGY THEORY OF THINKING

While the last objection was somewhat superficial, this one is very serious indeed. According to it, the analogy theory fails because it is basically circular, presupposing ideas it purports

to analyze. In attempting to elucidate the notion of silent thinking by reference to overt assertion, it tacitly assumes that the
latter could be adequately characterized without reference to
silent mental activities. Yet this assumption can be seen to be
false as soon as one focuses attention on the frame of mind appropriate to an assertion. As is clear from the argument of
Chapter IV, such a frame of mind will always involve a readiness to make further utterances if the speaker is asked, *hears,
and thus understands,* certain questions that might be put to
him in order to illuminate the actual claim that he has made.[17]
Also, if his assertion concerns the sensible qualities of a thing,
such as its color, his frame of mind will normally involve a
readiness to identify other instances of the qualities he is talking about. But to identify an object as having a certain color—
to *see* it *as being* of this color rather than that—is not just to
have a particular sense experience; it is to heed or notice something. This being so, it follows that if we are to specify everything involved in using language to make assertions, we shall
have to make some reference to silent mental activities—of
taking certain noises to be questions of this or that sort, and of
hearing, seeing, or feeling *that* such and such is so. Since the
notion of an overt assertion necessarily involves a reference to
such activities, it is clear that the analogy theorist's attempt
to analyze the latter by reference to the former is inescapably
circular. If we really had no conception of silent mental activities, we could not understand the explication the theory
provides.

In my view this objection is entirely successful in ruling out
any form of the analogy theory of thinking that rests its case on
an unqualified analogy between silent thought and intelligent,
audible speech *as we presently conceive the latter.* I fully agree
that our conception of such speech is now so rich and complicated that it enshrines the notion of silent thought already,
and thus could not be used to elucidate it. But even given this
admission, the spirit of the analogy theory of thinking can still
be saved by a fairly obvious amendment. One of the central

[17] See *Knowledge, Mind, and Nature,* ch. 4, sec. 1.

themes of preceding chapters was that the enrichment of a conceptual system tends to modify even the basic ideas of that system. If we take this theme seriously, an obvious strategy in meeting the above objection arises at once. The basic analogy must rather be drawn between silent thought and a protoversion of overt speech, a more primitive form of speech whose elucidation does not involve reference to, or mention of, silent mental activities.

It might naturally be thought that resorting to protoconcepts here is resorting to the fictional in order to make sense of the actual, or resorting to the dubious in order to clarify the obvious. But to take this attitude is to misunderstand the basic point that the analogy theorist is eager to make. What he wants to show, fundamentally, is that the notion of silent thinking can be seen as an essentially public one, derived from other notions that apply unproblematically to publicly accessible phenomena—in a sense of "derived" in which, say, the concept of lying can be said to be derived from a simpler notion of ingenuous assertion. Yet, when one considers such derived concepts, one sees that they can rarely be boiled down to the versions we now have of the concepts they presuppose. It is clear, for instance, that circularity would necessarily result if we tried to explicate "lying" in terms of "not" and "telling the truth," for our notion of telling the truth now involves the idea of having no intention to deceive. Still, there are very good reasons for thinking that *lying* is a very sophisticated concept, which is in some sense a development from a simpler notion of ingenuous assertion. This simpler notion would have to have the status of a protoconcept, since our present notion of an honest assertion is a good deal more sophisticated.

In order to exhibit the structure of the concept of silent thinking—to show that it can be understood as a derived one, built on notions that might naturally arise in a social context —we must then be able to outline the basic form of various protoconcepts. To do this is not to suggest that, in point of historical fact, we actually did operate with these protoconcepts at an early stage of our conceptual development. It is rather to indicate that the concept of thought we now have is understandable as a development from familiar principles that can apply, in a philosophically unproblematic way, entirely to overt linguistic phenomena.

Let us suppose, then, that at a certain stage of our intellectual development our linguistic resources were such as to permit us to ascribe very complicated dispositions to our fellows but not to allow reference to any inner *mental* phenomena.[18] With reference to such a linguistic framework, we can easily define less sophisticated counterparts to our familiar notions of seeing, trying, wanting, and the like. To protosee X, for instance, would involve training your eye on it, having some kind of sensory experience, and then uttering, or gaining a short-term disposition to utter, the words "X is. . . ." To prototry to secure Y would involve uttering, or being disposed to utter, the words "I want Y," moving in Y's direction, groping at it, and so on. A protoconcept of asserting could then be given in a fairly natural way. To proto-assert that *p* would be to utter appropriate sounds in a certain frame of mind. This frame of mind would be characterized entirely in prototerms: roughly, to follow up the one utterance with others of certain kinds, depending on what one protosees or protohears; to make movements of various sorts, depending on what one protosenses, and so on.[19]

Now, protoreasoning would be entirely overt reasoning, involving successive utterances. To the extent that one understands what this reasoning is, one will be in a position to see how one might act or speak as a result of a proto-inference from other information. Having this concept of inference will put one in a position to explain certain forms of intelligent, nonhabitual behavior. The notion of a covert protothought could then be constructed: an inner episode that is formally analogous to a proto-assertion.[20] The episode would be analo-

---

[18] Cf. Sellars' Jonesean myth in "Empiricism and the Philosophy of Mind," pp. 219 ff. of this volume.

[19] See *Knowledge, Mind, and Nature*, ch. 4, pp. 91f. and 100–101, for the basic idea of how this account would have to be filled out.

[20] This is, again, highly simplified, since not all silent thoughts would be assertive, even at this primitive stage.

gous to the assertion in the sense that its elements would be related to one another in the formal way in which the elements of the assertion are interrelated; in the terminology used earlier, the same abstract assertive role would be involved in both of them. Also, of course, each element of the protothought would be related to appropriate verbal and other dispositions, these being similar to those characteristic of the corresponding proto-assertion.

Once the concept of protothinking is in hand, virtually non-protoconcepts of seeing, hearing, and the like, could be developed, and so, stepwise, the notions of thinking and asserting as we presently understand them. This development would, I believe, meet the objection mentioned above; if the analogy is conceived as holding only between something like proto-asserting and silent thinking, then no circularity is involved whatever. But if, on the other hand, it is conceived in a less guarded way—as Geach, for instance, seems to conceive it[21]—then it runs into the objection head on.

7. AN OBJECTION BASED ON THE CONCEPT OF INTENTIONALITY

Although the amendment just given may be taken to save the analogy theory of thinking from one charge of circularity, it unfortunately prompts another such charge in its turn. This one is based on the consideration that the protoconcepts (of seeing, trying, saying) used to elucidate the notion of silent thinking are not as neutrally behavioral as they seem. Like the concept of silent thinking itself, they essentially involve the idea of intentionality. And when this idea is carefully examined (the objection goes), it can be shown to presuppose the concept of thought. Thus, in outlining a protoconcept of asserting, I tacitly introduced the feature of intentionality by reference to the language used. Yet language has intentionality (that is, reference) only because it is used to express thoughts. It is only

[21] See Geach, *Mental Acts*, ch. 17.

because people *mean things* by words that words can refer to things at all.[22]

To this objection, the immediate reaction of anyone sympathetic with the argument of preceding chapters ought to be that it is completely wrong-headed. If it is granted that the activity of thinking necessarily requires a conceptual scheme of some sort, the claim implicit in the objection would seem to be that human beings inherently possess a prelinguistic conceptual scheme in terms of which they can conceive the world in detail, and by reference to which they confer meaning on words of public languages. And surely all claims of this sort are false. We of course can define new words once we have a language, but we cannot, in point of fact, conceive our world in detail prior to having some language or other. On the contrary, as the later work of Wittgenstein and even the early work of the pragmatists has shown,[23] we develop our conception of the world in the very process of mastering a language that allows us to talk about it.

In spite of all this, there remains an important sense in which words can be said to possess meaning only *because* people mean things by them. If we look to the later work of Wittgenstein, we can easily develop this sense without commitment to prelinguistic concepts. Using his terminology, we can say that words have meaning because they are caught up in a system of characteristic human activities, because they play a special role in a certain "form of life."[24] The activity of using words in accordance with this "form of life" is what, in fact, using them to mean something seems to amount to. As I have shown, to use a word purposely in its conventional sense is to use it with intentions and expectations of a fairly specific sort; it is to use it when in a certain "frame of mind."[25] This frame of

[22] This is Chisholm's standard objection to linguistic theories of thinking. See *Perceiving*, ch. 11.

[23] I have C.S. Pierce especially in mind here. For a useful discussion of his views, which are often very similar to mine, see W.B. Gallie, *Pierce and Pragmatism* (Harmondsworth: Penguin Books, 1952).

[24] Wittgenstein, *Philosophical Investigations*, pt. 1, secs. 19, 23, *et passim*.

[25] See *Knowledge, Mind, and Nature*, ch. 4, and also p. 252 f. above.

mind will involve a complex state of readiness to respond in various ways (with movements, actions, other words) if this or that should eventuate. The peculiar pattern of these responses, being essential to a purposeful and understanding use of a conventional expression, is part of what a linguistic "form of life" consists in.

Although this favorable interpretation of the claim that words have meaning because people mean things by them would seem to be acceptable to a great many philosophers today, it unfortunately leaves us open to a modified form of the above objection, which is extremely difficult to refute. As so modified, the objection is best expressed negatively, as a challenge: "Can one, in point of fact, characterize the relevant form of life without using the intentional notions it is supposed to elucidate? It seems clear that this cannot be done. To make sense of a language-using form of life one will inevitably have to employ such intentional notions as human purpose, human need, and human interest—notions that immediately involve the phenomena of human consciousness. That notions of this sort will be essential should be no surprise; for as Brentano argued a long time ago, it is the phenomena of human consciousness that are the fundamental bearers of intentionality."

In my view this challenge is a profound and exciting mixture of insight and error. But to explore its force, it is necessary to have a more accurate grasp of the notion of intentionality. So far, this notion has been given only an intuitive significance, in terms of the reference of words and thoughts. Statements about mental phenomena have, however, certain logical peculiarities, which have been called "marks of intentionality."[26] It will be necessary to attend to these peculiarities in order to make an intelligent attack on the question at issue. What is in dispute is whether a language-using form of life can be adequately *described* without reference to inner, mental activities. If it should turn out, for example, that the logical peculiarities of mentalistic statements are *sui generis*, it would follow that

[26] Chisholm, *Perceiving*, ch. 11.

the mentalistic concepts involved in these statements could not be derivative from a more primitive concept of a language-using form of life.

8. THE MARKS OF INTENTIONALITY

According to Roderick Chisholm, a distinguished contemporary defender of Brentano's thesis, sentences used to describe psychological phenomena exhibit at least one of three distinctive marks of intentionality.[27] The *first mark* is exhibited by a simple declarative sentence if it contains a substantive expression (a name or descriptive phrase) in such a way that neither the sentence nor its negation implies either that there is or is not something to which the substantive expression truly applies. Chisholm's example of such a sentence is "Diogenes looked for an honest man." This sentence contrasts very sharply with such nonintentional (or extensional) sentences as "Jones rode a horse," which implies, if it is true, that there is a horse that Jones rode.

The *second mark* of intentionality is exhibited by noncompound sentences that contain a propositional clause in such a way that neither the sentence nor its negation implies either that the propositional clause is true or that it is false. An example of such a sentence is "Jones asserted that demons cause schizophrenia"; neither this sentence nor its negation implies either that demons do or do not cause schizophrenia. This

[27] Since writing his book *Perceiving* (see note 15), Chisholm has modified his views. He would still agree, I take it, that every psychological statement has one of the marks I mention, but he would deny that only psychological statements have such marks. In a recent letter to me he formulated the following sufficient condition for a more stringent sense of "intentionality," which I quote with his kind permission: "Consider the following two formulae: (1) (Ex) (Ey) $(y = a \& xRa)$, (2) (Ex)(Ey)$(y = a \& xRy)$. An expression which may occupy the place of 'R' [such as: 'believes that ——— is bald'] is intentional if there is an individual term which may occupy the place of 'a' with the result that (1) does not imply (2), and (2) does not imply (1)." I have not amended my discussion in the text to take account of this revised formulation because I believe that a discussion of his original marks is more illuminating. But see footnotes 38 and 43 below.

sentence plainly differs from such nonintentional sentences as "It is not the case that Jones died in India."

The *third mark* of intentionality is exhibited by sentences containing terms having what Frege called "indirect reference." Chisholm describes this mark of intentionality as follows:

> Suppose there are two names or descriptions which designate the same things and that E is a sentence obtained merely by separating these two names or descriptions by means of "is identical with" (or "are identical with" if the first word is plural). Suppose also that A is a sentence using one of these names or descriptions and that B is like A except that, where A uses the one, B uses the other. Let us say that A is intentional if the conjunction of A and E does not imply B.[28]

An example of a sentence of this sort would be "Tom believes that Cicero was a Roman"; from this sentence and the identity-statement "Tully is identical with Cicero" one cannot validly infer that Tom believes that Tully was a Roman.

As Chisholm points out, these marks of intentionality characterize sentences that are logically atomistic, that is, not truth-functional compounds of simpler sentences.[29] In order to get a more general criterion of intentionality, we can say that a sentence is intentional if and only if either it or one of its component sentences is intentional. Since it is possible to transform sentences with marks of intentionality into sentences which do not possess those marks by introducing technical terms (as we can transform the intentional "Jones is hunting demons" into "Jones is demonhunting," which lacks intentional marks), we shall say, following Chisholm, that a sentence is intentional even if, lacking intentional marks, it contains technical or indeed any other terms whose meaning can be given only by reference to sentences that are intentional.

In order to avoid confusion in what follows, note that in speaking of sentences as intentional, another sense of the word is being introduced. As used earlier in this chapter, it is states or episodes that are intentional. Whether mental or verbal, they

---

[28] Chisholm, *Perceiving*, p. 171.
[29] *Ibid.*, p. 172.

are intentional because they possess a reference or so-called intentional object. It is in the very different sense of possessing the "marks" Chisholm describes that sentences are intentional. This double use of the term "intentional" should cause no difficulty in subsequent discussion, for the sense in which it is used will always be clear from the context. There could in any case be little occasion to confuse a purely formal property of sentences with the "reference to an object" of a mental or verbal episode.

9. A NOTE ON SCIENTIFIC LANGUAGE

Before proceeding with the objection raised in section 7, it will be useful to unload a few more cards on the table, so that the scope of the problem at issue is more easily appreciated. According to distinctively empiricist philosophies of science,[30] the language of scientific theory is ideally extensional, with no room for expressions with Chisholm's marks. The ordinary language of scientific investigation is admittedly not extensional in this sense, for it contains modal locutions such as "it is possible that," "it is necessary that," and "If A were to happen, B would happen as well." But the empiricist philosopher typically argues that locutions of this sort ideally ought not to be there, either on the tough-minded ground that modalities are utter fictions, or on the more moderate ground that they reflect the structure of our language rather than (as their appearance in scientific language suggests) the structure of our world. According to this last alternative, modal terms are confusing "material mode" surrogates for metalinguistic expressions. The locution "It is physically necessary that p" is an unperspicuous representation of the metalinguistic fact that the statement that p is unconditionally assertable in virtue of some accepted theory; and the locution "If A were to happen, B

[30] The classic statement of this point of view is found in Ludwig Wittgenstein, *Tractatus Logico-Philosophicus*, trans. C.K. Ogden (New York: Harcourt, Brace, 1922). In this connection also see Rudolf Carnap, *The Logical Syntax of Language* (New York: Harcourt, Brace, 1937).

would happen as well" (which seems to point to "real connections" in nature) is an unperspicuous representation of the metalinguistic statement that, given normal conditions, "B happens" is inferrable from the premise "A happens."[31] On either of these alternatives the empiricist's claim is basically the same, however. It is that the language of science is ideally extensional. Anything essentially nonextensional is to be regarded as at best reflecting metalinguistic ideas, which apply to the structure of our linguistic framework rather than to the world itself.

It is not my purpose to defend this conception of science here, although I shall treat it sympathetically and even base certain remarks on the assumption that its fundamental claim is sound. But because it is a widely held conception among philosophers and methodologists of science, it follows that Chisholm's views about the irreducibly nonextensional character of discourse about psychological phenomena are extremely explosive. If the language of science *is* ideally extensional, Chisholm's view has the consequence that a science of psychology—to the extent that it is a science of the mental— is strictly impossible: there would be mental phenomena about which we could not construct a proper science. The shock value of this statement for any empiricist philosopher ought to be sufficient to indicate the magnitude of the issue that the mild words of Chisholm really generate.

10. AN IMPORTANT OBJECTION REFORMULATED

With this richer conception of the intentional, we are now in a position to probe the objection of section 7 with much greater

---

[31] This is only a very rough statement of the position in point. For a more adequate discussion, see Wilfrid Sellars, "Counterfactuals, Dispositions, and the Causal Modalities," in H. Feigl, M. Scriven, and G. Maxwell, eds., *Minnesota Studies in the Philosophy of Science*, vol. II (Minneapolis: Univ. of Minnesota Press, 1958), pp. 225–308. An earlier and less complicated approach along similar lines can be found in Gilbert Ryle, " 'If,' 'So', and 'Because'," in M. Black, ed., *Philosophical Analysis* (Ithaca, N.Y.: Cornell Univ. Press, 1950), pp. 323–340.

care. The point of that objection was that although we may be able to characterize the intentionality of language by reference to certain forms of life, these forms of life can in turn be adequately characterized only by reference to the thoughts, beliefs, intentions, and interests of the persons who live according to them. Consequently, any attempt to elucidate the intentionality of thought by reference to the intentionality of language is necessarily circular. Since the analogy theory of thinking is built on this circularity, it follows that it must be rejected as fundamentally unsound.

In order to assess the precise force of this objection, we must be entirely clear about two points: first, we must know which form of life is relevant to the issue; second, we must know which mental phenomena are supposed to be required for describing this form of life. Respecting this first point, the amendment to the analogy theory given in section 6 must be kept firmly in mind. According to this amendment, our ordinary mental concepts are to be elucidated by reference to a family of protoconcepts. This being so, it will have to be the protoconcepts whose intentionality is to be understood in relation to a particular language-using form of life. Plainly, the form of life relevant here will be an extremely primitive one; it will not be the one we actually have. The objection in question, therefore, must be that even this primitive form of life could not be adequately characterized without reference to mental phenomena.

Concerning the second point, it should be noted that proto-assertion is intentional both in the sense of referring to something and in the sense of satisfying Chisholm's marks.[32] Given this, it would appear that if Brentano was right about intentionality, proto-assertion is either mentalistic itself or else definable by reference to something that is mentalistic in some more exacting sense. Anyone backing the objection in question must obviously endorse the last of these alternatives, for the analogy theory of thinking as I have developed it is built on the

[32] More exactly, a given proto-assertion such as "Tom is tall" may be said to refer to Tom while a sentence such as "Bill proto-asserted that Tom is tall" will satisfy one of Chisholm's marks.

idea that proto-assertion may be an act of conscious thinking itself, and hence be as mental as any other act of thought. Accordingly, the basic questions at issue seem to boil down to the following: "Is not a proto-assertion intentional because, and only because, it is the manifestation of an *inner* mental process? And will not every attempt to account for this intentionality by reference to a primitive form of life be inescapably circular on the ground that an adequate description of the latter will require a reference to inner mental episodes?"

Set out in this way, the force of the objection is by no means obvious. A natural way of evaluating it—and of tracking down the ultimate source of intentionality—might be to consider just what minimal addition to a purely extensional account of the relevant form of life would be sufficient to account for the intentionality that clings to proto-assertion. If it turns out that this minimal addition will necessarily involve an explicit reference to an inner mental process, then the objection will have to be regarded as successful. But if nothing of this sort actually has to be mentioned, then the objection can reasonably be said to fail.

11. AN EXTENSIONAL COUNTERPART TO PROTO-ASSERTION

An extensional counterpart to the activity of proto-asserting is easily envisaged in view of previous discussion, particularly the discussion of Chapter IV. From an extensional point of view, what obviously happens in proto-assertion is that an agent emits certain noises while in a frame of mind (state of readiness) to emit certain other noises and perhaps to make certain physical movements, depending on what happens, occurs, or exists in his immediate sensory vicinity. Just what these further noises or movements would be in certain circumstances could be stated, of course, only by someone who knows the language the agent is speaking and is aware of the peculiar dispositions (the proto-interests) the agent has.[33] But assuming that this knowledge is

---

[33] Compare *Knowledge, Mind, and Nature*, ch. 2, pp. 213 ff.

in hand, it would not, in principle, be impossible to construct a complex description that would represent, in entirely extensional terms, the full pattern of movements and noises that characterize the man's frame of mind and, consequently, the empirical features of his entire speech-act.[34]

Assuming that we can imagine how such an extensional description could be constructed, the next step is to ask whether anything logically essential to proto-assertion has been left out. In approaching this question it is essential to see that from a strictly scientific point of view, nothing else need be involved in proto-assertion than what an extensional description of this sort would disclose.[35] An organism simply emits certain noises while disposed to emit a specific pattern of further noises and, perhaps, to make certain patterns of physical movements, depending on various external features of its immediate sensory environment. One might be inclined to dispute this contention very hotly on the grounds, first, that even an aseptic scientific account must allude to the linguistic rules the speaker is following and, second, that the conditional utterances and movements distinctive of his frame of mind will depend, not on what his future surroundings might actually be, but on what he will take them to be, or on how he will interpret them. But both of these points are implicitly accommodated by the extensional description envisaged above.

Consider the agent's linguistic rules. From an extensional point of view, such rules can affect a man's behavior in two logically distinct ways. Either his behavior is merely in accordance with them, as it would be if he were incapable of formulating them, or else it results from his conception of them, which means that he is able to formulate them and indeed "has them in mind" when he acts. The first alternative is obviously accommodated by the extensional description, since the entire range of the man's rule-conforming is laid out in detail.

---

[34] This, in effect, is what Willard Van Orman Quine tried to do in ch. 2 of his *Word and Object* (Cambridge, Mass.; London: John Wiley, M.I.T. Press, 1960).

[35] It is granted that patterns of neural activity might be held to be essential as well.

The second alternative is also accommodated, though perhaps less obviously, because by hypothesis the man belongs to a community of proto-asserters, who do all their thinking out loud. If, then, he acts on a formulated rule, his formulation of the rule will consist in overt verbal behavior, the connection with his other behavior (his other utterances and movements) being included in the extensional description.

The matter of the man's taking or interpreting his surroundings to be such-and-such and then acting appropriately is treated similarly. For he is, by hypothesis, capable only of prototaking —and this is nothing other than his describing, or being disposed to describe, his surroundings in such-and-such a way; which is to say, his applying, or being disposed to apply, certain words to them. And applying words to something is, from an extensional point of view, just a matter of uttering certain sounds in a particular set, in a particular state of readiness to utter certain other sounds and make other movements, depending for instance on the particular sensations one has.[36] Applying a word to an object is not to be understood, after all, as a matter of conceiving the object in some natural, innate conceptual system, and then relating this conception to the word one utters by *another* psychic act. If the presence of the object in one's immediate sensory vicinity is the occasion for a variety of language-entry moves that we know (on other grounds) to be appropriately related to the original move one made, then there is no longer any general question about whether one's original move did or did not involve a reference to that object —let alone a question that could be satisfactorily answered (in a public context!) by reference to a shadowy domain of psychic facts, incapable of inclusion in the above account.

But granting that from a narrowly scientific point of view the extensional account omits nothing strictly essential to proto-

---

[36] Here we must keep in mind a point developed in earlier chapters, namely that in order to elicit a response from a man in a certain mental set, these sensations need not be interpreted by him. The basic reason for this is that the response that in many cases is elicited by the sensation is the mental act that interprets it: for example, a throb of pain may elicit the response (given a certain set), "It hurts!"

assertion, is there not a broader sense in which something is very obviously missing from that account? The answer, of course, is "Yes." The specific interpretation that sounds get *as linguistic elements* is simply ignored. Anyone who did not know what the sounds picked out by the extensional description *mean* to the people who use them would not be able to tell, merely from scrutinizing that description, just what assertion a particular utterance was. Admittedly, if one happened to know the language in which the utterances are to count as tokens, one could very easily decide which assertions are normally made when they are uttered. But to one who did not have this information, the extensional description would provide no decoding of the utterances at all.

The basic reason for this last point is simply that no matter how complex the extensional description may be, it will provide nothing more than *de facto* correlations of different sound patterns with one another, with physical movements, and with certain features of reality that *we* might regard as occasioning them. But mere correlations cannot possibly capture the meaning of a linguistic element. Not only does the meaning of a word relate to possible as well as actual things or episodes, but its application cannot in any way be determined by the occasion, marked in terms of *our way* of viewing reality, on which it is uttered. This can be determined only by criteria internal to the conceptual scheme in point, criteria that define the relevant *kind* of thing or episode involved. Hence, to understand the application of various words we must have a conception of the *sort* of thing to which they can legitimately apply. Similarly with word-word relations. To know what certain noises mean in a certain linguistic community, we must know how they *may* be related so as to form intelligible utterances, descriptions, and so forth. And what *may* be done—what is legitimate, permissible, or sanctionable—cannot be boiled down to what *is* done.

This brings us to the heart of the matter. To know what proto-assertion is being made is to know, not just what is likely to follow upon its utterance (what noises, what movements), but what *may* be inferred from it, what *must* be the case if it is true, reasonable, or appropriate. These "mays" and "musts" are essential to the notion of an assertion because a linguistic move of this sort is possible only in relation to a system of linguistic norms or rules. It is, after all, norms or rules that specify the defining characteristics of assertions: that they have implications, denials; that they are clear, confused, consistent, self-contradictory, tautologous, and the like.

When this point is appreciated, it is easy to see why the extensional account can be, in its way, entirely complete and yet seem fearfully denuded of significant flesh. For to characterize an utterance, suitably produced, *as an assertion* is not to call attention to its empirical features—let alone describe it as springing from some arcane inner episode in connection with which such intentional objects as golden mountains "inexist." To characterize an utterance in this way is rather to subsume it under a network of essentially normative concepts. The description of an utterance as a proto-assertion is thus similar to the description of an arm-movement as a signal, a bodily interaction with a rubber sphere as a serve, and a peculiar adjustment of an ivory piece on a checkered board as a checkmate.[37] All these actions—asserting, signaling, serving, checkmating—have extensionally describable counterparts, but their identity as the specific acts that they are can be understood only with respect to the particular system of principles and rules that properly specify them.

But granting that this is so—granting that the notion of a proto-assertion, or indeed of any other linguistic act, is in part a normative one—just where does intentionality enter the picture? Is it perhaps just a feature of certain normative notions? And if it is, how exactly is this feature to be analyzed?

The answer to this first question seems to be obvious, once one considers it. As defined by Chisholm's three marks, intentionality is patently a feature of normative statements. Con-

---

[37] This is not to say, of course, that acts of signaling, serving, and checkmating do not have a mental component. This aspect of such acts will be discussed fully.

sider "It is right to repay one's debts," which exhibits his first mark, and "It is morally obligatory that men refrain from murder," which exhibits his second mark. Anyone who accepts the familiar principle that prescriptive discourse cannot be reduced to, or analyzed in terms of, "naturalistic" discourse should surely balk at the idea that the intentionality of these two statements is wholly due to naturalistic facts about minds, even if these facts are themselves mysteriously intentional in some way. The same is true of Chisholm's third mark: the fact that "It is a necessary truth that 9 is greater than 6" is intentional does not seem to require explanation or analysis in terms of facts about human minds, either. We could not know that this statement is true if we did not have minds, nor could we formulate it. But it does not follow that its truth, or even its meaning, must be analyzed by reference to inner mental processes.[38]

The fact that these three normative statements possess the marks of intentionality surely makes it tempting to trace the intentionality of proto-assertion to its essential normative features.[39] After all, a proto-assertion is not an inner mental process nor, as I have defined it, does it necessarily spring from such a process. But in order to evaluate this temptation it will be necessary to follow up the question raised a moment ago: "How, with respect to normative principles, is intentionality to be analyzed?" This is a very difficult question; and in order to work my way to what I believe is its answer for the special case of proto-assertion, I shall comment on another analogy, this time one holding between the normative activity of asserting and the rule-governed activity of playing chess. There

[38] According to Chisholm's most recent criterion, none of the three sentences I have mentioned is necessarily intentional (see footnote 27 above). This fact is not, however, damaging for the purposes of my argument in what follows. On the contrary, I shall attempt to show that the logical peculiarities of sentences satisfying his latest criterion are also due to essentially normative considerations, and thus will not support his commitment to "Brentano's thesis." See footnote 43 below.

[39] Here I adopt the position, characteristic of empiricism, that statements concerning what is necessarily true are essentially normative. Some of my reasons for taking this position will emerge in what follows.

are, of course, important limitations to this analogy,[40] but I shall base my discussion mainly on the noncontroversial assumption that both asserting and playing chess are rule-governed activities subject to evaluation by reference to appropriate norms.

12. INTENTIONALITY: ITS SOURCE AND ANALYSIS

Although we have standard material criteria of a reasonably definite kind for what is to count as the sort of piece that is to be moved in accordance with the standard rules of chess, it is not difficult to imagine other activities, associated with radically different pieces, which seem to involve the same basic rules, or at least involve rules that are formally analogous to the ones we happen to have. Thus, we might imagine (as Sellars suggests[41]) a game played in Texas, called "tess," in which automobiles of various makes are driven from county to county in accordance with rules that are the exact analogue of our rules for chess, which, let us suppose, is unknown in Texas. In view of the close formal analogy between the moves made in chess and tess, it would not be unreasonable to think of chess and tess as specific varieties of the same basic game, which might be called "bess." It would also be reasonable to think of chess and tess players coming into contact with one another, becoming aware of the close similarity between their games, deciding that they are both playing varieties of the same basic game, and proceeding to engage in correspondence games, in which each side plays with its own pieces, translating the reports of the opponent's moves into the idiom with which they themselves are familiar. Thus when, in such a game, a Texan moves a Volkswagen into

[40] The chief limitation is that while the rules of chess are sufficiently neat and tidy to make the identification of a certain move (as, say, a checkmate) virtually automatic, there is so much freedom and idiosyncrasy in language-using that the identification of the move that a man is actually making can often be done only with the greatest difficulty, and not just by perusal of grammars and dictionaries. On this limitation see footnote 6 above, and also ch. 6 of R.G. Collingwood's *The Principles of Art* (Oxford: Clarendon Press, 1937).

[41] Wilfrid Sellars, "Abstract Entities," *passim.*

a county adjacent to one occupied by a Rolls Royce, the opposing side represents this move on its board by a certain configuration of pawn and queen.

If we assume that these different bess players have not worked out a distinct bess vocabulary into which their own special bess moves (that is, their own tess or chess moves) are translatable, then they would have to represent bess moves by the juxtaposition of their own familiar pieces, relying sometimes, if they were new to the game, on translators to tell them which of their moves are the counterparts of certain moves of their opponents. In order, however, to distinguish bess games from chess or tess games, each side might find it useful to employ slightly different pieces for bess. Thus, in order to characterize a certain move as a bess move, without committing themselves to the specific form in which it was made (that is, whether by the pieces of chess or tess), they might flag the pieces and squares (or automobiles and counties) used to represent it. This practice would have an obvious similarity to our use of indirect discourse, as in "the assertion *that* it is raining," or to the use of Sellars' dot quotes: "the assertion ·It is raining·" to represent for English speakers a move that could be made by a standard use of either the English "It is raining" or the French "*Il pleut.*"[42]

Leaving the bess players for a moment, I want to consider a community of primitive speakers operating only with proto-concepts, who have no conception of a proto-assertion made in unfamiliar words and no means of indicating the gist of a proto-assertion in indirect discourse or in anything functionally similar. For them, to relate what anyone proto-asserted requires that one reproduce the actual words used. As they see it, only members of their community proto-assert, and a given proto-assertion can be conceived only as made in specific words. According to their conventions, only one linguistic form is available for the purpose of relating what someone proto-asserted, namely, "S proto-asserted 'P'." Since "P" may have

---

[42] For an explanation of this convention, see Sellars, "Abstract Entities" and "Notes on Intentionality."

considerable internal complexity, it is clear that "S proto-asserted . . ." locutions may exhibit something like all three of Chisholm's marks of intentionality, even though the speakers of the language have no conception of an inner mental process or of anything like it.

The fact that these locutions are virtually intentional—that they exhibit something like Chisholm's marks—is in no way mysterious. The reason we can infer neither " 'P' is true" nor " 'P' is false" from the statement "Jones proto-asserted 'P' " or from its negation is simply that while both " 'P' is true" and " 'P' is false" involve an *evaluation* of "P," no such evaluation is either implicit in or demanded by the claim or the denial of the claim that Jones proto-asserted "P." Again, from the dual claim "Jones proto-asserted 'a is f' and in fact $a = b$" we plainly cannot infer that Jones proto-asserted "b is f." The reason for this is simply that, given the primitive community in question, the identity of a proto-assertion is determined by the actual words used. Since "a is f" and "b is f" are different verbal forms, even the logical truth of "$a = b$" would not allow us to infer that if Jones proto-asserted the former, he proto-asserted the latter.

I said a moment ago that claims about what a man proto-asserted (in the sense just described) exhibit "something like" all three of Chisholm's marks of intentionality; I did not say that they actually did exhibit those marks. This qualification is essential because his marks were specified by reference to expressions used but not mentioned in the relevant sentences. In the examples just considered, however, the intentionality was traced to the presence of mentioned expressions, which occurred only within quotation marks. Since Chisholm's thesis evidently concerns the intentionality of sentences containing uses of terms and propositional clauses, the account of quasi-intentionality just given does not strictly affect his position at all.

The point of the chess analogy is that it suggests a way of extending the treatment just given to proto-assertions *of* "P" to the more general case of proto-assertion *that-p*, and thus of refuting the position Chisholm evidently wants to defend. I shall now try to effect this extension by returning to that analogy.

The proto-asserters just described correspond to chess players before they conceived the possibility of bess. For them, a check mate could be accomplished only by particular configurations of certain familiar pieces, just as the proto-asserters could conceive of assertions as made only by the utterance of certain noises with which they are familiar. But once these proto-asserters are able to conceive the possibility of empirically different linguistic elements being used in accordance with formally analogous norms, then the necessity will arise for introducing such empirically noncommittal means of representing proto-assertions as *oratio obliqua* or Sellars' dot quotes.

It is crucial to observe that the notion these proto-asserters might gain of a linguistic move that could in principle be made in countless verbal forms may still apply only to overt linguistic activities. Although they could conceive of asserting that p in any number of different languages, they need not thereby have anything like the concept of silent thinking or nonproto saying. Even without these concepts, however, they would have no trouble accounting for the intentionality of proto-assertion. In order to account for the intentionality of "Jones proto-asserted *that p*," they would only have to point out that to say it is true (or false) that p is to evaluate the move made in asserting that p, and then add that such an evaluation is neither implicit in nor derivable from the claim (or the denial of the claim) that Jones made such a move. Similar remarks could be made regarding sentences exhibiting Chisholm's other marks: to the extent that they involve only protoconcepts, their intentionality is easily accounted for without any reference to an *inner* mental process.[43]

[43] If Chisholm's latest criterion of intentionality makes sense (see footnote 27 above), it too would be satisfied by sentences about proto-assertions. Some of these sentences, such as "Tom proto-asserts that the king of France is bald," possess the same purely formal properties as sentences about a man's beliefs. I say "*if* it makes sense" because if I am right about belief (see *Knowledge, Mind, and Nature*, pp. 213–218), Chisholm's criterion actually involves an error of quantification. As I explain it, the clause "that p" is quasi-metalinguistic in such

Aside from undermining the evident position of Chisholm, these last considerations allow us to identify the key line of reasoning on which the analogy theory of thinking is built. The first step is to focus attention on the rule-governed activity of proto-asserting. In considering that the function of a given proto-assertion could in principle be accomplished by a proto-assertion with very different empirical features, one makes sense of the abstract notion of a linguistic role, which could be exemplified by the use of expressions in countless protolanguages. Then, by reference to the explanatory force of statements about linguistic activities such as calculation, one elucidates and justifies application of the notion of a covert or "mental" exemplification of a linguistic role. Once this notion is in hand, the protoconcepts of the original scheme are capable of promotion to nonproto status, which converts them into the concepts we now have, with their characteristic intentionality. On this accounting the key idea of the whole analogy theory is thus that of a linguistic role, whose exemplification in thought is conceived by analogy with its exemplification in observable speech acts.

---

contexts as "Tom believes that p." If I am right about this, the formula (2) of Chisholm's criterion involves an error of quantification analogous to that involved in "(Ex) (Tom said 'f(x)')."

14

Intentionality

GUSTAV BERGMANN

A book on botany mentions plants, but it need not mention botany. A zoological text mentions animals, but it need not contain the word zoology. Intentionality is like botany or zoology, not like plants or animals and their kinds. That is why I shall hardly mention it in this essay. The things I shall mention are awareness, meaning, truth, and, my method being what it is, inevitably also language, particularly language about language. Concerning my philosophical method and my views on some philosophical problems, I am in a quandary. I do not wish to proceed as if they were known and I do not quite know how to proceed without assuming that they are. So I shall compromise. I shall not explain once more either the notion of an ideal language, which is not really a language to be spoken, or how, speaking commonsensically about it and what it is about, one philosophizes. For the rest, I shall tell a connected story. I realize, though, that in order to grasp it fully some readers may have to turn to what I said elsewhere.[1] One device I shall employ to provide as many connections and as much context as I possibly can are some "historical" passages about the recent as well as about the more remote past. These should be taken

From *Semantica* (*Archivio di Filosofia*, Rome: Bocca, 1955), 177–216. Reprinted, with omission of Part II, by permission of the author and the editors of *Archivio di Filosofia*.

[1] A collection of eighteen of my essays has been published under the title *The Metaphysics of Logical Positivism* (New York, London, Toronto: Longmans, Green and Co., 1954). I shall quote these essays as MLP, followed by the number under which they appear in the volume.

structurally, not as excursions into scholarly history; for I do not pretend to be a scholar living in history. Only, I wouldn't know how to philosophize without the history, or the image of history, that lives in me. For another, I shall not be able to avoid the use of symbols; but I shall keep it at a minimum; nor do I wish to pretend that I could do much better. For, again, I am not a mathematician any more than I am a historian. Fortunately, certain matters can be left safely to the mathematicians, just as some others can be left to the historians. Every now and then, though, the philosopher who, since he is a philosopher, finds himself short of time and taste to emulate the achievements of these specialists, does need their services. Things would probably go more smoothly if those specialists were not all too often like miners who cannot tell the raw diamonds from the philosophical pebbles in the materials they bring to light. Some of the confusions I shall try to unravel can indeed be traced to the mathematical logicians. But, then, it may be fairer to lay them at the doorsteps of those philosophers who, admiring the mathematicians too much, knew too little of what they actually did.

Here is an outline of what I propose to do. *First*, I shall try to convince my readers that when we say, speaking as we ordinarily do, that *there are* awarenesses, what we say is true. If, then, there are awarenesses, one may ask whether they also *exist*, in the philosophical sense of 'exist'. (In its ordinary or commonsensical use 'exist' is expendable, since it can always be replaced by 'there are (is)'.) Awarenesses do exist. By this I mean three things. I mean, first, that instances of awareness are particulars in exactly the same sense in which a tone is a particular. I mean, second, that there are certain characters, among them at least one that is simple, which are in fact exemplified by those and only those particulars I call awarenesses, in exactly the same sense in which the simple characters called pitches are in fact exemplified by those and only those particulars that are called tones. An awareness may, for instance, be *a* remembering, i.e., an instance of remembering, just as a tone may be *a* middle *c*, i.e., an instance of middle *c*. The third thing which I mean I shall mention presently. Like everybody else, philosophers are sometimes aware of their awarenesses. Many philosophers nevertheless deny that they exist. One very important one, Ludwig Wittgenstein, spent the second half of his life trying painfully to convince himself, not only that they do not exist, but even that there are none. Such persistent refusals to admit the obvious are so strange that one must try to explain them. That will be my *second* step. Philosophers did not see how they could consistently hold that there are awarenesses without also holding that there are interacting minds, i.e., mental particulars causally interacting with physical objects in exactly the same sense in which the latter interact among each other. Thus, when the belief in interacting minds became less and less tenable, some philosophers denied, with the intellectual violence that is so characteristic of all of us, that there are awarenesses. This is the story of the classical act and its later vicissitudes. In its final stages one kind of concern with language came to the fore. Another kind lies at the root of all analytical philosophy. I shall turn in my *third* step to some aspects of this second concern with language. Each of the two different concerns produced some confusions; there were also some illegitimate fusions between the two. The fusions and confusions support each other. To clear up the latter and to undo the former is one half of the analysis which vindicates awareness. *Fourth*, I shall propose what I believe is the correct form of those sentences in the ideal language that mention awarenesses. This is the other half of the analysis which, in the nature of things, involves the analysis of meaning and truth. It is also the heart of the essay. All the latter amounts to, in a sense, is therefore a proposal for transcribing such sentences as 'I see that this is green' in the ideal language. The transcription will show that awarenesses and, in fact, only awarenesses exemplify certain peculiar characters, which I call *propositions*. (This is the third thing I mean when I say that awarenesses exist.) Because of these characters statements about awarenesses are, loosely and ambiguously speaking, statements about statements. To tighten the looseness and to eliminate the ambiguity is virtually the same thing as to clear up the confusions and to undo the fusions of which I just spoke. This is the reason for my expository strategy.[2]

I

I stand in front of a tree, look at it, and see it. As we ordinarily speak, we say that the situation has three constituents, myself, the tree, and the seeing. Ordinarily we let it go at that. Upon a little reflection, still safely within commonsense, we notice that 'myself', 'tree', and 'seeing' may be taken in either of two contexts. In one of these, the first two words, 'myself' and 'tree', refer to two physical objects, namely, my body and the tree, while the third, 'seeing', refers, not to a third physical object, but to a relation between such, namely, the relation exemplified whenever one says truly that someone sees something. About this very complex relation physicists, physiologists, and behavioristic psychologists know a good deal. In the other context, 'seeing' refers to something mental, as we ordinarily use 'mental', and this mental something is again distinct from myself, the seer, as well as from what is seen, the tree. This seeing is an awareness. An awareness is thus something mental, distinct from what, if anything, is aware as well as from what it is aware of. That much is evident and to that much I commit myself therefore without hesitation. To three other beliefs one is, I think, not committed by commonsense. I, for one, hold all three to be false. One of them is crucial. Whether the other two are, in fact, false makes no difference for what I intend to say. Even so, I shall briefly mention all three; for it is well to grasp clearly what does and what does not depend on what.

I do not believe that an instance of seeing, or of any other awareness, is merely the exemplification of a relation, or of any other character, between two "things," as indeed the physical

---

[2] The fundamental ideas of this essay are first stated, very badly, in two papers that appeared over a decade ago: "Pure Semantics, Sentences, and Propositions," *Mind*, 53 (1944), 238–57; "A Positivistic Metaphysics of Consciousness," *Mind*, 54 (1945), 193–226.

seeing is. I believe, instead, that an awareness is itself a "thing." I say thing rather than particular because it makes no difference for what I want to say right now whether or not the other two terms do or do not refer to particulars. (Presently we shall see that the content of an awareness could not possibly be a particular.) This is crucial. The second belief which I hold to be false is that there is a mental thing referred to by 'myself'. To make it quite clear that nothing I shall say depends on whether or not this belief is in fact false, I shall eventually transcribe, not 'I see that this is green' but, instead, '(It is) seen (by me) that this is green' without paying any attention to the problems connected with the two words in the second parenthesis. Third. Some philosophers believe that the object or, as one also says, the content of an awareness is, in some cases, a physical object. According to these philosophers, my illustration is such a case; the content in question is of course the tree, or, perhaps more accurately, something that is in some sense a part of its surface. To these philosophers I grant that when we use 'see' as we ordinarily do in such situations, we certainly mean to mention a physical object. Some other philosophers insist that the content of an awareness is always a mental object; in my illustration, a tree percept. To these philosophers I grant that there is a perfectly plain sense of 'directly apprehending' or of 'being directly acquainted with' such that what we directly apprehend, even in a so-called perceptual situation, is a mental object. But, again, nothing I shall say depends on which side one takes on this issue, even though at one place I shall *seem* to side with the second view. (To dispel the appearance of this seeming is one of the things I cannot take time to do in this essay.)

Sometimes I shall find it convenient to speak of an awareness as a mental state of the person who, as one ordinarily says, has it or owns it. In fact, I do not know what one could possibly mean when, speaking literally, one says that someone has or is in a certain mental state if not that he has an awareness of a certain kind. But I shall ordinarily not call an awareness a mental content. The reasons for this caution as well as for the qualification, ordinarily, are, I think, fairly obvious. Since I shall use 'content' to refer to what an awareness is the awareness of, and since I have committed myself to the distinction between the two, it is prudent to avoid expressions that may tend to blur it. The reason for the qualification is that one awareness is sometimes the content of another. (How would we otherwise know that there are any?) When I am aware of something, then I am aware of this thing, not of the awareness through which I am aware of it. But I may also, either at the same time or at some other time, be aware of that awareness. In this event the first awareness is the content of the second. Notice, though, that the second awareness is not, either directly or indirectly, an awareness of the content of the first, just as it is not, if I may so express myself, aware of itself.

Perceiving is one kind of awareness; directly apprehending, remembering, doubting whether, thinking of, wondering are others. The analysis of some of these kinds is very complex. For what I intend to do I can, happily, limit myself to direct apprehension. When I speak in the rest of this essay without further qualification of awareness I should therefore be taken to speak of direct apprehension. Similarly, when I speak of *an* awareness, I should be understood to speak of an instance of directly apprehending. Again, the difference really makes no difference. But I wish to make as clear as I can which problems I shall not discuss without, however, either belittling them or denying that they are problems.

Ordinarily we say 'I see this tree' but we also say 'I am aware of this being a tree', 'I know that this is a tree', 'I wonder whether this is a tree', and so on. If we choose, we can rephrase the first of these sentences: 'I see that this is a tree'. A statement mentioning an awareness can always be so rephrased that its content is referred to by a sentence. Grammatically this sentence appears in our language either as a dependent clause or as a participial phrase (e.g., 'this being a tree'). This is what I mean by the formula: *The content of every awareness is propositional.* If, for instance, I see (or directly apprehend, or remember; the difference makes no difference) a red spot, the content of my awareness is a state of affairs or fact, namely, a certain particular being red.

If one asks the proper question of one who has an awareness while he has it, one elicits a certain answer. If, for instance, somebody points at the tree while I am looking at it and asks me what it is, I shall say "This is a tree." This statement is the *text* of my awareness. This and only this sort of thing is what I mean by the text of an awareness. In many cases it is not easy to hit upon the right question or to be sure that the answer one receives is what one was asking for. In some cases the difficulties are very great. But, no matter how formidable they may be, they lie always within the limits of commonsense and its long arm, science; in no case are they philosophical difficulties. The notion of a text is therefore itself entirely commonsensical. Three things about texts are worth noticing, though. Notice first that the text of an awareness states its content and only its content, without mentioning the awareness itself. This jibes well with what I said in the second to the last paragraph. Notice next that this is the first time I mention language in a certain way. More precisely, this is the first time I mention linguistic behavior as such. Notice, third, that the connection I thus establish between an awareness and its text is purely *external*. This means, first, that I am not dealing with the awarenesses one may have of the words he utters or hears uttered; and it means, second, that I am at this point not concerned with the question whether or not and in what sense one's inner speech is a "part" of his awarenesses. (These comments lay the ground for the unraveling of some of the fusions and confusions I mentioned in the outline.)

Let us return to my awareness of the red spot. The situation involves two particulars, the spot and the particular awareness. It also involves *at least* two states of affairs or facts, referred to by statements, namely, first, the spot being an instance of red and, second, the awareness being an instance of perceiving or, perhaps, of directly apprehending. The first of the two states of affairs is the content of this particular awareness. I said at least because the analysis is patently still incomplete. What it omits to mention is, in fact, the very crux of the matter, namely, that the one particular, the awareness, is an awareness *of* the state of affairs of which the other particular, the spot, is an ingredient. This third constituent fact of the situation is, I submit, not (1) that two particulars exemplify a relation, nor (2) that the one particular, the awareness, and the state of affairs of which the other is an ingredient exemplify a pseudorelation, but, (3) that the awareness exemplifies another nonrelational character, of the sort I call a proposition, which I shall specify in good time when I shall state my proposal. Alternatives (1) and (2) bring us to the classical act and thus to the development I wish to consider in my second step. In this development the difference between (1) and (2) was not always clearly seen. Nor shall I bother to distinguish between them in my quasi-historical account of it. However, we shall need the distinction later on, in the fourth step; so I shall state it now. A (binary) relation obtains between what is referred to by two terms. A (binary) pseudorelation obtains either between what is referred to by a term and what is referred to by a sentence or between what is referred to by two sentences. Symbolically, in the usual notation: 'xRy', 'xPp', 'pPq'. Connectives are, of course, not pseudorelations but truth tables. Logical atomism is the thesis that the ideal language contains no pseudorelations.

. . . .

III

Linguistic events, whether they are mental or noises, are events among events. Linguistic things, such as marks on paper, are things among things. Talking about either, one talks about language as part of the world. This is the way scientists talk about it. Philosophers look at language as a pattern, that is, as a picture of the world rather than as a part of it. Event *vs.* pattern, part *vs.* picture; the formula is suggestive. That is why I begin with it. Yet, like all formulas, it needs unpacking. The following three propositions and five comments state what is sound in it.

Propositions and comments are both very succinct. If I went into detail, I would do what I said I would not do, namely, explain once more the method of philosophizing by means of an ideal language.

There is of course nothing that is not part of the world. Clearly, then, the negative half of the metaphor must not be taken literally. The following propositions unpack it. (1) The construction of the ideal language L proceeds syntactically, i.e., as a study in geometrical design, without any reference to its interpretation. A schema so constructed is as such not a language; it becomes one, at least in principle, only by interpretation. (2) The philosopher interprets L by coordinating to awarenesses not their actual texts but ideal texts, i.e., sentences of L. (3) Having so interpreted L, he can, by speaking about both it and what it refers to or speaks about, first reconstruct and then answer the philosophical questions. This is the meaning of the positive half of the picture metaphor, according to which the ideal language is a picture, or, in the classical phrase, a logical picture of the world. These are the three propositions. Now for the five comments. (a) Notice that in (2) 'sentence' is used proleptically. Only by interpretation of L do certain of its designs become "sentences." (b) The connection between an awareness and its ideal text is as external as that between it and its actual text. (c) In coordinating his ideal texts to awarenesses the linguistic philosopher acknowledges in his own way the Cartesian turn. (d) The text of an awareness refers to its content. Some texts, whether actual or ideal, refer therefore to awarenesses. But a text does not refer to an awareness merely because it is coordinated to one. (e) Familiarity with the traditional dialectic shows that the undefined descriptive constants of L must refer to what we are directly acquainted with, in the sense in which the classical phenomenalists maintained that we are not directly acquainted with physical objects.[3]

The picture metaphor also misled some, among them the

---

[3] This is the issue mentioned earlier on which I *seem* to side with the classical phenomenalists. The appearance is dispelled in MLP.

Wittgenstein of the *Tractatus*. One of the several errors[4] it
caused is the belief that the ideal language cannot "speak about
itself." Let me first show how this confused idea came to seem
plausible. Change the metaphor slightly, introducing a mirror
instead of a picture. Take an object and let it stand for the
world. The mirror may mirror the object; it does not and cannot
mirror its own mirroring it. One may, of course, place a second
mirror so that it mirrors the object, the first mirror, and the
latter's mirroring of the former. But now one who understood
what was said before might remark that when this is done then
the first mirror and its mirroring have themselves become part
of the world (of the second mirror). The remark is not yet the
analysis, but it points at the crucial spot. The source of the
confusion is an unnoticed ambiguity of 'about'. This ambiguity
is not likely to be noticed unless one distinguishes clearly between the two ways of looking at language, once as part of the
world, once as its picture.

Commonsensically we say that a sentence (or a word) *refers*
to, or is *about*, a state of affairs (or a thing). This makes
sense if and only if what is said to refer to something, or to be
or speak about something, is a linguistic event or a kind of such.
Notice, first, that in the two comments (d) and (e) above I
myself used 'refer' and 'about' in this sense. In fact, I never use
them otherwise, for I do not understand any other use of them.
Notice, furthermore, how well all this fits with what was said
earlier. What a linguistic event or a kind of such refers to is also
its meaning, in one of the two commonsensical and scientific
meanings of 'meaning'. And when scientists speak about language they speak of course always about linguistic events. What
one asserts, then, when one asserts, with this meaning of 'about',
that language cannot "speak about itself" is that there cannot
be kinds of noises which, as we use them, refer to other kinds
of noises. The assertion is so implausible that I hardly know
how to argue against it. The best one can do if one wishes to
dispose of it as thoroughly as possible is what I am doing in

[4] For an analysis of some others see MLP 3.

this section, namely, analyze the major sources of the illusion. But let me first dispose of what is even more obvious. If we use a language in which reference is not univocal, we will eventually get into trouble. This is just commonsense. Thus, if we use a certain kind of noise to refer to a certain kind of animals, say, dogs, we had better not also use it to refer to something else and, in particular, not to itself, i.e., to this particular kind of noise. Any adequate language will therefore distinguish between the two kinds of design on the next line:

dog   'dog'.

This is the origin of the quoting device. In any language that is not on grounds of sheer commonsense foredoomed, the linguistic events about linguistic events, or, if you please, the part of the language that is "about itself" are therefore those and only those that contain single quotes or their equivalents, e.g., the phrase 'the word dog'.

What, if anything, could be meant by saying that the ideal language speaks about itself? Every awareness has an ideal text. Let 'b' be the name of (refer to) an awareness and let '$gr(a)$' ('This is green') be its text. From what was said earlier we know that the name of an awareness, in this case 'b', could not possibly occur in its text, in this case '$gr(a)$'; for the text of an awareness refers to its content, which is always distinct from the awareness itself. But consider now another awareness, c, whose content contains b. Since c is about b, its text contains at least one clause that predicates some character of b; for otherwise it wouldn't be about b. Let '. . . (b)' be this clause, with the dots marking the place of the name of that character. Assume next that L contains as the name of the character the predicate expression ' '$gr(a)$' '.[5] Then the text of c *contains* ' '$gr(a)$' (b)'. L, therefore, contains an expression of its own between single quotes. This is the exact point at which the illusion arises that the ideal language may speak about itself *in the same sense* in which language as event may do so. Or, to put the same thing differently, this is the only clear sense in

[5] These are not double quotes but one pair of single quotes within another.

which the ideal language as a pattern could be said to "speak about itself." Moreover, this is, as we now see, *not* the sense in which language as a part of the world may speak about itself. After one has seen that, one may if one wishes continue to use the phrase, as I occasionally shall, and say that *in this sense* the ideal language may and must "speak about itself." Only, and this is my real point, or, rather, this is the point that matters most for my story, there is again no reason whatsoever why in this sense the ideal language should not or could not "speak about itself." Again, the assertion is not even plausible. One of two apparent reasons that made it seem plausible is, if I may so express myself, the grammar of the picture metaphor. This, I believe, is the reason why Wittgenstein propounded the dogma in the *Tractatus*. The other reason, which probably did not sway Wittgenstein but which seemed a good reason to some others, is that the mathematicians proclaimed they had proved that language cannot both be consistent and say certain things "about itself." The mathematicians had indeed proved something. They usually do. Only, what they had proved was not by any stretch of the imagination what they mistook it for. It took indeed all the philosophical clumsiness and insensitivity which mathematicians sometimes display to make this mistake, just as it took the wrong kind of awe in which some philosophers hold mathematics to believe them. In the rest of this section I shall analyze the mistake; partly in order to dispose of the strange dogma as thoroughly as I possibly can; mainly because this is the best place to introduce the notion of *truth* into the story. For the philosophical analyses of awareness, meaning, and truth belong together.

The mathematicians thought they had proved that a schema syntactically constructed cannot (a) be consistent[6] and upon interpretation contain (b) arithmetic as well as (c) a predicate with the literal meaning of 'true'. To be a plausible candidate for the role of ideal language, a schema must obviously satisfy conditions (a) and (b). As to (c), one of the things one would

---

[6] Consistency can be defined syntactically.

naturally want to say in a language that "speaks about itself" is that its sentences are true or false (not true), as the case may be. Thus, if the mathematicians had proved what they thought they proved, there would be a difficulty. In fact, they proved that no schema can simultaneously fulfill (a), (b), and a third condition, (c'), which they mistook for (c).

In order to fix the ideas I speak for the time being about language as part of the world. Sentences, then, are kinds of linguistic events (or things). Literally, only sentences are true or false. Explicitly, 'true' is therefore a linguistic predicate in the sense that it is truly predicated only of the names of certain linguistic kinds. This, by the way, is the only meaning of 'linguistic' that is clear and does not stand in need of explication. Implicitly, truth involves more than the linguistic events themselves. *A sentence is true if and only if what it refers to (means) is the case.* Let me call this sentence (A). It is a truism: yet, firmly grasped, it has three important consequences. *First.* Some linguistic properties are syntactical properties. In the case of marks on paper, for instance, a property of a sentence or of any other expression is syntactical if and only if it is defined in terms of the shapes and the arrangement of its signs and of nothing else. Truth is obviously not a syntactical property of sentences. *Second.* Introducing 'true' into a schema means two things. It means (α) introducing into the schema a sentence which upon interpretation becomes (A). It means (β) that this sentence ought to be a "linguistic truth," in a sense of the phrase, linguistic truth, which is by no means clear and must therefore be explicated. It follows, *third*, that if all this is to be achieved, the schema must contain certain expressions, one which can be interpreted as 'refer' and others that can be interpreted as names of sentences. In the nature of things, these expressions must be descriptive.

The property mentioned in (c') *is a syntactical property of sentences; truth, the linguistic property mentioned in* (c), *is not.* Not to have seen this is the mathematicians' major mistake. They also made two subsidiary ones. One of these is that, accurately speaking, the property mentioned in (c′) is not even a syntactical property.

Goedel, who did not make any of these mistakes, invented a method that allowed him to use arithmetic in speaking commonsensically about an uninterpreted schema. Specifically, he invented a rule by which to each expression of the schema[7] one and only one integer is coordinated in a manner that depends only on the shapes and the arrangement of the signs in the expression itself. (This is, in fact, the least achievement of that great mathematician.) In speaking commonsensically about the schema we can therefore use the number (n_A) which by the rule corresponds to an expression 'A' as the "name" of this expression. By the same rule, a class of integers corresponds to every syntactical property, namely, the class of all the integers coordinated to expressions which have the property. The name of a class of integers is called an arithmetical predicate. (E.g., 'square' is the name of the class [1, 4, 9, . . .].) Now remember (b). By assumption our schema contains number-signs (not numbers!), i.e., expressions we intend to interpret as referring to integers, and arithmetical-predicate-expressions, i.e., expressions we intend to interpret as referring to classes of integers. Assume now that one of these latter expressions, '*pr*', upon interpretation becomes an arithmetical predicate that is coordinated to a syntactical property. In this case the mathematicians say that the schema contains the "name" of the syntactical property, just as they say that in the number-signs it contains the "names" of its own expressions. This use of 'name' is inaccurate. For one, an uninterpreted schema does not contain the name (or the "name") of anything. For another, in the intended interpretation '*pr*' obviously refers to a class of integers and not to a syntactical property just as the number-signs refer to integers and not to expressions. Assume, third, that we actually use the (interpreted) schema as a language. We could not *in* it state what

---

[7] More precisely, the rule works only for schemata of a certain kind; all plausible candidates for the role of ideal language belong to that kind. This is but one of the many omissions I shall permit myself on more technical matters.

the mathematicians say *about* it unless it contained further expressions, namely, those which upon interpretation become the names of expressions and of their syntactical properties, and, in addition, the means to state *in* the schema the rules by which, speaking *about* it, we make integers and classes of integers the "names" of linguistic things and characters. This is the reason why, as we shall presently see, the property mentioned in (c′) is, accurately speaking, not even a syntactical property. Not to have seen that is one of the two subsidiary mistakes. Its root is the mathematicians' special use of 'name'. For their own special purposes it is, as it happens, quite harmless. Philosophically, it is disastrous to believe that one can state *in* the interpreted schema what can only be stated *about* it. Why this is so is obvious. The one and only schema which interests the philosopher is that which upon interpretation becomes L, the ideal language. And in L one must in principle be able to say everything nonphilosophical.

I am ready to state what the mathematicians did prove. Let 'A' be a sentence of a schema that satisfies (a) and (b) as well as some other conditions, of a purely technical nature, which every plausible candidate for the role of L must satisfy. Let n_A be the number we have coordinated to 'A'; let 'N_A' be the number-sign of the schema which upon interpretation transcribes n_A; let finally 'pr' be an arithmetical-predicate-expression. What has been proved is this.[8] The schema contains no 'pr' such that

(T) $$pr\,(N_A) \equiv A$$

is *demonstrable* for all (closed) sentences of the schema. But I see that I must again explain, first, what demonstrability is, then why anybody should think that (T) ought to be demonstrable.

Analyticity is a syntactical property of sentences. More precisely, what philosophers mean by 'analytic' can and must be explicated by means of a syntactical property. Demonstrability

---

[8] D. Hilbert and P. Bernays, *Grundlagen der Mathematik* (Berlin: Springer, 1939), vol. II, pp. 245 f.

is another syntactical property of sentences. Every demonstrable sentence is analytic, though not conversely. (The second half is one of Goedel's celebrated results.) Thus, while there is no 'pr' for which (T) is demonstrable, there could conceivably be one for which it is analytic. That there actually is none is a purely mathematical matter which does not interest me here at all. The question that interests me is: Why should one who believes, however mistakenly, that an arithmetical-predicate-expression could ever transcribe 'true', also believe that the transcription is adequate only if (T) is demonstrable? The answer is instructive. Remember the condition (β), which requires that (A) be a "linguistic truth." (T) was mistaken for the transcription of (A); demonstrability was implicitly offered as the explication of the problematic notion of linguistic truth. This is the second subsidiary mistake. It is a mistake because in the light of Goedel's result demonstrability is not at all a plausible explication of 'linguistic truth'. Analyticity might be. In the next section I shall propose what I believe to be the correct transcription of (A) in *L*; and I shall show that this transcription is analytic.

IV

The sentence I proposed to transcribe in the ideal language is 'I see that this is green'; or, rather, in order to sidestep the issue of the self, '(It is) seen (by me) that this is green'; or, still more precisely, since I wish to limit myself to the indubitably simple character of direct acquaintance, 'direct acquaintance with this being green'. Let the undefined descriptive constants '*a*', '*aw*', '*gr*' name a particular and two simple characters, direct acquaintance and greenness, respectively. Consider '$aw(gr(a))$'; call it (1). On first thought one might hit upon (1) as the transcription of our sentence. A little reflection shows that for at least two reasons we are already committed to reject (1).

To be a direct acquaintance, or an imagining, and so on, are, as we saw, characters of particular awarenesses. Let '*b*' be the name of the awareness whose text I wish to transcribe. '*aw*' must then be predicated of '*b*' and not, as in (1), of '*gr(a)*', which refers to the content of *b*. This is the first reason why we must reject (1). '*gr*' and '*a*' refer to a character and a particular with both of which I am directly acquainted. Speaking as we ordinarily do, what they refer to is thus called mental. (This is my "point of contact" with the phenomenalists.) Change the example; consider '*kn* (p_1)'; call it (1'); let '*kn*' and 'p_1' stand for 'known that' and 'This stone is heavy' respectively. 'p_1' refers to a physical state of affairs; to say that it refers to anything mental is to fall into the absurdities of the phenomenalists. '*kn*', on the other hand, names a character which, speaking as we ordinarily do, we specifically and characteristically call mental.[9] It follows that (1') mixes the physical and the mental in the manner that leads to the interactionist catastrophe. Perhaps this becomes even clearer if for a moment I write, relationally, '*aw*(*self*, p_1)', which is of course the pattern of the classical act. However, the difference between the relational and nonrelational alternatives makes no real difference so far as mixing the physical and the mental goes. This is the second reason why we must reject (1). But now a critic might insist that when somebody knows or sees something there is indeed a transaction[10] between what is known or seen and the knower or seer. Quite so. Only, this transaction is properly spoken of as the scientists speak about it, that is, in principle, behavioristically. (This is my "point of contact' with materialists and epiphenomenalists.) Notice that, in spite of the "phenomenalistic" feature of my ideal language, I can say all this and even find it necessary to say it. This alone should go a long way toward convincing anyone that I avoid the absurdities of the various classical positions.

Let us take stock. Negatively, we understand why (1) cannot be the transcription. Positively, we see that the transcription

[9] So used, 'knowing' refers to a character of awarenesses. To insist on that one need not deny that there are other uses of the word, e.g., those of which Ryle now makes too much.

[10] I use this clumsy word in order to avoid 'relation', which would be syntactically false since the "transaction" is a pseudorelation.

must contain the clause '$aw(b)$'. In this clause, by the way, 'aw' is a predicate and therefore, strictly speaking, the name of a character. In (1) it is a nonrelational pseudopredicate and therefore, as I use 'character', not really the name of a character. Of this presently. For the moment we notice that '$aw(b)$' could not possibly be the whole ideal text of our sentence since it does not say what b is an awareness *of*. Thus, there must be at least one more clause. To provide it, I make use of an idea I introduced before. That an awareness is an awareness of something I represent in the ideal language by a character of this awareness which is *in some sense* (I shall presently explicate it) a simple character; in our instance, call this character ''$gr(a)$' '[11]; generally, I call it ''p_1'', where 'p_1' refers to the content of the awareness or, what amounts to the same thing, is its (ideal) text. The transcription of our sentence becomes then

(2) $aw \ (b) \cdot 'gr(a)'(b)$

Undoubtedly there is something peculiar about ''$gr(a)$''. For one, the expression itself is very complex, even though it names a character that is simple. For another, the expression is not, as a syntactically introduced, undefined, descriptive predicate ought to be, wholly innocent of its interpretation. One can, of course, as I presently shall, syntactically construct a schema that contains it. But that in itself means nothing. Even so, ''$gr(a)$'' is innocent of the intended interpretation in that (a) it remains fully indeterminate as long as 'gr' and 'a' are. But it is not so innocent in that (β), after 'gr' and 'a' have been interpreted, if I am to achieve my purposes, ''$gr(a)$'' must be interpreted as the name of the character which an awareness possesses if and only if it is an awareness of what '$gr(a)$' refers to. On the other hand, we would like to say that (β) is "merely a linguistic matter" or, as I once put it, that to be an awareness of a certain kind and to have a certain content (and, therefore, text) is one thing and not two. Let there be no illusion. In so speaking we ourselves use 'linguistic' philosophically, i.e., in a

[11] Again, these are not double quotes but one pair of single quotes within another.

problematic way that needs explication. The point is that what I am saying in this section is, among other things, the explication. The following are three salient points of it. (a) I introduce into the ideal language the sentence ' '$gr(a)$'$Mgr(a)$' as the transcription of what we *sometimes* mean when we say that the proposition (or sentence) this is green *means* that this is green. (b) I so extend the notion of a logical sign that 'M' becomes logical and not descriptive. (c) I so extend the notion of analyticity that ' '$gr(a)$'$Mgr(a)$' and all similar sentences become analytic.

Sometimes, when we assert such things as, say, that the proposition (or sentence) this is green means that this is green, we would be dissatisfied if we were told that in asserting it we use 'means' in the sense of either reference or context. The cause of the dissatisfaction is that we feel, however confusedly, that we did not say anything, or did not want to say anything, about linguistic events. Or, if you please, we feel that what we really wanted to say is something "linguistic" in some other sense of this problematic term. 'M' transcribes this meaning of 'means'. I am tempted to call it the hidden or philosophical meaning; hidden, because it got lost in the development I described in the second section; philosophical, because I believe that it is what the philosophers who were not sidetracked by that development groped for. However, I ordinarily call meanings (or uses) philosophical if and only if, remaining unexplicated, they produce philosophical puzzlement. So I shall resist the temptation and call this third meaning, transcribed by 'M', the *intentional* meaning of 'means'.

This is as good a place as any to introduce a fourth meaning of 'means' (and 'meaning'). This I call the *logical* meaning. But first for two comments that might help to forestall some misunderstandings. (a) I have mentioned four meanings of 'means'. Two of them, reference and context, I called scientific; one I call logical; another I was at least tempted to call philosophical. There are good reasons for choosing these names; but one must not let the names obscure the fact that *'means' occurs with each of these four meanings in ordinary discourse*, sometimes with

the one, sometimes with the other, sometimes with some combination. As long as one speaks commonsensically one does
not get into trouble. As soon as one begins to philosophize in
the traditional way about "meaning," the fourfold ambiguity
begins to produce the traditional philosophical troubles. (b)
There are quite a few further meanings of 'meaning'. They
occur in moral, esthetic, and scientific discourse and in discourse about such discourse. I know this as well as the next
man, even if that man should hail from Oxford. The four
meanings I single out are nevertheless those which through
fusion and confusion have produced one of the major tangles
of first philosophy. Compared with the task of untying this fourfold knot the explication of the other meanings of 'meaning' is
not very difficult.

Logicians often say that two sentences of a schema, 'p_1' and
'p_2', have the same meaning if and only if '$p_1 \equiv p_2$' is analytic.
This is the logical meaning of 'means'. In logic the idea is important; hence the adjective, logical. Nor is there any doubt
that it explicates *one* of the ordinary uses of 'means'. Technically, the basic notion in this case is not meaning but having-
the-same-meaning; so the former must be explicated in terms
of the latter, say, as the class of all sentences having the same
meaning. These, however, are mere technicalities with which
we need not bother.

I am ready to put the last touch to my main proposal. One
may wonder whether

(2') $aw(b) \cdot {}'gr(a)'(b) \cdot {}'gr(a)'Mgr(a)$

is not preferable to (2). (2') has the advantage that, since its
third clause mentions the content of the awareness whose text
it transcribes, one can be quite sure of what in the case of (2)
one may conceivably doubt, namely, that nothing essential has
been omitted. Interestingly, one need not choose. The third
clause of (2'), the one which makes the difference between it
and (2), is, as I mentioned before, analytic. (2) and (2') are
thus like 'p_1' and '$p_1 \cdot p_2$', where 'p_2' is analytic. In this case
'$p_1 \equiv p_1 \cdot p_2$' is also analytic. (2) and (2') have therefore the

same logical meaning. The meaning transcription must preserve is logical meaning. It follows that the difference between (2) and (2′) makes no real difference.

Consider everything I have said so far in this section as preliminary, merely an exposition of the main ideas, to be followed by the more formal presentation and argument on which I am about to embark. First, though, I want to attend to two related matters.

The predicates of the ideal language L which I form by surrounding sentences of L with single quotes name those characters which I call *propositions*. Propositions are therefore not kinds of linguistic things or events in the sense in which certain marks on paper, certain sounds, and certain visual and auditory "contents" are linguistic things or events. And this latter sense is, as we know, the only clear and unproblematic sense of 'linguistic'. It is therefore a mistake, or, at least, it is confusing to say that what I call a proposition is a linguistic character. If a qualifying adjective must be used at all, I would rather say that propositions are mental or psychological characters. But then again, it would be another mistake to think that I propose what is traditionally called a psychological theory of propositions. To understand why it is a mistake one merely has to remember that, as the term is traditionally used in philosophy, propositions are a peculiar kind of entity of which some philosophers claim they are the real contents of awarenesses. I do not believe that there are propositions in this sense. So I would not propose a theory, either psychological or otherwise, to provide some status for these chimeras. Why then, one may wonder, use a word that invites mistakes and confusions. I hold no brief for the word. I needed a name. This one came to mind. It is, I think, as good as any other. Also, I welcome the opportunity it provides to cast new light on certain kinds of mistakes and confusions. This is one of the two matters to which I wanted to attend.

Some particulars are tones. This does not imply that L must contain an undefined predicate interpreted as 'tone'. If, for instance, L contains the undefined names of the various

pitches, middle *c*, *c* sharp, *d*, and so on, one could try, in *L*, to
define a tone as anything that exemplifies a pitch. The technicalities of this business need not concern us here.[12] Similarly,
since awarenesses are in fact those particulars which exemplify
propositional characters, one may wonder whether *L* must contain undefined descriptive predicates, such as '*aw*', which are
interpreted as the names of different modes of awareness, in the
sense in which direct acquaintance, wondering, remembering,
doubting, and so on, are modes of awareness. There are undoubtedly such modes, just as there are shapes, tones, smells,
and so on. The only question is whether, omitting from *L* all
undefined names for any of them, one can in *L* still account for
the differences among them; i.e., whether one can in principle
account for these differences in term of "content" and of "content" alone. I have pondered the question for years. (Hume
threw out a casual suggestion concerning it when he distinguished "ideas" from "impressions" by their "faintness.") I
am not sure what the answer is, though I am now inclined to
believe that it is negative. That is why I proceed as if it were
negative. But it is also important to see clearly that whatever it
is does not make much difference for anything else I have said
and shall still say in this essay. The only difference is that if the
answer were positive then propositions would be the only characters that are in fact exemplified by awarenesses alone. This is
the other matter to which I wanted to attend.

Russell and the Wittgenstein of the *Tractatus* were the first
who practiced the method of philosophizing by means of an
ideal language. Since then quite a few philosophers, whether
they knew it or not, have more or less consistently employed
this method. With two exceptions, they all proposed essentially
the same syntactical schema. This schema, I shall call it the
conventional schema or L_c, is of the *Principia Mathematica*
type. The New Nominalists are one exception; the other, for
over a decade now, has been myself. The New Nominalists,
who do not belong in our story, believe that *L* must be syn-

[12] See also MLP 12 and "Undefined Descriptive Predicates," *Philosophy and
Phenomenological Research*, 8 (1947), 55–82.

tactically poorer than L_c.[13] I believe that L_c is in one respect and in one respect only not rich enough to serve as L. My reason should now be obvious. I do believe that L_c can serve as a clarified language to be spoken, in principle, about everything which, as one usually says, is an object of mind—including mind itself, as long as we speak about it scientifically, that is, in principle, behavioristically. But I also believe that L_c does not provide adequate transcriptions for many statements we make about minds or mental things when we speak commonsensically. It follows, on my conception of philosophy, that one cannot, by talking about L_c and what it talks about, solve some of the philosophical problems concerning mind and its place in nature.[14] Hence L_c cannot be the ideal language. Positively, I believe that L_c becomes the ideal language if it is supplemented by two further primitive signs, namely

$$M \text{ and } `\cdot \cdot \cdot \cdot \cdot \cdot \cdot',$$

i.e., the relational pseudopredicate which I interpret as the intentional 'means' and the quoting operator. We have incidentally come upon another reason why the question whether L must contain '*aw*' and other undefined names for the several modes of awareness is not as fundamental as it might seem. '*aw*' and its cognates are predicates; thus they exemplify a syntactical category provided by L_c. '*M*' is a pseudopredicate. Thus it belongs to a syntactical category unknown to L. As it happens, it is also the only primitive sign that represents this category in L_c. And what holds for '*M*' in these two respects also holds for the quoting operator. Presently I shall make much of these points. But I see that I am once more illuminating basic ideas when the ground for a more formal presentation has already been laid. So I shall proceed as follows. *First*, I shall very concisely describe those features of L_c that matter most for my purpose. *Second*, I shall construct syntactically the

---

[13] For an analysis of the New Nominalism see MLP 4, MLP 5, and "Particularity and the New Nominalism," *Methodos*, 6 (1954), 131–47, and also pp. 91–105 of my book, *Meaning and Existence* (Madison, Wis.: Univ. of Wisconsin Press, 1960).

[14] See also MLP 6.

schema I believe to be L. It contains the two syntactical categories represented by 'M' and by the quoting operator. This feature requires a redefinition of the syntactical notions of *logical sign* and *analyticity*. The two new notions are broader than the conventional ones in that every primitive sign logical in L_c and every sentence analytic in L_c are also in L logical and analytic respectively, but not conversely. *Third*, I shall state explicitly what is implicit in this essay as a whole, namely, that the enriched schema can be made to bear the burden of the philosophy of mind.

The primitive signs of L_c fall into two classes, logical and descriptive. The logical signs are of two kinds. There are, first, two signs, each individually specified, each belonging to a syntactical category of its own, each the only primitive representative of its category in L_c. These two signs are, of course, a connective and a quantifier, interpreted in the familiar fashion as, say, 'neither-nor' and 'all'.[15] The second kind of logical signs, not individually specified, consists of an indefinite number of variables of each of the several types. Each type is a syntactical category; but they are all categories of "terms." The essence of a term is that it combines with terms to form sentences. L_c contains no pseudoterms, i.e., no category (except connectives) whose members combine either with sentences or with terms and sentences to form sentences. The primitive descriptive signs or, as one also says, the undefined descriptive constants of L_c are distributed over the various types of "terms." If a sentence S of L_c contains descriptive terms, then replace them all according to certain rules by variables. Call the resulting sentence the "form" of S. The syntactical definition of analyticity is so constructed that whether or not a sentence is analytic depends only on its "form." The syntactical significance of the distinction between the two kinds of signs lies thus in the role it plays in the syntactical definition of analyticity. The philosophical significance of the latter, and thus of both syntactical distinc-

---

[15] If, as strictly speaking one must, one is to dispense with definitions, then a third logical primitive, the abstraction operator, is necessary. This is another of the omissions and simplifications for which I must take the responsibility.

tions, lies in the circumstance that *in all cases but one* it can serve as the explication of what philosophers mean when they say that a sentence is "analytic," or a "formal" truth, or a "linguistic" truth. The exception where the conventional definition of analyticity is not adequate for this purpose is, as one might expect, the case of such sentences as "The sentence (proposition) this is green means that this is green," when 'means' is used intentionally.

That the definition of analyticity in L_c achieves its philosophical purpose depends of course on its details; they are specified in what is technically known as validity theory. I cannot here state the definition accurately; but I shall recall its nature by means of two elementary illustrations. Take the two forms '$p \vee \sim p$' and '$(x)f(x) \supset (\exists x)f(x)$'. The first is analytic because its truth table is tautological; the second is analytic because if '$f(x)$' is read 'x is a member of f', then it becomes a set-theoretical truth for all subsets of all nonempty sets. The definition of analyticity (validity) is thus combinatorial; arithmetical in the simplest case, set-theoretical in all others. What makes it philosophically significant is, first, the combinatorial feature, and, second, the circumstance that as far as we know all analytical statements are in fact true.[16]

Technically, validity theory is a branch of mathematics with many difficult problems. So it is perhaps not surprising that it, too, provided the philosophers with an opportunity to be misled by the mathematicians. The following two comments will show what I have in mind. (a) For all philosophical purposes (with the one notorious exception) our definition is an adequate explication of what philosophers mean by 'analytic'. Mathematically, it is not as interesting. It would be, if we knew a procedure which, applied to *any* sentence S of L_c, after a finite number of steps yielded an answer to the question whether S is analytic. There is and there can be no such procedure. (That there can be none even if one restricts S to the so-called lower functional calculus is the famous result of Church.) This is the

[16] See also MLP 4, MLP 14.

reason why mathematicians are not very interested in validity; unfortunately, their lack of interest has blinded some philosophers to the philosophical significance of this explication of 'analytic'. (b) In speaking about a schema we always speak commonsensically. In framing the explication of analyticity in terms of validity we use set theory "commonsensically." Yet it is a matter of record that "commonsensical" set theory itself got into difficulties that had to be straightened out by the construction of schemata. Mathematicians may therefore feel that the explication of analyticity in terms of validity uncritically takes for granted what is in fact uncertain and problematic. For some mathematical purposes that may indeed be so. Yet, we must not allow the mathematicians to persuade us that we, as philosophers, ought to strive for certainty, or constructivity, or decidability, in the sense in which the finitists among them do. We seek, not certainty of any peculiar noncommonsensical kind, but, rather, the clarity achieved by explications framed in terms of commonsense, that commonsense of which science and (nonformalized) mathematics are but the long arm. If yesterday's "commonsense" got us into trouble that had to be straightened out by the construction of schemata, we shall today still use this "amended commonsense" to construct "commonsensically" the schemata of today. And if tomorrow we should get into trouble again, we shall start all over again. For what else could we possibly do?

One more feature of L_c must be mentioned. Let 'F_1' and 'F_2' be predicate expressions of any type, 'X' a variable of its subject type, '$\Phi(F_1)$' any sentence containing 'F_1', '$\Phi(F_2)$' a sentence made out of '$\Phi(F_1)$' by replacing at least one occurrence of 'F_1' by 'F_2'. It is a consequence of our definition of analyticity that

$$(E) \qquad (X)[F_1(X) \equiv F_2(X)] \supset [\Phi(F_1) \equiv \Phi(F_2)]$$

is analytic. Thus, if the antecedent of (E) is true, so is the consequent; and if the antecedent is analytic, so is the consequent. This feature is called the extensionality of L_c. I turn to the syntactical description of L. With the qualification entailed by 1 it contains L_c.

1. Only closed expressions are sentences of L. (This is merely a technical detail, necessary to avoid undesirable consequences of the quantification rules for expressions containing 'M'.)

2. L contains sentential variables. (Since L contains no primitive sentential constants, this modification has, upon my conception of ontology,[17] no untoward ontological consequences.)

3. L contains two additional primitive signs, the relational pseudopredicate 'M' and the quoting operator, with the following formation rules:

a. Every sentence of L surrounded by quotes becomes a nonrelational first-order predicate (type:f) with all the syntactical properties of a primitive descriptive predicate.

b. Every sentence of the form 'fMp' is well formed. Call these sentences the simple clauses of 'M'.

These are the formation rules of L. Now for the definition of analyticity.

4a. Every sentence analytic according to L_c is analytic.

4b. Every simple clause of 'M' is either analytic or it is contradictory, i.e., its negation is analytic. It is analytic if and only if the predicate to the left of 'M' is formed by the quoting operator from the sentence to the right of 'M'.

The part of L that contains 'M' is not extensional. To see that, let 'A' be a constant of the same type as 'X' and assume that '$(X)[F_1(X) \equiv F_2(X)]$' is true. If L were extensional, then ' '$F_1(A)$'$MF_1(A)$' $\equiv$ ' '$F_1(A)$'$MF_2(A)$' would have to be true. In fact, this sentence is not only false, it is contradictory, for by *4b* its left side is analytic and its right side is contradictory.

I call 'M' and the quoting operator, together with the two primitive logical signs of L_c, the four primitive logical signs of L. But then, one may ask, are the two new signs "really"

[17] See also MLP 4, MLP 13, and "Particularity and the New Nominalism."

logical? I can of course call them so. Yet, obviously, I do not wish to argue merely about words. The only real argument consists in stating clearly the similarities and the differences between the old and the new "logical" signs. I shall present this argument or, as I had better say, these reflections in three steps. *First.* Each of the four signs, both old and new, is individually specified. Each of the four signs, both old and new, belongs to a syntactical category of its own. Each of the four signs, both old and new, is the only primitive member of the syntactical category to which it belongs. These similarities are impressive. Nor is that all. *Second.* Consider the role the four signs play in the definition of analyticity. If in view of the three similarities just mentioned one accepts the two new signs as logical, then one can in view of 4a and 4b again say that whether a sentence of L is analytic depends only on its "form." This similarity, too, is impressive. But there is also a difference with respect to analyticity which I do not at all intend to minimize. For philosophy, as I understand it, is not advocacy, least of all advocacy of uses of words, but accurate description. The difference is that 4b is not a combinatorial criterion in the sense in which 4a is one. On the other hand, though, the "new" analytic sentences, i.e., those which are analytic by 4b, have a unique feature which in its own way is just as sweeping as any combinatorial one. They are all simple clauses of 'M' and each of these clauses is either analytic or contradictory. *Third.* Sentences which are analytic in the "old" sense of L_o (or 4a) are also called "formal" or "linguistic" truths. These are of course philosophical and therefore problematic uses of 'formal' and 'linguistic'. Analyticity in the old sense is their explication. Now we know that such sentences as "The sentence (proposition) this is green means that this is green" are sometimes also called "linguistic" truths and that this use of 'linguistic' is equally problematic. L transcribes these sentences into those that are analytic by 4b. Our "new" notion of analyticity thus clarifies two of the problematic uses of 'formal' and 'linguistic'; it exhibits accurately both the similarities and the differences between them; and it does not tear asunder what in the structural history of philosophical thought belongs together.

I have not, I shall not, and I could not in this essay show that L is the ideal language. What I have shown is merely this. *If* $(a)L_c$ is an adequately clarified language which one can in principle speak about everything except minds, and if $(\beta)L$ provides in principle adequate transcriptions for what we say, commonsensically and not behavioristically, about minds, *then* L is the ideal language. Furthermore, I have shown (β) by showing, at the beginning of this section, that L contains adequate transcriptions of such sentences as 'direct awareness of this being green'. With this I have accomplished the main task I set myself in this essay. Again, if this is so, then the differences between L and L_c must provide us with the accurate description, or, in the classical phrase, with the logical picture of the nature of minds and their place in the world. Let us see. In the world of L_c there are tones, shapes, colors, and so on. That is, there are particulars such that *in fact* they and they alone exemplify certain simple characters, say, in the case of tones, the pitches. In the world of L there are in addition also awarenesses. That is, there are particulars such that *in fact* they and they alone exemplify certain additional simple characters, those I called propositions and, probably, also some among those I called modes of awareness. These, to be sure, are important differences; yet they are not as radical as the one I saved quite deliberately for the end of the list. This difference is that L requires two new logical primitives. For what novelty, I ask, could possibly be more radical than one which cannot be spoken about without new syntactical categories. Notice, finally, that the two new primitives determine *in a minimal fashion* that part of L which is, in a technical sense I explained, nonextensional. So far I have avoided the use of 'intentional' for 'nonextensional'. Now we might as well remember that philosophers, speaking philosophically, have insisted that "intentionality" is the differentiating characteristic of minds. Since they spoke philosophically, one cannot be completely certain what they meant. Yet, I am confident that my analysis is the explication of what they reasonably could have meant.

It will pay to reflect briefly on why I used the phrase 'in fact' at the two italicized places above. Interpret '*bl*' and '*a*' as 'blue' and as the name of a particular which is a tone. Let ''p_1'' stand for the name of a propositional character. Both '$bl(a)$' and ''p_1'(a_1)' are well-formed sentences; all one can say is that they are *in fact* false. To say anything else, such as, for instance, that they are ill-formed or, even, that they are contradictory, amounts to accepting some form of the synthetic *a priori* and, probably, also some form of substantialism. I, for one, accept neither.[18]

In the third section I told one half of the story of truth. I am now ready to tell the other half. Then I shall be done.

In an unforgettable metaphor G.E. Moore once called awareness diaphanous or transparent. What he wanted to call attention to was that, because we are so prone to attend to their contents, the awarenesses themselves easily elude us. Intentional meaning is, as we now understand, closely connected with awareness. Not surprisingly, then, it is similarly elusive. That is why, when I first mentioned it, I proceeded negatively, as it were. Remember what I did. I selected a sentence to serve as illustration: "The sentence (proposition) this is green means that this is green." Then I insisted that we sometimes so use such sentences that we do not speak about either the contexts or the referents of linguistic events, in the only clear sense of 'linguistic event'; but, rather, about something "linguistic" in a sense of 'linguistic' which is problematic and therefore in need of explication. The explication, as we now know, is this. (a) The sentence is transcribed by ''$gr(a)$'$Mgr(a)$', which is analytic. (b) ''$gr(a)$'' refers to or names a proposition, i.e., a character of awarenesses. (c) '$gr(a)$' refers to a state of affairs. (d) 'M', being a logical sign, does not refer to or name anything in the sense in which descriptive expressions refer to something. (a) and (d) are the source of the problematic use

18 See MLP 3, MLP 8, MLP 11.

of 'linguistic'. (b) and (c) show that intentional meaning is a logical pseudorelation between a propositional character and a state of affairs; they also show accurately in which respects it makes no sense whatsoever to say that intentional meaning is "linguistic."

When I spoke in the third section about truth, I spoke about language as event—with some reservation, or, as I put it, merely in order to fix the ideas. The reason for the reservation was that 'true', like 'means', has an intentional meaning. Or, to say what corresponds exactly to what I said before and just repeated in the case of 'means', sometimes, when we say "The sentence (proposition) this is green is true if and only if this is green," we speak neither about the contexts nor about the referents of linguistic events but, rather, "linguistically" in a problematic sense of 'linguistic'. I shall now explicate this sense by first proposing a definition of 'true' in L and then commenting on it.

A defined sign or expression is logical if and only if all the primitive signs in its definition are logical. Defined logical signs, like primitive ones, do not refer to anything in the sense in which descriptive ones do. 'True', as I explicate it, is a defined logical predicate of the second type with a nonrelational argument. Thus 'true', or, as I shall write, 'Tr', like 'M', does not refer to anything in the sense in which 'a', 'gr', ' '$gr(a)$' ', and '$gr(a)$' all do. The idea is, as one might expect, to define 'Tr' in terms of 'M' and of other logical signs, i.e., variables, quantifiers, and connectives. The actual definition is

(D) '$Tr(f)$' for '$(\exists p)[fMp \cdot p]$'.

Notice that although 'Tr' can be truly predicated only of the names of characters which are propositions, its definition is nevertheless in terms of the variable of the appropriate type. '$Tr(gr)$', for instance, though it is false, is therefore well formed. To proceed otherwise amounts to accepting some version of the synthetic a priori. This is the same point I made before. Now for four comments to establish that (D) is in fact an adequate transcription of the intentional meaning of 'true'.

I. Remember the sentence I called (A): A sentence is true if and only if what it refers to (means) is the case. Since we are now dealing with intentions, I had better amend it to (A′): *A proposition is true if and only if what it means is the case.* Consider next that in view of (D)

(D′) $Tr(f) \equiv (\exists p)[fMp \cdot p]$

is analytic; for our notion of analyticity is of course so arranged that every sentence stands to a definition in the relation in which (D′) stands to (D) is analytic. (This is just one of the many details I skipped.) Now read (D′) in words: Something is true if and only if there is a state of affairs such that it means this state of affairs and this state of affairs is the case. The only verbal discrepancies between this sentence and (A′) are due to the greater precision which the formalism forces upon us. We must say 'something' instead of 'proposition'; and we must make the existential quantification explicit. (D′), being analytic, is thus an adequate transcription of (A′). A little reflection shows that 'Tr' is and is not "linguistic" in exactly the same senses in which 'M' is. I don't think I need to repeat the distinctions I just made under (a), (b), (c), and (d).

II. Ordinarily we think of true and false as contradictories. I define 'Fs', to be interpreted as 'false', by

'$Fs(f)$' for '$(\exists p)[fMp \cdot \sim p]$.

It follows that 'Fs' and 'Tr' are not contradictories, or, what amounts to the same thing, that '$(f)[Tr(f) \lor Fs(f)]$' cannot be shown to be analytic. On first thought this may make our transcription look less than adequate. Closer examination reveals that we have come across one of its strengths. We do not really want to say that "everything" is either true or false. What we want to say is, rather, that "every sentence" is either true or false. Technically, this means that '$Tr('p_1') \lor Fs('p_1')$' ought to be analytic for every proposition. And that this is so is easily shown. For those who care for this sort of detail I write down the steps of the demonstration: '$p_1'Mp_1$; '$p_1'Mp \cdot (p_1 \lor \sim p_1)$; $(\exists p)['p_1'Mp \cdot (p \lor \sim p)]$; $(\exists p)['p_1'Mp \cdot p)] \lor [(\exists p)'p_1' Mp \cdot \sim p]$.

III. We are in a position to dispose of a question over which recently more ink has been spilled than it deserves. Do 'p_1' and '$Tr('p_1')$' have the same meaning? To ask this question is, as we know, to ask four. With respect to *context*, we do not care and we need not bother. Take the two sentences 'Peter died' and 'It is true that Peter died'; and assume that a person hears once the one and once the other. Whether what he does is the same and whether his mental states are the same on the two occasions is a question for psychologists and psychologists only. As a matter of commonsense, though, the answer will vary, depending on many circumstances, from sentence to sentence, from person to person, and, for the same person, from occasion to occasion. The attempt to answer this question by constructing schemata and trying to discern in them something that corresponds to this meaning of having-the-same-meaning is thus patently absurd. Unhappily, Carnap and some of his students have recently spent a good deal of time and effort on this goose chase. With respect to *reference* the answer is obvious. The two sentences do not refer to the same thing. The same holds for *intentional* meaning. To see that, one merely has to consider that while ''p_1'Mp_1' and ''$Tr('p_1')$'$MTr('p_1')$' are analytic, ''p_1'$MTr('p_1')$' and ''$Tr('p_1')$'Mp_1' are contradictory. There remains *logical* meaning, or, what amounts to the same thing, there remains the question whether '$p_1 \equiv Tr('p_1')$' is analytic for every proposition. This, I believe, is the question which most of those who recently dealt with the issue wanted to discuss. The answer is affirmative. Upon our broader conception of analyticity the sentence is analytic. Some will probably consider that another strength of our transcription. For those who care for this sort of thing I again write down the steps of the demonstration. For the proof that '$p_1 \supset Tr('p_1')$' is analytic they are: p_1; 'p_1'$Mp_1 \cdot p_1$; $(\exists p)$ ['p_1'$Mp \cdot p$]; $p_1 \supset (\exists p)$ ['p_1'$Mp \cdot p$]. To prove that '$Tr('p_1') \supset p_1$' is analytic, the definition of analyticity in L must be technically implemented with what is intuitively obvious. I add then to 4a and 4b a third clause 4c: If '$\Phi(p)$' is an expression such that when a sentence of L is substituted for the variable the sentence it becomes is analytic for every sentence of L, then '$(p)\Phi(p)$' is analytic. Now the proof proceeds as follows. '$(\text{'}p_1\text{'}Mp_1 \cdot p_1)$ $\supset p_1$' is obviously analytic. For every other p_i, '$(\text{'}p_1\text{'}Mp_i \cdot p_i)$ $\supset p_1$' is analytic because the first factor in the antecedent is contradictory. Hence, by 4c, '$(p)[\text{'}p_1\text{'}Mp \cdot p) \supset p_1]$' is analytic. This sentence is equivalent to the one to be proved.

IV. Everybody is familiar with the Liar paradox, that is, with the difficulties one can produce by supposing that a sentence "says about itself" that it is false. When the mathematicians proved what I explained in the third section, they drew part of their inspiration from this conundrum. Assume 'pr' to be an arithmetical-predicate-expression that can be interpreted as 'false'. We know this assumption to be absurd; but that is not the point now. If there is such a predicate expression then one can be using Goedel's ideas show that there is an integer, n, such that if 'N' is the number-sign interpreted as n, the number coordinated to '$pr(N)$' is n. Speaking as inaccurately as the mathematicians do, one could then say that '$pr(N)$' says about itself that it is false. That is why, by a pattern taken from the Liar paradox, the mathematicians drew their conclusions from this sentence. Under the circumstances it is worth noticing that L could not possibly contain a sentence which literally "says about itself" that it is false, or, for that matter, anything else. Assume that S is such a sentence and that, written down, it is a sequence of, say, 17 primitive signs. Its name is then a sequence of 18 primitive signs, the 17 original ones and the quoting operator. Since this name is a predicate and not itself a sentence, any sentence containing it is a sequence of at least 19 primitive signs. S, which is a sequence of only 17 primitive signs, cannot be such a sentence and can therefore not literally say anything about itself. It follows that no sentence of a clarified language can literally say anything about itself.[19] The belief that there are such sentences is one of the illusions created by the logical deficiencies of our natural language.

[19] As I recently discovered, this idea can be read into prop. 3.333 of the *Tractatus*. Wittgenstein made an essential mistake, though. He omitted the quotes.

15

Notes on Intentionality

WILFRID SELLARS

My aim in this paper is to develop, in fairly short compass, some central themes pertaining to intentionality. Since I do not have the space for discussing usefully even a few of the major approaches to this complex topic, I shall limit myself to sketching the kind of position I am inclined to hold, and contrasting it with a carefully worked out alternative which belongs in the same philosophical neighborhood.

I shall assume that there are inner conceptual episodes proper ("thoughts") which are expressed by candid overt speech. These episodes can be referred to as "mental acts" provided that one is careful not to confuse 'act' with 'action' in the sense of "piece of conduct." Thoughts are acts in the sense of *actualities* (as contrasted with dispositions or propensities).[1]

I shall not attempt to botanize the varieties of mental act. Their diversity corresponds to the diversity of the linguistic utterances in which, in candid or uncontrived speech, they find their natural culmination. I shall focus my attention on such thoughts as are expressed by subject-predicate empirical statements, and make use where possible of the tidy forms of PMese.

I said above that candid meaningful linguistic utterances ex-

From *The Journal of Philosophy*, 61 (1964), 655–665. Reprinted by permission of the author and the editors. Presented in a symposium on "Intentionality" at the sixty-first annual meeting of the American Philosophical Association, Eastern Division, December 29, 1964.

[1] This is not to say that there are no such things as mental actions in the conduct sense, but are more complex in structure.

press thoughts. Here it is essential to note that the term 'express', indeed the phrase 'express a thought', is radically ambiguous. In one sense, to say of an utterance that it expresses a thought is to say, roughly, that a thought episode *causes* the utterance.[2] But there is another and radically different sense in which an utterance can be said to express a thought. This is the sense in which the utterance expresses a proposition, i.e., a thought in Frege's sense (*Gedanke*)—an "abstract entity" rather than a mental episode. Let me distinguish between these two senses of 'express' as the 'causal' and the 'logical', and between the two senses of 'thought' by referring to *thinkings* and *propositions*. These distinctions are represented by the following diagram:

$$* \text{ proposition that-}p$$

$$\text{Thinking that-}p * \rightarrow * \text{ speaking that-}p$$

This diagram obviously raises the question: What is the relation between the thinking that-*p* and the proposition that-*p*? One possible move is to treat the relation between the speaking and the proposition as the logical product of the causal relation between the speaking and the thinking and a relation between the *thinking* and the proposition; thus:

$$* \text{ proposition that-}p$$

$$\text{Thinking that-}p * \rightarrow * \text{ speaking that-}p$$

(Roughly: for a speaking to mean that-*p* is for it to be caused by a thinking that-*p*.)

Another possible move is to treat the relation between the thinking and the proposition as the logical product of the causal relation between the speaking and the thinking and a relation between the *speaking* and the proposition, a situation which the first diagram can also be used to represent. (Roughly: to be a

---

[2] I say "roughly," because the word 'cause' is a dangerous one unless used with proper care. Here it means that the occurrence of the thought explains (on certain assumptions about the context) the occurrence of the utterance.

thinking that-*p* is to be an episode of a sort that causes speakings that express the proposition that-*p*.)

I propose, instead, to work with the following more complex
framework in which the idea that thinkings belong to "inner
speech" is taken seriously, and is combined with the idea that
expressions in different languages can stand for (express in the
logical sense) the same proposition. This can be represented,
at least initially, by the following diagram:

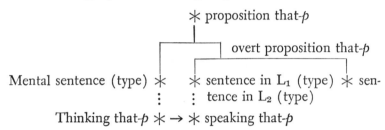

According to this account, neither the relation of the speaking
to the proposition nor the relation of the thinking to the
proposition is to be analyzed as a logical product along the lines
of the last two paragraphs. This claim is intended to be compatible with the idea that there is an internal relation between
the idea of a speaking's expressing a certain proposition and the
idea of a speaking's being caused, *ceteris paribus*, by a thinking
that expresses the same proposition.[3]

The structure of the above diagram can perhaps be clarified
by pointing out that, according to the position I am defending,
the framework of thinkings is an analogical one, the *funda-
mentum* of which is meaningful overt speech, i.e., speech understood in terms of the uniformities and propensities that connect
utterances (a) with other utterances (at the same or a different
level of language), (b) with the perceptible environment, and
(c) with courses of action (including linguistic behavior). I

---

[3] It is important to distinguish between two senses of 'meaningless utterance':
(a) An utterance is meaningless if it does not token a properly formed expression in a language. (b) An utterance is meaningless if it is uttered parrotingly
by one who does not know the language. It is worth reflecting on the idea of a
meaningless mental utterance. We might not call it a thinking, but it would
stand to thinkings as meaningless utterances stand to "saying something."

say uniformities, but the uniformities are not *mere* uniformities, for they are grounded in rules in a way most difficult to analyze, but which involves the causal efficacy of rule expressions.[4]

Thus the concept of a proposition as something that can be expressed by sentences in both Mentalese and, say, English is an analogical extension of the concept of a proposition as something that can be expressed by sentences in both English and German. My next move, therefore, will be to explore what it is for a token of a sentence in, for example, German to express a proposition.

Instead, however, of dealing with this topic directly, I shall ask the closely related question, What is it for a German noun, say 'Himmel', to express a concept: the concept sky?[5]

I have written on a number of occasions[6] that "meaning is not a relation," although statements about what expressions mean "convey" information that would be directly expressed by statements among which would be relational ones. I want now to make additional payments on these promissory notes. Let me begin by acknowledging that there is a perfectly good sense in which

'Himmel' (in German) expresses the concept sky

is a relational statement. Hence, if

'Himmel' (in German) means sky

had the same sense, it, too, would be a relational statement. But the former is a relational statement only in the special way in which

[4] See Wittgenstein, *Philosophical Investigations,* § 198 ff.; also my essay "Some Reflections on Language Games," *Philosophy of Science,* 21 (1954), reprinted in a revised version as ch. 11 of Sellars, *Science, Perception and Reality* (New York: Humanities Press, 1963).

[5] In Fregean terminology, both concepts and propositions, as I am using these terms, are *senses,* and I am exploring what it is for a sentence to express a sense, by asking the parallel question about less complex expressions.

[6] Cf. "Empiricism and Abstract Entities," in Paul Schilpp, ed., *The Philosophy of Rudolf Carnap* (LaSalle, Ill.: Open Court, 1964), pp. 431–468, especially pp. 464 ff.; also "Empiricism and the Philosophy of Mind" in *Minnesota Studies in the Philosophy of Science,* vol. I (Minneapolis: Univ. of Minnesota Press, 1956), pp. 253–329 (reprinted as ch. 5, *Science, Perception and Reality*), especially § 31.

Lions are members of the class of animals

is a relational statement. The special character of the latter consists in the fact that it is a second-level relational counterpart of the first-level nonrelational statement:

Lions are animals

The original statement (let me rewrite it, for reasons which will shortly emerge):

'Himmel's (in German) express the concept sky

has as its nonrelational counterpart

'Himmel's (in German) are ·sky·s

where "·sky·" is a common noun which applies to items in any language that play the role played in our language by the sign design that occurs between the dot quotes. The hypothesis I wish to propose, therefore, is that

'Himmel' (in German) means sky

is, at bottom, the PMese statement.

'Himmel' (in German) ⊂ ·sky·

Here the specific word 'means' serves to indicate that the context is linguistic and to remind us that, in order for the statement to do its job directly, the unique common-noun–forming convention must be understood, and the sign design *sky* must be present in the active vocabulary of the person to whom the statement is made, playing there the role played by 'Himmel' in German.

To characterize a statement of the form

$$A \subset B$$

as "relational" is a mistake of the same nature as characterizing

$$p \text{ or } q$$

as a relational form, or

$$\sim p$$

as predicating negation of a state of affairs. The first of these statements is equivalent by definition to

$$\hat{x}(x\epsilon A) \subset \hat{x}(x\epsilon B)$$

and, ultimately, to

$$(x)\ x\epsilon A \supset x\epsilon B$$

The expressions 'A' and 'B' that appear in 'A ⊂ B' are no more to be construed as proper names than these same expressions as they appear in its unpacked equivalents. As a first approximation we can say that 'A ⊂ B' preserves the predicative character of these expressions which is explicit in the latter statement.[7] 'A ⊂ B' must not be confused with its higher-order counterpart:

The class of As (or A-kind) is included in the class of Bs (or B-kind)

which is, in its way, a relational statement. This distinction is closely parallel to that between the nonrelational statement form '*fa*' and its higher-order counterpart '*a* exemplifies *f*-ness' which is discussed in the next paragraph.

It will probably be objected that the above account simply disguises the relational character of meaning. For surely, it will be said, the role played by the design *sky* in our language is that of expressing the concept sky, and, consequently, I have no more shown that meaning is nonrelational than I would have shown that largeness is nonrelational by pointing out that

New York is large

has the nonrelational form

$$f(x)$$

To come to grips with this challenge I must say an additional word or so about the relational character of

[7] Strictly speaking, of course, the predicates in the latter are 'ϵA' and 'ϵB' in which the 'A' and 'B' are the differentiating components and the 'ϵ' serves (like 'is a' in English) as a syncategorematic component which expresses the classificatory rather than adjectival character of the predicates. See "Classes as Abstract Entities and the Russell Paradox," *Review of Metaphysics* 17 (1963), 67–90, specifically, pp. 67–69; also "Counterfactuals, Dispositions and the Causal Modalities" in *Minnesota Studies in the Philosophy of Science*, vol. II, pp. 225–308, especially pp. 252–266.

'Himmel's (in German) express the concept sky

and, in general, about the connection between those special relational statements which have nonrelational counterparts and these counterparts. I gave above as an example the pair:

> Lions are members of the class of animals
> Lions are animals

What is the relation between these two statements? It is, at bottom, that of

> Socrates exemplifies wisdom

to

> Socrates is wise

where, to tip my hand, the former can be replaced by

> Wisdom is true of Socrates

In other words, as I see it, to claim that 'Himmel' (in German) means sky *because* 'Himmel' (in German) expresses the concept sky is analogous to claiming that Socrates is wise *because* wisdom is true of Socrates. To see that the latter claim would be a mistake one needs only reflect that it would be akin to claiming that Socrates is wise *because* that Socrates is wise is true.[8]

I have argued elsewhere[9] that the truth of statements in a language is to be defined in terms of the truth of propositions. In the framework sketched above, the definition can be represented by the following schema:

$$S \text{ (in L) is true} =_{\text{df}} (\exists p) \, S \text{ (in L) means } p, \text{ and that-}p \text{ is true}$$

[8] Of course, I may know that Socrates is wise because Plato tells me so, and I know that what Plato says is true. But not everything that explains how one knows something to be so explains why it is so. For an elaboration of this interpretation of the relational counterparts of nonrelational statements see "Grammer and Existence: A Preface to Ontology," *Mind*, 69 (1960), 499–533, reprinted as ch. 8, *Science, Perception and Reality*; also "Abstract Entities," *Review of Metaphysics*, 16 (1963), 627–671.

[9] "Truth and 'Correspondence'," *The Journal of Philosophy*, 59 (1962), 29–56, reprinted as ch. 6, *Science, Perception and Reality*.

If we extend these considerations to the case of sentential expressions, we see that

'Es regnet' (in German) means *it is raining*

is, at bottom, the nonrelational PMese statement:

'Es regnet' (in German) ⊂ ·it is raining·

and that, although it has a relational counterpart, namely,

'Es regnet' (in German) expresses the proposition that it is raining

the existence of the latter does not point to a relational analysis of meaning-statements.

If this seems to involve a conflation of two radically different variables, viz. 'that-p' and 'p', the appearance is an illusion, for the propositional expression 'that-p' is related to 'p', as it occurs in the context 'S (in L) means p', as 'the bishop' in

The bishop is a diagonal mover

to 'bishop' in

Bishop ⊂ diagonal movers

They are at the same level of language,[10] and hence no fallacy of treating expressions at different levels as values of the same variable is involved in the above definition. Explicated, it now becomes

S (in L) is true $=_{\text{df}}$ (Ⅎ·p·) S (in L) ⊂ ·p· and ·p· ⊂ true

Once one makes the move of accounting for the truth of statements in language in terms of the propositions they express, the philosophical problem of truth becomes that of explaining how statements like

That it is raining is true

are related to their lower-level counterparts, here

It is raining

[10] This point is elaborated and defended in "Abstract Entities," *Review of Metaphysics*, 16 (1963), 627–671.

Gustav Bergmann, in an important essay on intentionality,[11] makes an interesting use of the structure of Carnap's definition of 'true sentence in L' in which he applies it to the truth of the propositions; thus,

$$\text{That-}p \text{ is true} =_{\text{df}} (\exists q) \text{ that-}p \text{ means } q \text{ and } q$$

Bergmann argues that statements of the form

$$\text{—means ...}$$

are either "analytic" or "self-contradictory" according to extended applications of these terms which he finds to be justified by the fact that these applications bring together things that belong together. (Just what the *intension* is which is supposed to be common to the original and extended applications is left somewhat obscure—a matter of being decidable on purely quasi-linguistic grounds.)

The initial effect of this approach is to make it appear that Bergmann is assimilating the way in which 'it is raining' occurs on the right-hand side of

$$\text{That it is raining means it is raining}$$

to the way in which it occurs in analytic extensional contexts, thus on the right-hand side of

$$\text{Not (it is raining) or it is raining}$$

To switch the metaphor, his logically atomistic left hand works on the principle that 'p' can occur in sentential contexts only if the latter are truth-functional, so that, in order for 'it is raining' to occur in 'that it is raining means it is raining', the latter statement must be analyzable in terms of truth-functional connectives in such a way that the apparently *predicative* character of 'means' disappears.[12] His equally agile right hand,

[11] "Intentionality" (included in this volume).
[12] In his *Introduction to Semantics*, Carnap so introduces 'designates (in L)' that 'it is raining' does occur on the right-hand side of

$$S \text{ designates (in L) it is raining}$$

by defining it in terms of disjunction, conjunction, and identity construed as a PMese connective. This generates *at best* the "telephone directory" account of however, works on the principle that 'means' functions as a predicate. Is it a predicate? It cannot, he assures us, be analyzed into the familiar connectives. Yet its character as predicate is *somehow* bogus. At this juncture, Bergmann simply tells us that 'means' is a *unique* connective. His purpose is clear. If 'means' is a *connective*, then 'it is raining' genuinely occurs on the right-hand side of '. . . means it is raining', while if it is *not reducible* to the familiar connectives, it must be *added* to PMese to capture the unique character of mentalistic discourse. Bergmann is on to something important, but his formulations strike me, to use a Russellian metaphor, as light-fingered.

What is the alternative? As I see it the correct move is not to introduce a new "connective," but to explore in greater detail the unique way in which 'it is raining' occurs in 'it is true that it is raining'. But before doing so, let me note that, on the view sketched earlier in this paper, as well as on Bergmann's view,

> That it is raining means it is raining

is analytic. On my view, however, it is analytic in a straight-forward sense, for it amounts to nothing more than

> ·It is raining· ⊂ ·it is raining·

The crucial difference between our two accounts concerns the concept of a proposition. On my view, it is essential to distinguish between a proposition and the mental sentence directly tokened by mental acts or thinkings. Bergmann runs these two together, with, as I see it, disastrous consequences to his whole philosophy of mind. For this running together, when combined with the insight that it is just as appropriate to speak of what mental sentences mean, leads him to the mistaken conclusion that statements about *the meanings of propositions* are basic to the theory of mind, meaning, and truth.

Thus, if we use '⟨⟨Es regnet⟩⟩' to stand for the kind of mental act that occurs in the minds of German-speaking people and

---

meaning and truth correctly satirized by Max Black in his well-known paper on the "Semantical Definition of Truth," *Analysis*, 7 (1947).

finds its overt expression in candid utterances of 'Es regnet', then it makes as good sense to say

⟨⟨Es regnet⟩⟩ (in the minds of German speakers) means it is raining[13]

as it does to say

'Es regnet' (in German) means it is raining

That there is a close connection between these statements is clear, but it is not such as to make (using a corresponding convention)

⟨⟨It is raining⟩⟩ (in the minds of English speakers) means it is raining

as trifling as

That it is raining means it is raining

Inner sentence episodes can differ in their descriptive character and yet express the same proposition, just as can overt sentence episodes.[14] And just as the generically specified character of the shapes and motions and relative locations demanded of chess pieces *must* have determinate embodiment in actual games, so the generically specified character of pieces, positions, and moves which is common to determinate ways of playing the same conceptual game must be determinately embodied in the natural order. In other words, although a mental act that expresses the proposition that it is raining is *ipso facto* an ·it is raining·, it must also belong to a specific variety of ·it is raining·, just as a token of the corresponding English sentence not only is an ·it is raining· but has the specific empirical character by virtue of which it sounds (or reads) like *that*.

The fact that conceptual "pieces" or "role-players" *must*

---

[13] I pointed out above (p. 323 f.) that the concept of a proposition as expressed by mental and overt sentences is an analogical extension of the concept of a proposition as something expressed by overt sentences. Roughly, to be a that-*p* item in the more inclusive sense is to be an item of a kind that plays a role in *either* thinkings *or* overt speakings similar in relevant respects to that played in our overt speech by the design represented by '*p*'.

[14] There is indeed, every reason to suppose that Japanese inner speech differs systematically from English inner speech in a way which reflects the differences between these two languages.

have determinate *factual* character, even though we don't know what that character is, save in the most general way, is the hidden strength of the view that identifies mental acts with neurophysiological episodes.[15]

If the foregoing remarks are correct, then, whereas the truth of mental statements must, like that of overt statements, be defined in terms of the truth of propositions, according to the schema

$$S \text{ (in L) is true} =_{df} (\exists \cdot p \cdot) S \text{ (in L) means } \cdot p \cdot \text{ and } \cdot p \cdot \subset \text{ true}$$

the truth of propositions is *not* to be so defined, but requires a radically different treatment.

I shall limit my positive account to the truth of empirical propositions and to the bare bones of that. The central theme is that the "inference" represented by the sequence

> It is true that Tom is tall
> Tom is tall

differs radically from the inference represented by the sequence

> Tom is tall and wise
> Tom is tall

or

> There is lightning
> It will thunder

In the latter two examples, the sequences are authorized by principles that do not themselves belong in the sequences. In the first example, however, the inscribing of ·Tom is tall· is a performance which has as its authority a statement inscribed above it.

It might be thought that I am offering something like the "warranted assertability" theory of truth, according to which the first sequence has the form

---

[15] This point is elaborated in Sellars, "Philosophy and the Scientific Image of Man" in Robert Colodny, ed., *Frontiers of Science and Philosophy* (Pittsburgh: Univ. of Pittsburgh Press, 1962), reprinted as ch. 1 in *Science, Perception, and Reality*; see especially sec. 6.

> The tokening of ·Tom is tall· is warranted
> Tom is tall

But to make this move is to confuse truth with probability, for, presumably, to be warranted is to be warranted by evidence. There is, indeed, a close connection between truth and probability, but it is not so simple as that.

What is the basic job of empirical statements? The answer is, in essence, that of the *Tractatus*, i.e., to compete for places in a picture of how things are, in accordance with a complex manner of projection. Just how such a manner of projection is to be described is a difficult topic in its own right.[16] The important thing for our purposes is that the relation between conceptual picture and objects pictured is a factual relation. Thus, whereas an item in the picture is, say, an ·*fa*·, and the concept of an ·*fa*· ultimately involves (as does the concept of a pawn) the concept of what it is to satisfy a norm or a standard, the point of the norms or standards pertaining to conceptual "pieces" is to bring it about that *as items in the natural order* they picture the way things are.

To say of a basic empirical proposition, e.g., that-*fa*, that it is true is to say that an ·*fa*· belongs in a telling of the world story that it is the business of empirical inquiry to construct. And the statement:

> An ·fa· belongs to the story

makes sense even where one neither knows nor has good reason to think an ·*fa*· belongs in the story. If, however, one constructs two columns, a right-hand column purporting to be a fragment of the story, and to the left a fragmentary list of statements about what belongs in the story, then it is clear that to inscribe ·An ·*fa*· belongs in the story· in the latter left-hand column is to be committed to the inscribing of an ·*fa*· in the right-hand column, and vice versa. If we represent this commitment by

---

[16] I have explored this topic in the paper on "Truth and 'Correspondence' " referred to in footnote 9 above.

That-*fa* is true: *fa*

then we can say that the implication statements:

That-*fa* is true implies that-*fa*
That-*fa* implies that that-*fa* is true

are derivative from the former in that the latter are vindicated by pointing out that the pair of inscriptions referred to above can be regarded as a special case of both of the kinds of sequence represented by:

That-*fa* is true   *fa*
fa   That-*fa* is true

which it would be the point of the implication statements to authorize.

Notice, in conclusion, that the *practical* connection between inscribing ·That-*fa* is true· and inscribing ·*fa*· is a special case of a family of practical connections. Another example is that which relates ·that-*fa* implies that-*ga*· to world-story telling. Commitment to ·that-*fa* is true· picks for further consideration out of all constructible world stories those which include an ·*fa*·. Commitment to ·that-*fa* implies that-*ga*· picks out those which do not include an ·*fa*· unless they include a ·*ga*·, nor a competitor of ·*ga*· unless it includes a competitor of ·*fa*·.

PART IV

Intentionality, Logic, and Semantics

16

On Sense and Nominatum

GOTTLOB FREGE

The idea of sameness[1] challenges reflection. It raises questions which are not quite easily answered. Is sameness a relation? A relation between objects? Or between names or signs of objects? I assumed the latter alternative in my *Begriffsschrift*. The reasons that speak in its favor are the following: "a = a" and "a = b" are sentences of obviously different cognitive significance: "a = a" is valid *a priori* and according to Kant is to be called analytic, whereas sentences of the form "a = b" often contain very valuable extensions of our knowledge and cannot always be justified in an *a priori* manner. The discovery that it is not a different and novel sun which rises every morning, but that it is the very same, certainly was one of the most consequential ones in astronomy. Even nowadays the recognition (identification) of a planetoid or a comet is not always a matter of self-evidence. If we wished to view identity as a relation between the objects designated by the names 'a' and 'b' then "a = b" and "a = a" would not seem different if "a = b" is true. This would express a relation of a thing to itself, namely, a relation such that it holds between every

From *Readings in Philosophical Analysis*, eds. H. Feigl and W. Sellars. Copyright 1949 by Appleton-Century-Crofts, Inc. Reprinted by permission of the publisher. Translated by H. Feigl from the article "Ueber Sinn und Bedeutung," *Zeitschr. f. Philos. und Philos. Kritik*, 100 (1892). The terminology adopted is largely that used by R. Carnap in *Meaning and Necessity* (Chicago: Univ. of Chicago Press, 1947).

[1] I use this word in the sense of identity and understand "a = b" in the sense of "a is the same as b" or "a and b coincide."

thing and itself but never between one thing and another. What one wishes to express with "a = b" seems to be that the signs of names 'a' and 'b' name the same thing; and in that case we would be dealing with those signs: a relation between them would be asserted. But this relation could hold only inasmuch as they name or designate something. The relation, as it were, is mediated through the connection of each sign with the same nominatum. This connection, however, is arbitrary. You cannot forbid the use of an arbitrarily produced process or object as a sign for something else. Hence, a sentence like "a = b" would no longer refer to a matter of fact but rather to our manner of designation; no genuine knowledge would be expressed by it. But this is just what we do want to express in many cases. If the sign 'a' differs from the sign 'b' only as an object (here by its shape) but not by its role as a sign, that is to say, not in the manner in which it designates anything, then the cognitive significance of "a = a" would be essentially the same as that of "a = b," if "a = b" is true. A difference could arise only if the difference of the signs corresponds to a difference in the way in which the designated objects are given. Let a, b, c be straight lines which connect the corners of a triangle with the midpoints of the opposite sides. The point of intersection of a and b is then the same as that of b and c. Thus we have different designations of the same point and these names ('intersection of a and b', 'intersection of b and c') indicate also the manner in which these points are presented. Therefore the sentence expresses a genuine cognition.

Now it is plausible to connect with a sign (name, word combination, expression) not only the designated object, which may be called the nominatum of the sign, but also the sense (connotation, meaning) of the sign in which is contained the manner and context of presentation. Accordingly, in our examples the *nominata* of the expressions 'the point of intersection of a and b' and 'the point of intersection of b and c' would be the same, but not their senses. The nominata of 'evening star' and 'morning star' are the same but not their senses.

From what has been said it is clear that I here understand by 'sign' or 'name' any expression which functions as a proper name, whose nominatum accordingly is a definite object (in the widest sense of this word). But no concept or relation is under consideration here. These matters are to be dealt with in another essay. The designation of a single object may consist of several words or various signs. For brevity's sake, any such designation will be considered as a proper name.

The sense of a proper name is grasped by everyone who knows the language or the totality of designations of which the proper name is a part;[2] this, however, illuminates the nominatum, if there is any, in a very one-sided fashion. A complete knowledge of the nominatum would require that we could tell immediately in the case of any given sense whether it belongs to the nominatum. This we shall never be able to do.

The regular connection between a sign, its sense, and its nominatum is such that there corresponds a definite sense to the sign and to this sense there corresponds again a definite nominatum; whereas not one sign only belongs to one nominatum (object). In different languages, and even in one language, the same sense is represented by different expressions. It is true, there are exceptions to this rule. Certainly there should be a definite sense to each expression in the complete configuration of signs, but the natural languages in many ways fall short of this requirement. We must be satisfied if the same word, at least in the same context, has the same sense. It can perhaps be granted that an expression has a sense if it is formed in a grammatically correct manner and stands for a proper name. But as to whether there is a denotation corresponding to the connotation is hereby not decided. The words 'the heavenly body which has the greatest distance from the

[2] In the case of genuinely proper names like 'Aristotle' opinions as regards their sense may diverge. As such may be suggested, e.g., Plato's disciple and the teacher of Alexander the Great. Whoever accepts this sense will interpret the meaning of the statement "Aristotle was born in Stagira" differently from one who interpreted the sense of 'Aristotle' as the Stagirite teacher of Alexander the Great. As long as the nominatum remains the same, these fluctuations in sense are tolerable. But they should be avoided in the system of a demonstrative science and should not appear in a perfect language.

earth' have a sense; but it is very doubtful as to whether they have a nominatum. The expression 'the series with the least convergence' has a sense; but it can be proved that it has no nominatum, since for any given convergent series, one can find another one that is less convergent. Therefore the grasping of a sense does not with certainty warrant a corresponding nominatum.

When words are used in the customary manner then what is talked about are their nominata. But it may happen that one wishes to speak about the words themselves or about their senses. The first case occurs when one quotes someone else's words in direct (ordinary) discourse. In this case one's own words immediately name (denote) the words of the other person and only the latter words have the usual nominata. We thus have signs of signs. In writing we make use of quotes enclosing the word-icons. A word-icon in quotes must therefore not be taken in the customary manner.

If we wish to speak of the sense of an expression 'A' we can do this simply through the locution 'the sense of the expression 'A''. In indirect (oblique) discourse we speak of the sense, e.g., of the words of someone else. From this it becomes clear that also in indirect discourse, words do not have their customary nominata; they here name what customarily would be their sense. In order to formulate this succinctly we shall say: words in indirect discourse are used *indirectly*, or have *indirect* nominata. Thus we distinguish the *customary* from the *indirect* nominatum of a word; and similarly, its *customary* sense from its *indirect* sense. The indirect nominatum of a word is therefore its customary sense. Such exceptions must be kept in mind if one wishes correctly to comprehend the manner of connection between signs, senses, and nominata in any given case.

Both the nominatum and the sense of a sign must be distinguished from the associated image. If the nominatum of a sign is an object of sense perception, my image of the latter is an inner picture[3] arisen from memories of sense impressions and

---

[3] With the images we can align also the percepts in which the sense impressions and activities themselves take the place of those traces left in the mind.

activities of mine, internal or external. Frequently this image is suffused with feelings; the definiteness of its various parts may vary and fluctuate. Even with the same person the same sense is not always accompanied by the same image. The image is subjective; the image of one person is not that of another. Hence, the various differences between the images connected with one and the same sense. A painter, a rider, a zoologist probably connect very different images with the name 'Bucephalus'. The image thereby differs essentially from the connotation of a sign, which latter may well be common property of many and is therefore not a part or mode of the single person's mind; for it cannot well be denied that mankind possesses a common treasure of thoughts which is transmitted from generation to generation.[4]

While, accordingly, there is no objection to speak without qualification of the sense in regard to images, we must, to be precise, add *whose* images they are and at what time they occur. One might say: just as words are connected with different images in two different persons, the same holds of the senses also. Yet this difference would consist merely in the manner of association. It does not prevent both from apprehending the same sense, but they cannot have the image. *Si duo idem faciunt, non est idem.* When two persons imagine the same thing, each still has his own image. It is true, occasionally we can detect differences in the images or even in the sensations of different persons. But an accurate comparison is impossible because these images cannot be had together in one consciousness.

The nominatum of a proper name is the object itself which is designated thereby; the image which we may have along with it is quite subjective; the sense lies in between, not subjective as is the image, but not the object either. The following simile

---

For our purposes the difference is unimportant, especially since besides sensations and activities recollections of such help in completing the intuitive presentation. 'Percept' may also be understood as the object, inasmuch as it is spatial or capable of sensory apprehension.

[4] It is therefore inexpedient to designate fundamentally different things by the one word 'image' (or 'idea').

may help in elucidating these relationships. Someone observes the moon through a telescope. The moon is comparable with the nominatum; it is the object of the observation which is mediated through the real image projected by the object lens into the interior of the telescope, and through the retinal image of the observer. The first may be compared with the sense, the second with the presentation (or image in the psychological sense). The real image inside the telescope, however, is relative; it depends upon the standpoint; yet, it is objective in that it can serve several observers. Arrangements could be made such that several observers could utilize it. But every one of them would have only his own retinal image. Because of the different structures of the eyes not even geometrical congruence could be attained; a real coincidence would in any case be impossible. One could elaborate the simile by assuming that the retinal image of A could be made visible to B; or A could see his own retinal image in a mirror. In this manner one could possibly show how a presentation itself can be made into an object; but even so, it would never be to the (outside) observer what it is to the one who possesses the image. However, these lines of thought lead too far afield.

We can now recognize three levels of differences of words, expressions, and complete sentences. The difference may concern at most the imagery, or else the sense but not the nominatum, or finally also the nominatum. In regard to the first level, we must note that, owing to the uncertain correlation of images with words, a difference may exist for one person that another does not discover. The difference of a translation from the original should properly not go beyond the first level. Among the differences possible in this connection we mention the shadings and colorings which poetry seeks to impart to the senses. These shadings and colorings are not objective. Every listener or reader has to add them in accordance with the hints of the poet or speaker. Surely, art would be impossible without some kinship among human imageries; but just how far the intentions of the poet are realized can never be exactly ascertained.

We shall henceforth no longer refer to the images and

picturizations; they were discussed only lest the image evoked by a word be confused with its sense or its nominatum.

In order to facilitate brief and precise expression we may lay down the following formulations:

A proper name (word, sign, sign-compound, expression) expresses its sense, and designates or signifies its nominatum. We let a *sign express* its sense and *designate* its nominatum.

Perhaps the following objection, coming from idealistic or skeptical quarters, has been kept in abeyance for some time: "You have been speaking without hesitation of the moon as an object; but how do you know that the name 'the moon' has in fact a nominatum? How do you know that anything at all has a nominatum?" I reply that it is not our intention to speak of the image of the moon, nor would we be satisfied with the sense when we say 'the moon'; instead, we presuppose a nominatum here. We should miss the meaning altogether if we assumed we had reference to images in the sentence "the moon is smaller than the earth". Were this intended we would use some such locution as 'my image of the moon'. Of course, we may be in error as regards that assumption, and such errors have occurred on occasion. However, the question whether we could possibly always be mistaken in this respect may here remain unanswered; it will suffice for the moment to refer to our intention in speaking and thinking in order to justify our reference to the nominatum of a sign; even if we have to make the proviso: if there is such a nominatum.

Thus far we have considered sense and nominatum only of such expressions, words, and signs which we called proper names. We are now going to inquire into the sense and the nominatum of a whole declarative sentence. Such a sentence contains a proposition.[5] Is this thought to be regarded as the sense or the nominatum of the sentence? Let us for the moment assume that the sentence has a nominatum! If we then substitute a word in it by another word with the same nominatum

[5] By 'proposition' I do not refer to the subjective activity of thinking but rather to its objective content which is capable of being the common property of many.

but with a different sense, then this substitution cannot affect the nominatum of the sentence. But we realize that in such cases the proposition is changed; e.g., the proposition of the sentence "the morning star is a body illuminated by the sun" is different from that of "the evening star is a body illuminated by the sun". Someone who did not know that the evening star is the same as the morning star could consider the one proposition true and the other false. The proposition can therefore not be the nominatum of the sentence; it will instead have to be regarded as its sense. But what about the nominatum? Can we even ask this question? A sentence as a whole has perhaps only sense and no nominatum? It may in any case be expected that there are such sentences, just as there are constituents of sentences which do have sense but no nominatum. Certainly, sentences containing proper names without nominata must be of this type. The sentence "Odysseus deeply asleep was disembarked at Ithaca" obviously has a sense. But since it is doubtful as to whether the name 'Odysseus' occurring in this sentence has a nominatum, so it is also doubtful that the whole sentence has one. However, it is certain that whoever seriously regards the sentence either as true or as false also attributes to the name 'Odysseus' a nominatum, not only a sense; for it is obviously the nominatum of this name to which the predicate is either ascribed or denied. He who does not acknowledge the nominatum cannot ascribe or deny a predicate to it. It might be urged that the consideration of the nominatum of the name is going farther than is necessary; one could be satisfied with the sense, if one stayed with the proposition. If all that mattered were only the sense of the sentence (i.e., the proposition) then it would be unnecessary to be concerned with the nominata of the sentence-components, for only the sense of the components can be relevant for the sense of the sentence. The proposition remains the same, no matter whether or not the name 'Odysseus' has a nominatum. The fact that we are at all concerned about the nominatum of a sentence-component indicates that we generally acknowledge or postulate a nominatum for the sentence itself. The proposition loses in interest as soon as we recognize that one of its parts is lacking a nominatum. We may therefore be justified to ask for a nominatum of a sentence, in addition to its sense. But why do we wish that every proper name have not only a sense but also a nominatum? Why is the proposition alone not sufficient? We answer: because what matters to us is the truth-value. This, however, is not always the case. In listening to an epic, for example, we are fascinated by the euphony of the language and also by the sense of the sentences and by the images and emotions evoked. In turning to the question of truth we disregard the artistic appreciation and pursue scientific considerations. Whether the name 'Odysseus' has a nominatum is therefore immaterial to us as long as we accept the poem as a work of art.[6] Thus, it is the striving for truth which urges us to penetrate beyond the sense to the nominatum.

We have realized that we are to look for the nominatum of a sentence whenever the nominata of the sentence-components are the thing that matters; and that is the case whenever and only when we ask for the truth value.

Thus we find ourselves persuaded to accept the *truth-value* of a sentence as its nominatum. By the truth-value of a sentence I mean the circumstance of its being true or false. There are no other truth-values. For brevity's sake I shall call the one the true and the other the false. Every declarative sentence, in which what matters are the nominata of the words, is therefore to be considered as a proper name; and its nominatum, if there is any, is either the true or the false. These two objects are recognized, even if only tacitly, by everyone who at all makes judgments, holds anything as true, thus even by the sceptic. To designate truth-values as objects may thus far appear as a capricious idea or as a mere play on words, from which no important conclusion should be drawn. What I call an object can be discussed only in connection with the nature of concepts and relations. That I will reserve for another essay. But this

---

[6] It would be desirable to have an expression for signs which have sense only. If we call them 'icons' then the words of an actor on the stage would be icons; even the actor himself would be an icon.

might be clear even here: in every judgment[7]—no matter how obvious—a step is made from the level of propositions to the level of the nominata (the objective facts)

It may be tempting to regard the relation of a proposition to the true not as that of sense to nominatum but as that of the subject to the predicate. One could virtually say: "the proposition that 5 is a prime number is true". But on closer examination one notices that this does not say any more than is said in the simple sentence "5 is a prime number". This makes clear that the relation of a proposition to the true must not be compared with the relation of subject and predicate. Subject and predicate (interpreted logically) are, after all, components of a proposition; they are on the same level as regards cognition. By joining subject and predicate we always arrive only at a proposition; in this way we never move from a sense to a nominatum or from a proposition to its truth-value. We remain on the same level and never proceed from it to the next one. Just as the sun cannot be part of a proposition, so the truth-value, because it is not the sense, but an object, cannot be either.

If our conjecture (that the nominatum of a sentence is its truth value) is correct, then the truth-value must remain unchanged if a sentence-component is replaced by an expression with the same nominatum but with a different sense. Indeed, Leibnitz declares: *"Eadem sunt, quae sibi mutuo substitui possunt, salva veritate"*. What else, except the truth-value, could be found, which quite generally belongs to every sentence and regarding which the nominata of the components are relevant and which would remain invariant for substitutions of the type indicated?

Now if the truth-value of a sentence is its nominatum, then all true sentences have the same nominatum, and likewise all false ones. This implies that all detail has been blurred in the nominatum of a sentence. What interests us can therefore never be merely the nominatum; but the proposition alone does not give knowledge; only the proposition together with its nomi-

---

[7] A judgment is not merely the apprehension of a thought or proposition but the acknowledgment of its truth.

natum, i.e., its truth-value, does. Judging may be viewed as a movement from a proposition to its nominatum, i.e., its truth-value. Of course this is not intended as a definition. Judging is indeed something peculiar and unique. One might say that judging consists in the discerning of parts within the truth-value. This discernment occurs through recourse to the proposition. Every sense that belongs to a truth-value would correspond in its own manner to the analysis. I have, however, used the word 'part' in a particular manner here: I have transferred the relation of whole and part from the sentence to its nominatum. This I did by viewing the nominatum of a word as part of the nominatum of a sentence, when the word itself is part of the sentence. True enough, this way of putting things is objectionable since as regards the nominatum the whole and one part of it does not determine the other part; and also because the word 'part' in reference to bodies has a different customary usage. A special expression should be coined for what has been suggested above.

We shall now further examine the conjecture that the truth-value of a sentence is its nominatum. We have found that the truth-value of a sentence remains unaltered if an expression within the sentence is replaced by a synonymous one. But we have as yet not considered the case in which the expression-to-be-replaced is itself a sentence. If our view is correct, then the truth-value of a sentence, which contains another sentence as a part, must remain unaltered when we substitute for the part another of the same truth-value. Exceptions are to be expected if the whole or the part are either in direct or indirect discourse; for as we have seen, in that case the nominata of the words are not the usual ones. A sentence in direct discourse nominates again a sentence but in indirect discourse it nominates a proposition.

Our attention is thus directed to subordinate sentences (i.e., dependent clauses). These present themselves of course as parts of a sentence-structure which from a logical point of view appears also as a sentence, and indeed as if it were a main clause. But here we face the question whether in the case of dependent clauses it also holds that their nominata are truth-values. We know already that this is not the case with sentences in indirect discourse. The grammarians view clauses as representatives of sentence-parts and divide them accordingly into subjective, relative, and adverbial clauses. This might suggest that the nominatum of a clause is not a truth-value but rather that it is of similar nature as that of a noun or of an adjective or of an adverb; in short, of a sentence-part whose sense is not a proposition but only part thereof. Only a thorough investigation can provide clarity in this matter. We shall herein not follow strictly along grammatical lines, but rather group together what is logically of comparable type. Let us first seek out such instances in which, as we just surmised, the sense of a clause is not a self-sufficient proposition.

Among the abstract clauses beginning with 'that' there is also the indirect discourse, of which we have seen that in it the words have their indirect (oblique) nominata which coincide with what are ordinarily their senses. In this case then the clause has as its nominatum a proposition, not a truth-value; its sense is not a proposition but it is the sense of the words 'the proposition that . . .', which is only a part of the proposition corresponding to the total sentence-structure. This occurs in connection with 'to say', 'to hear', 'to opine', 'to be convinced', 'to infer', and similar words.[8] The situation is different, and rather complicated in connection with such words as 'to recognize', 'to know', 'to believe', a matter to be considered later.

One can see that in these cases the nominatum of the clause indeed consists in the proposition, because whether that proposition is true or false is immaterial for the truth of the whole sentence. Compare, e.g., the following two sentences: "Copernicus believed that the planetary orbits are circles" and "Copernicus believed that the appearance of the sun's motion is produced by the real motion of the earth". Here the one clause can be substituted for the other without affecting the

---

[8] In "A lied, that he had seen B" the clause denotes a proposition of which it is said, first, that A asserted it as true, and second, that A was convinced of its falsity.

truth. The sense of the principal sentence together with the clause is the single proposition; and the truth of the whole implies neither the truth nor the falsity of the clause. In cases of this type it is not permissible to replace in the clause one expression by another of the same nominatum. Such replacement may be made only by expressions of the same indirect nominatum, i.e., of the same customary sense. If one were to infer: the nominatum of a sentence is not its truth-value ("because then a sentence could always be replaced by another with the same truth-value"), he would prove too much; one could just as well maintain that the nominatum of the word 'morning star' is not Venus, for one cannot always substitute 'Venus' for 'morning star'. The only correct conclusion is that the nominatum of a sentence is *not always* its truth-value, and that 'morning star' does not always nominate the planet Venus, for this is indeed not the case when the word is used with its indirect nominatum. Such an exceptional case is before us in the clauses just considered, whose nominatum is a proposition.

When we say "it seems that . . ." then we mean to say "it seems to me that . . ." or "I opine that . . .". This is the same case over again. Similarly with expressions such as: 'to be glad', 'to regret', 'to approve', 'to disapprove', 'to hope', 'to fear'. When Wellington, toward the end of the battle of Belle-Alliance, was glad that the Prussians were coming, the ground of his rejoicing was a conviction. Had he actually been deceived, he would not have been less glad, as long as his belief persisted; and before he arrived at the conviction that the Prussians were coming he could not have been glad about it, even if in fact they were already approaching.

Just as a conviction or a belief may be the ground of a sentiment, so it can also be the ground of another conviction such as in inference. In the sentence "Columbus inferred from the roundness of the earth that he could, traveling westward, reach India" we have, as nominata of its parts, two propositions: that the earth is round and that Columbus traveling westward could reach India. What matters here is only that Columbus was convinced of the one as well as of the other and that the one conviction furnishes the ground for the other. It is irrelevant for the truth of our sentence whether the earth is really round and whether Columbus could have reached India in the manner he fancied. But it is not irrelevant whether for 'the earth' we substitute 'the planet accompanied by one satellite whose diameter is larger than one-fourth of its own diameter'. Here also we deal with the indirect nominata of the words.

Adverbial clauses of purpose with 'so that', likewise belong here; obviously the purpose is a proposition; therefore: indirect nominata of the words, expressed in subjunctive form.

The clause with 'that' after 'to command', 'to request', 'to forbid' would appear in imperative form in direct discourse. Imperatives have no nominata; they have only sense. It is true, commands or requests are not propositions, but they are of the same type as propositions. Therefore the words in the dependent clauses after 'to command', 'to request', etc., have indirect nominata. The nominatum of such a sentence is thus not a truth-value but a command, a request, and the like.

We meet a similar situation in the case of dependent questions in phrases like 'to doubt if', 'not to know what'. It is easy to see that the words, here too, have to be interpreted in terms of their indirect nominata. The dependent interrogatory clauses containing 'who', 'what', 'where', 'when', 'how', 'whereby', etc., often apparently approximate closely adverbial clauses in which the words have their ordinary nominata. These cases are linguistically distinguished through the mode of the verb. In the subjunctive we have a dependent question and the indirect nominata of the words, so that a proper name cannot generally be replaced by another of the same object.

In the instances thus far considered the words in the clause had indirect nominata; this made it intelligible that the nominatum of the clause itself is indirect, i.e., not a truth-value, but a proposition, a command, a request, a question. The clause could be taken as a noun; one might even say, as a proper name of that proposition, command, etc., in whose role it functions in the context of the sentence-structure.

We are now going to consider clauses of another type, in

which the words do have their customary nominata although there does not appear a proposition as the sense or a truth-value as the nominatum. How this is possible will best be elucidated by examples.

He who discovered the elliptical shape of the planetary orbits, died in misery.

If, in this example, the sense of the clause were a proposition, it would have to be expressible also in a principal sentence. But this cannot be done because the grammatical subject 'he who' has no independent sense. It merely mediates the relations to the second part of the sentence: 'died in misery'. Therefore the sense of the clause is not a complete proposition and its nominatum is not a truth-value, but Kepler. It might be objected that the sense of the whole does include a proposition as it part; namely, that there was someone who first recognized the elliptical shape of the planetary orbits; for if we accept the whole as true we cannot deny this part. Indubitably so; but only because otherwise the clause "he who discovered the elliptical shape, etc." would have no nominatum. Whenever something is asserted then the presupposition taken for granted is that the employed proper names, simple or compound, have nominata. Thus, if we assert "Kepler died in misery", it is presupposed that the name 'Kepler' designates something. However, the proposition that the name 'Kepler' designates something is, the foregoing notwithstanding, not contained in the sense of the sentence "Kepler died in misery". If that were the case the denial would not read "Kepler did not die in misery" but "Kepler did not die in misery, or the name 'Kepler' is without nominatum". That the name 'Kepler' designates something is rather the presupposition of the assertion "Kepler died in misery" as well as of its denial. Now, it is a defect of languages that expressions are possible within them, which, in their grammatical form, seemingly determined to designate an object, nevertheless do not fulfill this condition in special cases; because this depends on the truth of the sentence. Thus it depends upon the truth of the sentence "there was someone who discovered the ellipticity of the orbits" whether the clause 'he

who discovered the elliplicity of the orbits' really designates an
object, or else merely evokes the appearance thereof, while
indeed being without nominatum. Thus it may seem as if our
clause, as part of its sense, contained the proposition that there
existed someone who discovered the ellipticity of the orbits. If
this were so, then the denial would have to read "he who first
recognized the ellipticity of the orbits did not die in misery, or
there was no one who discovered the ellipticity of the orbits."
This, it is obvious, hinges upon an imperfection of language of
which, by the way, even the symbolic language of analysis is
not entirely free; there, also, sign compounds may occur which
appear as if they designated something, but which at least
hitherto are without nominatum, e.g., divergent infinite series.
This can be avoided, e.g., through the special convention that
the nominatum of divergent infinite series be the number o. It
is to be demanded that in a logically perfect language (logical
symbolism) every expression constructed as a proper name in
a grammatically correct manner out of already introduced
symbols in fact designate an object; and that no symbol be
introduced as a proper name without assurance that it have a
nominatum. It is customary in logic texts to warn against the
ambiguity of expressions as a source of fallacies. I deem it at
least as appropriate to issue a warning against apparent proper
names that have no nominata. The history of mathematics has
many a tale to tell of errors which originated from this source.
The demagogic misuse is close (perhaps closer) at hand as
in the case of ambiguous expressions. 'The will of the people'
may serve as an example in this regard, for it is easily established that there is no generally accepted nominatum of that
expression. Thus it is obviously not without importance to ob-
struct once and for all the source of these errors, at least as
regards their occurrence in science. Then such objections as the
one discussed above will become impossible, for then it will
be seen that whether a proper name has a nominatum can never
depend upon the truth of a proposition.

Our considerations may be extended from these subjective
clauses to the logically related relative and adverbial clauses.

Relative clauses, too, are employed in the formation of compound proper names—even if, in contradistinction to subjective clauses, they are not sufficient by themselves for this purpose. These relative clauses may be regarded as equivalent to appositions. Instead of 'the square root of 4 which is smaller than 0' we can also say 'the negative square root of 4'. We have here a case in which out of a conceptual expression a compound proper name is formed, with the help of the definite article in the singular. This is at any rate permissible when one and only one object is comprised by the concept.[9] Conceptual expressions can be formed in such a fashion that their characteristics are indicated through relative clauses as in our example through the clause 'which is smaller than 0'. Obviously, such relative clauses, just as the subjective clauses above, do not refer to a proposition as their sense nor to a truth-value as their nominatum. Their sense is only a part of a proposition, which in many cases, can be expressed by a simple apposition. As in the subjective clauses an independent subject is missing and it is therefore impossible to represent the sense of the clause in an independent principal sentence.

Places, dates, and time-intervals are objects from a logical point of view; the linguistic symbol of a definite place, moment, or span of time must therefore be viewed as a proper name. Adverbial clauses of space or time can then be used in the formation of such proper names in a fashion analogous to the one we have just remarked in the case of subjective and relative clauses. Similarly, expressions for concepts which comprise places, etc., can be formed. Here too, it is to be remarked, the sense of the subordinate clauses cannot be rendered in a principal clause, because an essential constituent, namely the determination of place and time, is missing and only alluded to by a relative pronoun or a conjunction.[10]

---

[9] According to our previous remarks such an expression should always be assured of a nominatum, e.g., through the special convention that the nominatum be the number 0 if there is no object or more than one object denoted by the expression.

[10] Regarding these sentences, however, several interpretations are easily conceivable. The sense of the sentence "after Schleswig-Holstein was torn away

In conditional clauses also, there is, just as we have realized in the case of subjective, relative, and adverbial clauses, a constituent with indeterminate indication corresponding to which there is a similar one in the concluding clause. In referring to one another the two clauses combine into a whole which expresses, as a rule, only one proposition. In the sentence "if a number is smaller than 1 and greater than 0, then its square also is smaller than 1 and greater than 0" this constituent in the conditional clause is 'a number' and in the concluding clause it is 'its'. Just through this indeterminacy the sense acquires the universal character which one expects of a law. But it is in this way also that it comes about that the conditional clause alone does not possess a complete proposition as its sense, and that together with the concluding clause it expresses a single proposition whose parts are no longer propositions. It is not generally the case that a hypothetical judgment correlates two judgments. Putting it in that (or a similar) manner would amount to using the word 'judgment' in the same sense that I have attributed to the word 'proposition'. In that case I would have to say: in a hypothetical proposition two propositions are related to each other. But this could be the case only if an indeterminately denoting constitutent were absent;[11] but then universality would also be missing.

---

from Denmark, Prussia and Austria fell out with one another" could also be rendered by "after the separation of Schl.-H. from Denmark, Prussia and Austria fell out with one another." In this formulation it is sufficiently clear that we should not regard it as part of this sense that Schleswig-Holstein once was separated from Denmark; but rather that this is the necessary presupposition for the very existence of a nominatum of the expression "after the separation of Schl.-H. from D." Yet, our sentence could also be interpreted to the effect that Schl.-H. was once separated from D. This case will be considered later. In order to grasp the difference more clearly, let us identify ourselves with the mind of a Chinese who, with his trifling knowledge of European history, regards it as false that Schl.-H. ever was separated from D. This Chinese will regard as neither true nor false the sentence as interpreted in the first manner. He would deny to it any nominatum because the dependent clause would be lacking a nominatum. The dependent clause would only apparently indicate a temporal determination. But if the Chinese interprets our sentence in the second manner, then he will find it expressing a proposition which he would consider false, in addition to a component which, for him, would be without nominatum.

[11] Occasionally there is no explicit linguistic indication and the interpretation has to depend upon the total context.

If a time point is to be indeterminately indicated in a conditional and a concluding clause, then this is not infrequently effected by *tempus praesens* of the verb, which in this case does not connote the present time. It is this grammatical form which takes the place of the indeterminately indicating constituent in the main and the dependent clause. "When the sun is at the Tropic of Cancer, the northern hemisphere has its longest day" is an example. Here, too, it is impossible to express the sense of the dependent clause in a main clause. For this sense is not a complete proposition; if we said: "the sun is at the Tropic of Cancer" we would be referring to the present time and thereby alter the sense. Similarly, the sense of the main clause is not a proposition either; only the whole consisting of main and dependent clauses contains a proposition. Further, it may occur that several constituents common to conditional and concluding clauses are indeterminately indicated.

It is obvious that subjective clauses containing 'who, 'what', and adverbial clauses with 'where', 'when', 'wherever', 'whenever' are frequently to be interpreted, inasmuch as their sense is concerned, as conditional sentences; e.g., "He who touches pitch soils himself".

Conditional clauses can also be replaced by relative clauses. The sense of the previously mentioned sentence can also be rendered by "the square of a number which is smaller than 1 and larger than 0, is smaller than 1 and larger than 0."

Quite different is the case in which the common constituent of main and dependent clause is represented by a proper name. In the sentence: "Napoleon who recognized the danger to his right flank, personally led his troops against the enemy's position" there are expressed two propositions:

1. Napoleon recognized the danger to his right flank.
2. Napoleon personally led his troops against the enemy's position.

When and where this happened can indeed be known only from the context, but is to be viewed as thereby determined. If we pronounce our whole sentence as an assertion we thereby assert simultaneously its two component sentences. If one of the components is false the whole is false. Here we have a case in which the dependent clause by itself has a sense in a complete proposition (if supplemented by temporal and spatial indications). The nominatum of such a clause is therefore a truth-value. We may therefore expect that we can replace it by a sentence of the same truth-value without altering the truth of the whole. This is indeed the case; but it must be kept in mind that for a purely grammatical reason, its subject must be 'Napoleon'; because only then can the sentence be rendered in the form of a relative clause attaching to 'Napoleon'. If the demand to render it in this form and if the conjunction with 'and' is admitted, then this limitation falls away.

Likewise, in dependent clauses with 'although' complete propositions are expressed. This conjunction really has no sense and does not affect the sense of the sentence; rather, it illuminates it in a peculiar fashion.[12] Without affecting the truth of the whole the implicate may be replaced by one of the same truth value; but the illumination might then easily appear inappropriate, just as if one were to sing a song of sad content in a cheerful manner.

In these last instances the truth of the whole implied the truth of the component sentences. The situation is different if a conditional sentence expresses a complete proposition; namely, when in doing so it contains instead of a merely indicating constituent a proper name or something deemed equivalent to a proper name. In the sentence: "if the sun has already risen by now, the sky is heavily overcast", the tense is the present—therefore determinate. The place also is to be considered determinate. Here we can say that a relation is posited such that the case does not arise in which the antecedent sentence nominates the true and the consequent sentence nominates the false. Accordingly, the given (whole) sentence is true if the sun has not as yet risen (no matter whether or not the sky be heavily overcast), and also if the sun has risen and the sky is heavily

---

[12] Similarly in the cases of 'but', 'yet'.

overcast. Since all that matters are only the truth-values, each of the component sentences can be replaced by another one of the same truth-value, without altering the truth-value of the whole sentence. In this case also, the illumination would usually seem inappropriate; the proposition could easily appear absurd; but this has nothing to do with the truth-value of the sentence. It must always be remembered that associated thoughts are evoked on the side; but these are not really expressed and must therefore not be taken account of; their truth-values cannot be relevant.[13]

We may hope we have considered the simple types of sentences. Let us now review what we have found out!

The sense of a subordinate clause is usually not a proposition but only part of one. Its nominatum is therefore not a truth-value. The reason for this is *either*: that the words in the subordinate clause have only indirect nominata, so that the nominatum, not the sense, of the clause is a proposition, *or*, that the clause, because of a contained indeterminately indicating constituent, is incomplete, such that only together with the principal clause does it express a proposition. However, there are also instances in which the sense of the dependent clause is a complete proposition, and in this case it can be replaced by another clause of the same truth-value without altering the truth-value of the whole; that is, inasmuch as there are no grammatical obstacles in the way.

In a survey of the various occurrent clauses one will readily encounter some which will not properly fit within any of the considered divisions. As far as I can see, the reason for that is that these clauses do not have quite so simple a sense. It seems that almost always we connect associated propositions with the main proposition which we express; these associated propositions, even if unexpressed, are associated with our words according to psychological laws also by the listener. And because they appear as associated automatically with our words (as in

---

[13] The proposition of the sentence could also be formulated thus: "either the sun has not as yet risen or the sky is heavily overcast." This shows how to interpret this type of compound sentence.

the case of the main proposition) we seem to wish, after all, to express such associated propositions along with the main propositions. The sense of the sentence thereby becomes richer and it may well happen that we may have more simple propositions than sentences. In some cases the sentence may be interpreted in this way; in others, it may be doubtful whether the associated proposition belongs to the sense of the sentence or whether it merely accompanies it.[14] One might find that in the sentence: "Napoleon, who recognized the danger to his right flank, personally led his troops against the enemy's position" there are not only the previously specified two propositions, but also the proposition that the recognition of the danger was the reason why he led his troops against the enemy. One may indeed wonder whether this proposition is merely lightly suggested or actually expressed. Consider the question whether our sentence would be false if Napoleon's resolution had been formed before the recognition of the danger. If our sentence were true even despite this, then the associated proposition should not be regarded as part of the sense of the sentence. In the alternative case the situation is rather complicated: we should then have more simple propositions than sentences. Now if we replaced the sentence: "Napoleon recognized the danger for his right flank" by another sentence of the same truth-value, e.g., by: "Napoleon was over 45 years old", this would change not only our first but also our third proposition; and this might thereby change also the truth-value of the third proposition—namely, if his age was not the reason for his resolution to lead the troops against the enemy. Hence, it is clear that in such instances sentences of the same truth-value cannot always be substituted for one another. The sentence merely by virtue of its connection with another expresses something more than it would by itself alone.

Let us now consider cases in which this occurs regularly. In the sentence: "Bebel imagines that France's desire for vengeance could be assuaged by the restitution of Alsace-Lorraine"

---

[14] This may be of importance in the question as to whether a given assertion be a lie, an oath, or a perjury.

there are expressed two propositions, which, however, do not correspond to the main and the dependent clause—namely:

1. Bebel believes that France's desire for vengeance could be assuaged by the restitution of Alsace-Lorraine;
2. France's desire for vengeance cannot be assuaged by the restitution of Alsace-Lorraine.

In the expression of the first proposition the words of the dependent clause have indirect nominata; while the same words, in the expression of the second proposition, have their usual nominata. Hence, we see that the dependent clause of our original sentence really is to be interpreted in a twofold way, i.e., with different nominata, one of which is a proposition and the other a truth-value. An analogous situation prevails with expressions like 'to know', 'to recognize', 'it is known'.

A condition clause and its related main clause express several propositions which, however, do not correspond one-to-one to the clauses. The sentence: "Since ice is specifically lighter than water, it floats on water" asserts:

1. Ice is specifically lighter than water.
2. If something is specifically lighter than water, it floats on water.
3. Ice floats on water.

The third proposition, being implied by the first two, would perhaps not have to be mentioned expressly. However, neither the first and the third, nor the second and the third together would completely render the sense of our sentence. Thus we see that the dependent clause 'since ice is specifically lighter than water' expresses both our first proposition and part of the second. Hence, our clause cannot be replaced by another of the same truth-value; for thereby we are apt to alter our second proposition and could easily affect its truth-value.

A similar situation holds in the case of the sentence: "If iron were lighter than water it would float on water". Here we have the two propositions that iron is not lighter than water and that whatever is lighter than water floats on water. The clause again expresses the one proposition and part of the other. If we interpret the previously discussed sentence: "After Schleswig-Holstein was separated from Denmark, Prussia and Austria fell out with one another" as containing the proposition that Schleswig-Holstein once was separated from Denmark, then we have: first, this proposition, second, the proposition that, at a time more precisely determined by the dependent clause, Prussia and Austria fell out with one another. Here, too, the dependent clause expresses not only one proposition but also part of another. Therefore, it may not generally be replaced by another clause of the same truth-value.

It is difficult to exhaust all possibilities that present themselves in language; but I hope, in essence at least, to have disclosed the reasons why, in view of the invariance of the truth of a whole sentence, a clause cannot always be replaced by another of the same truth-value. These reasons are:

1. that the clause does not denote a truth-value in that it expresses only a part of a proposition;
2. that the clause, while it does denote a truth-value, is not restricted to this function in that its sense comprises, beside one proposition, also a part of another.

The first case holds

a. with the indirect nominata of the words;
b. if a part of the sentence indicates only indirectly without being a proper name.

In the second case the clause is to be interpreted in a twofold manner; namely, once with its usual nominatum; the other time with its indirect nominatum; or else, the sense of a part of the clause may simultaneously be a constituent of another proposition which, together with the sense expressed in the dependent clause, amounts to the total sense of the main and the dependent clause.

This makes it sufficiently plausible that instances in which a clause is not replaceable by another of the same truth-value do not disprove our view that the nominatum of a sentence is its truth-value and its sense a proposition.

Let us return to our point of departure now.

When we discerned generally a difference in cognitive significance between "a = a" and "a = b" then this is now explained by the fact that for the cognitive significance of a sentence the sense (the proposition expressed) is no less relevant than its nominatum (the truth-value). If a = b, then the nominatum of 'a' and of 'b' is indeed the same and therefore also the truth-value of "a = b" is the same as that of "a = a". Nevertheless, the sense of 'b' may differ from the sense of 'a'; and therefore the proposition expressed by "a = b" may differ from the proposition expressed by "a = a"; in that case the two sentences do not have the same cognitive significance. Thus, if, as above, we mean by 'judgment' the transition from a proposition to its truth-value, then we can also say that the judgments differ from one another.

17

On Denoting

BERTRAND RUSSELL

By a "denoting phrase" I mean a phrase such as any one of the following: a man, some man, any man, every man, all men, the present King of England, the present King of France, the center of mass of the solar system at the first instant of the twentieth century, the revolution of the earth round the sun, the revolution of the sun round the earth. Thus a phrase is denoting solely in virtue of its *form*. We may distinguish three cases: (1) A phrase may be denoting, and yet not denote anything; e.g., "the present King of France". (2) A phrase may denote one definite object; e.g., "the present King of England" denotes a certain man. (3) A phrase may denote ambiguously; e.g., "a man" denotes not many men, but an ambiguous man. The interpretation of such phrases is a matter of considerable difficulty; indeed, it is very hard to frame any theory not susceptible of formal refutation. All the difficulties with which I am acquainted are met, so far as I can discover, by the theory which I am about to explain.

The subject of denoting is of very great importance not only in logic and mathematics, but also in theory of knowledge. For example, we know that the center of mass of the solar system at a definite instant is some definite point, and we can affirm a number of propositions about it; but we have no immediate *acquaintance* with this point, which is only known to us by description. The distinction between *acquaintance* and *knowl-*

Reprinted from *Mind*, 14 (1905), by kind permission of the editors.

edge about is the distinction between the things we have presentations of, and the things we only reach by means of denoting phrases. It often happens that we know that a certain phrase denotes unambiguously, although we have no acquaintance with what it denotes; this occurs in the above case of the center of mass. In perception we have acquaintance with the objects of perception, and in thought we have acquaintance with objects of a more abstract logical character but we do not necessarily have acquaintance with the objects denoted by phrases composed of words with whose meanings we are acquainted. To take a very important instance: There seems no reason to believe that we are ever acquainted with other people's minds, seeing that these are not directly perceived; hence what we know about them is obtained through denoting. All thinking has to start from acquaintance; but it succeeds in thinking *about* many things with which we have no acquaintance.

The course of my argument will be as follows. I shall begin by stating the theory I intend to advocate;[1] I shall then discuss the theories of Frege and Meinong, showing why neither of them satisfies me; then I shall give the grounds in favor of my theory; and finally I shall briefly indicate the philosophical consequences of my theory.

My theory, briefly, is as follows. I take the notion of the *variable* as fundamental; I use "C (x)" to mean a proposition[2] in which x is a constituent, where x, the variable, is essentially and wholly undetermined. Then we can consider the two notions "C (x) is always true" and "C (x) is sometimes true".[3] Then *everything* and *nothing* and *something* (which are the most primitive of denoting phrases) are to be interpreted as follows:

C (everything) means "C (x) is always true;"
C (nothing) means " 'C(x) is false' is always true;"

[1] I have discussed this subject in *Principles of Mathematics*, ch. 5, and § 476. The theory there advocated is very nearly the same as Frege's, and is quite different from the theory to be advocated in what follows.

[2] More exactly, a propositional function.

[3] The second of these can be defined by means of the first, if we take it to mean, "It is not true that 'C(x) is false' is always true."

C (something) means "It is false that 'C (x) is false' is always true".[4]

Here the notion "C (x) is always true" is taken as ultimate and indefinable, and the others are defined by means of it. *Everything, nothing,* and *something,* are not assumed to have any meaning in isolation, but a meaning is assigned to *every* proposition in which they occur. This is the principle of the theory of denoting I wish to advocate: that denoting phrases never have any meaning in themselves, but that every proposition in whose verbal expression they occur has a meaning. The difficulties concerning denoting are, I believe, all the result of a wrong analysis of propositions whose verbal expressions contain denoting phrases. The proper analysis, if I am not mistaken, may be further set forth as follows.

Suppose now we wish to interpret the proposition, "I met a man". If this is true, I met some definite man; but that is not what I affirm. What I affirm is, according to the theory I advocate:

"'I met x, and x is human' is not always false".

Generally, defining the class of men as the class of objects having the predicate *human,* we say that:

"C (a man)" means "'C (x) and x is human' is not always false".

This leaves "a man," by itself, wholly destitute of meaning, but gives a meaning to every proposition in whose verbal expression "a man" occurs.

Consider next the proposition "all men are mortal". This proposition[5] is really hypothetical and states that *if* anything is a man, it is mortal. That is, it states that if x is a man, x is mortal, whatever x may be. Hence, substituting 'x is human' for 'x is a man,' we find:

---

[4] I shall sometimes use, instead of this complicated phrase, the phrase "C(x) is not always false," or "C(x) is sometimes true," supposed *defined* to mean the same as the complicated phrase.

[5] As has been ably argued in Mr. Bradley's *Logic,* bk. I, ch. 2.

"All men are mortal" means " 'If x is human, x is mortal' is always true".

This is what is expressed in symbolic logic by saying that "all men are mortal" means " 'x is human' implies 'x is mortal' for all values of x". More generally, we say:

"C (all men)" means " 'If x is human, then C (x) is true' is always true."

Similarly

"C (no men)" means " 'If x is human, then C (x) is false' is always true".

"C (some men)" will mean the same as "C (a man)",[6] and

"C (a man)" means "It is false that 'C (x) and x is human' is always false".

"C (every man)" will mean the same as "C (all men)."

It remains to interpret phrases containing *the*. These are by far the most interesting and difficult of denoting phrases. Take as an instance "the father of Charles II was executed". This asserts that there was an x who was the father of Charles II and was executed. Now *the*, when it is strictly used, involves uniqueness; we do, it is true, speak of "*the* son of So-and-so" even when So-and-so has several sons, but it would be more correct to say "*a* son of So-and-so". Thus for our purposes we take *the* as involving uniqueness. Thus when we say "x was *the* father of Charles II" we not only assert that x had a certain relation to Charles II, but also that nothing else had this relation. The relation in question, without the assumption of uniqueness, and without any denoting phrases, is expressed by "x *begat Charles II*". To get an equivalent of "x was the father of Charles II," we must add, "If y is other than x, y did not beget Charles II," or, what is equivalent, "If y begat Charles II, y is identical with x". Hence "x is the father of Charles II" becomes "x begat

---

[6] Psychologically "C (a man)" has a suggestion of *only one*, and "C (some men)" has a suggestion of *more than one*; but we may neglect these suggestions in a preliminary sketch.

Charles II; and 'if *y* begat Charles II, *y* is identical with *x*' is always true of *y*".

Thus "the father of Charles II was executed" becomes:

> "It is not always false of *x* that *x* begat Charles II and that *x* was executed and that 'if *y* begat Charles II, *y* is identical with *x*' is always true of *y*."

This may seem a somewhat incredible interpretation; but I am not at present giving reasons, I am merely *stating* the theory.

To interpret "C (the father of Charles II)," where C stands for any statement about him, we have only to substitute C (*x*) for "*x* was executed" in the above. Observe that, according to the above interpretation, whatever statement C may be, "C (the father of Charles II)" implies:

> "It is not always false of *x* that 'if *y* begat Charles II, *y* is identical with *x*' is always true of *y*,"

which is what is expressed in common language by "Charles II had one father and no more". Consequently if this condition fails, *every* proposition of the form "C (the father of Charles II)" is false. Thus, e.g., every proposition of the form "C (the present King of France)" is false. This is a great advantage in the present theory. I shall show later that it is not contrary to the law of contradiction, as might be at first supposed.

The above gives a reduction of all propositions in which denoting phrases occur to forms in which no such phrases occur. Why it is imperative to effect such a reduction, the subsequent discussion will endeavor to show.

The evidence for the above theory is derived from the difficulties which seem unavoidable if we regard denoting phrases as standing for genuine constituents of the propositions in whose verbal expressions they occur. Of the possible theories which admit such constituents the simplest is that of Meinong.[7] This theory regards any grammatically correct denoting phrase as standing for an *object*. Thus "the present King of France,"

---

[7] See *Untersuchungen zur Gegenstandstheorie und Psychologie* (Leipzig, 1904), the first three articles (by Meinong, Ameseder, and Mally, respectively).

"the round square," etc., are supposed to be genuine objects. It is admitted that such objects do not *subsist*, but nevertheless they are supposed to be objects. This is in itself a difficult view; but the chief objection is that such objects, admittedly, are apt to infringe the law of contradiction. It is contended, for example, that the existent present King of France exists, and also does not exist; that the round square is round, and also not round; etc. But this is intolerable; and if any theory can be found to avoid this result, it is surely to be preferred.

The above breach of the law of contradiction is avoided by Frege's theory. He distinguishes, in a denoting phrase, two elements, which we may call the *meaning* and the *denotation*.[8] Thus "the center of mass of the solar system at the beginning of the twentieth century" is highly complex in *meaning*, but its *denotation* is a certain point, which is simple. The solar system, the twentieth century, etc., are constituents of the *meaning*; but the *denotation* has no constituents at all.[9] One advantage of this distinction is that it shows why it is often worthwhile to assert identity. If we say "Scott is the author of *Waverley*," we assert an identity of denotation with a difference of meaning. I shall, however, not repeat the grounds in favor of this theory, as I have urged its claims elsewhere (*loc. cit.*), and am now concerned to dispute those claims.

One of the first difficulties that confront us, when we adopt the view that denoting phrases *express* a meaning and *denote* a denotation,[10] concerns the cases in which the denotation appears to be absent. If we say "the King of England is bald," that is, it would seem, not a statement about the complex *meaning*

[8] See his "On Sense and Nominatum," in this volume.
[9] Frege distinguishes the two elements of meaning and denotation everywhere, and not only in complex denoting phrases. Thus it is the *meanings* of the constituents of a denoting complex that enter into its *meaning*, not their *denotation*. In the proposition "Mont Blanc is over 1,000 metres high," it is, according to him, the *meaning* of "Mont Blanc," not the actual mountain, that is a constituent of the *meaning* of the proposition.
[10] In this theory, we shall say that the denoting phrase *expresses* a meaning; and we shall say both of the phrase and of the meaning that they *denote* a denotation. In the other theory, which I advocate, there is no *meaning*, and only sometimes a *denotation*.

"the King of England," but about the actual man denoted by the meaning. But now consider "the King of France is bald". By parity of form, this also ought to be about the denotation of the phrase "the King of France". But this phrase, though it has a *meaning* provided "the King of England" has a meaning, has certainly no denotation, at least in any obvious sense. Hence one would suppose that "the King of France is bald" ought to be nonsense; but it is not nonsense, since it is plainly false. Or again consider such a proposition as the following: "If u is a class which has only one member, then that one member is a member of u," or, as we may state it, "If u is a unit class, *the u is a u*". This proposition ought to be *always* true, since the conclusion is true whenever the hypothesis is true. But "the u" is a denoting phrase, and it is the denotation, not the meaning, that is said to be a u. Now if u is *not* a unit class, "the u" seems to denote nothing; hence our proposition would seem to become nonsense as soon as u is not a unit class.

Now it is plain that such propositions do *not* become nonsense merely because their hypotheses are false. The King in "The Tempest" might say, "If Ferdinand is not drowned, Ferdinand is my only son". Now "my only son" is a denoting phrase, which, on the face of it, has a denotation when, and only when, I have exactly one son. But the above statement would nevertheless have remained true if Ferdinand had been in fact drowned. Thus we must either provide a denotation in cases in which it is at first sight absent, or we must abandon the view that the denotation is what is concerned in propositions which contain denoting phrases. The latter is the course that I advocate. The former course may be taken, as by Meinong, by admitting objects which do not subsist, and denying that they obey the law of contradiction; this, however, is to be avoided if possible. Another way of taking the same course (so far as our present alternative is concerned) is adopted by Frege, who provides by definition some purely conventional denotation for the cases in which otherwise there would be none. Thus "the King of France," is to denote the null-class; "the only son of Mr. So-and-so" (who has a fine family of ten), is to denote the class of all his sons; and so on. But this procedure, though it may not lead to actual logical error, is plainly artificial, and does not give an exact analysis of the matter. Thus if we allow that denoting phrases, in general, have the two sides of meaning and denotation, the cases where there seems to be no denotation cause difficulties both on the assumption that there really is a denotation and on the assumption that there really is none.

A logical theory may be tested by its capacity for dealing with puzzles, and it is a wholesome plan, in thinking about logic, to stock the mind with as many puzzles as possible, since these serve much the same purpose as is served by experiments in physical science. I shall therefore state three puzzles which a theory as to denoting ought to be able to solve; and I shall show later that my theory solves them.

1. If *a* is identical with *b*, whatever is true of the one is true of the other, and either may be substituted for the other in any proposition without altering the truth or falsehood of that proposition. Now George IV wished to know whether Scott was the author of *Waverley*; and in fact Scott *was* the author of *Waverley*. Hence we may substitute *Scott* for *the author of "Waverley,"* and thereby prove that George IV wished to know whether Scott was Scott. Yet an interest in the law of identity can hardly be attributed to the first gentleman of Europe.

2. By the law of excluded middle, either "A is B" or "A is not B" must be true. Hence either "the present King of France is bald" or "the present King of France is not bald" must be true. Yet if we enumerated the things that are bald, and then the things that are not bald, we should not find the present King of France in either list. Hegelians, who love a synthesis, will probably conclude that he wears a wig.

3. Consider the proposition "A differs from B". If this is true, there is a difference between A and B, which fact may be expressed in the form "the difference between A and B subsists". But if it is false that A differs from B, then there is no difference between A and B, which fact may be expressed in the form "the difference between A and B does not subsist". But

how can a nonentity be the subject of a proposition? "I think, therefore I am" is no more evident than "I am the subject of a proposition, therefore I am," provided "I am" is taken to assert subsistence or being,[11] not existence. Hence, it would appear, it must always be self-contradictory to deny the being of anything; but we have seen, in connection with Meinong, that to admit being also sometimes leads to contradictions. Thus if A and B do not differ, to suppose either that there is, or that there is not, such an object as "the difference between A and B" seems equally impossible.

The relation of the meaning to the denotation involves certain rather curious difficulties, which seem in themselves sufficient to prove that the theory which leads to such difficulties must be wrong.

When we wish to speak about the *meaning* of a denoting phrase, as opposed to its *denotation*, the natural mode of doing so is by inverted commas. Thus we say:

> The center of mass of the solar system is a point, not a denoting complex;
> "The center of mass of the solar system" is a denoting complex, not a point.

Or again,

> The first line of Gray's *Elegy* states a proposition.
> "The first line of Gray's *Elegy*" does not state a proposition.

Thus taking any denoting phrase, say C, we wish to consider the relation between C and "C", where the difference of the two is of the kind exemplified in the above two instances.

We say, to begin with, that when C occurs it is the *denotation* that we are speaking about; but when "C" occurs, it is the *meaning*. Now the relation of meaning and denotation is not merely linguistic through the phrase: there must be a logical relation involved, which we express by saying that the meaning denotes the denotation. But the difficulty which confronts us is that we cannot succeed in *both* preserving the connection of

[11] I use these as synonyms.

meaning and denotation *and* preventing them from being one and the same; also that the meaning cannot be got at except by means of denoting phrases. This happens as follows.

The one phrase C was to have both meaning and denotation. But if we speak of "the meaning of C", that gives us the meaning (if any) of the denotation. "The meaning of the first line of Gray's *Elegy*" is the same as "The meaning of 'The curfew tolls the knell of parting day'," and is not the same as "The meaning of 'the first line of Gray's *Elegy*'". Thus in order to get the meaning we want, we must speak not of "the meaning of C", but of "the meaning of 'C'," which is the same as "C" by itself. Similarly "the denotation of C" does not mean the denotation we want, but means something which, if it denotes at all, denotes what is denoted by the denotation we want. For example, let "C" be "the denoting complex occurring in the second of the above instances". Then C = "the first line of Gray's *Elegy*", and the denotation of C = The curfew tolls the knell of parting day. But what we *meant* to have as the denotation was "the first line of Gray's *Elegy*". Thus we have failed to get what we wanted.

The difficulty in speaking of the meaning of a denoting complex may be stated thus: The moment we put the complex in a proposition, the proposition is about the denotation; and if we make a proposition in which the subject is "the meaning of C", then the subject is the meaning (if any) of the denotation, which was not intended. This leads us to say that, when we distinguish meaning and denotation, we must be dealing with the meaning: the meaning has denotation and is a complex, and there is not something other than the meaning, which can be called the complex, and be said to *have* both meaning and denotation. The right phrase, on the view in question, is that some meanings have denotations.

But this only makes our difficulty in speaking of meanings more evident. For suppose C is our complex; then we are to say that C *is* the meaning of the complex. Nevertheless, whenever C occurs without inverted commas, what is said is not true of the meaning, but only of the denotation, as when we say: The center of mass of the solar system is a point. Thus to speak of C itself, i.e., to make a proposition about the meaning, our subject must not be C, but something which denotes C. Thus "C", which is what we use when we want to speak of the meaning, must be not the meaning, but something which denotes the meaning. And C must not be a constituent of this complex (as it is of "the meaning of C"); for if C occurs in the complex, it will be its denotation, not its meaning, that will occur, and there is no backward road from denotations to meanings, because every object can be denoted by an infinite number of different denoting phrases.

Thus it would seem that "C" and C are different entities, such that "C" denotes C; but this cannot be an explanation, because the relation of "C" to C remains wholly mysterious; and where are we to find the denoting complex "C" which is to denote C? Moreover, when C occurs in a proposition, it is not *only* the denotation that occurs (as we shall see in the next paragraph); yet, on the view in question, C is only the denotation, the meaning being wholly relegated to "C". This is an inextricable tangle, and seems to prove that the whole distinction of meaning and denotation has been wrongly conceived.

That the meaning is relevant when a denoting phrase occurs in a proposition is formally proved by the puzzle about the author of *Waverley*. The proposition "Scott was the author of *Waverley*" has a property not possessed by "Scott was Scott," namely the property that George IV wished to know whether it was true. Thus the two are not identical propositions; hence the meaning of "the author of *Waverley*" must be relevant as well as the denotation, if we adhere to the point of view to which this distinction belongs. Yet, as we have just seen, so long as we adhere to this point of view, we are compelled to hold that only the denotation can be relevant. Thus the point of view in question must be abandoned.

It remains to show how all the puzzles we have been considering are solved by the theory explained at the beginning of this article.

According to the view which I advocate, a denoting phrase

is essentially *part* of a sentence, and does not, like most single words, have any significance on its own account. If I say "Scott was a man," that is a statement of the form "*x* was a man," and it has "Scott" for its subject. But if I say "the author of *Waverley* was a man," that is not a statement of the form "*x* was a man," and does not have "the author of *Waverley*" for its subject. Abbreviating the statement made at the beginning of this article, we may put, in place of "the author of *Waverley* was a man," the following: "One and only one entity wrote *Waverley*, and that one was a man". (This is not so strictly what is meant as what was said earlier; but it is easier to follow.) And speaking generally, suppose we wish to say that the author of *Waverley* had the property φ, what we wish to say is equivalent to "One and only one entity wrote *Waverley*, and that one had the property φ".

The explanation of *denotation* is now as follows. Every proposition in which "the author of *Waverley*" occurs being explained as above, the proposition "Scott was the author of *Waverley*" (i.e., "Scott was identical with the author of *Waverley*") becomes "One and only one entity wrote *Waverley*, and Scott was identical with that one"; or, reverting to the wholly explicit form: "It is not always false of *x* that *x* wrote *Waverley*, that it is always true of *y* that if *y* wrote *Waverley* *y* is identical with *x*, and that Scott is identical with *x*". Thus if "C" is a denoting phrase, it may happen that there is one entity *x* (there cannot be more than one) for which the proposition "*x* is identical with C" is true, this proposition being interpreted as above. We may then say that the entity *x* is the denotation of the phrase "C". Thus Scott is the denotation of "the author of *Waverley*". The "C" in inverted commas will be merely the *phrase*, not anything that can be called the *meaning*. The phrase *per se* has no meaning, because in any proposition in which it occurs the proposition, fully expressed, does not contain the phrase, which has been broken up.

The puzzle about George IV's curiosity is now seen to have a very simple solution. The proposition "Scott was the author of *Waverley*," which was written out in its unabbreviated form in the preceding paragraph, does not contain any constituent "the author of *Waverley*" for which we could substitute "Scott". This does not interfere with the truth of inferences resulting from making what is *verbally* the substitution of "Scott" for "the author of *Waverley*," so long as "the author of *Waverley*" has what I call a *primary* occurrence in the proposition considered. The difference of primary and secondary occurrences of denoting phrases is as follows:

When we say: "George IV wished to know whether so-and-so," or when we say "So-and-so is surprising" or "So-and-so is true," etc., the "so-and-so" must be a proposition. Suppose now that "so-and-so" contains a denoting phrase. We may either eliminate this denoting phrase from the subordinate proposition "so-and-so," or from the whole proposition in which "so-and-so" is a mere constituent. Different propositions result according to which we do. I have heard of a touchy owner of a yacht to whom a guest, on first seeing it, remarked, "I thought your yacht was larger than it is"; and the owner replied, "No, my yacht is not larger than it is". What the guest meant was, "The size that I thought your yacht was is greater than the size your yacht is"; the meaning attributed to him is, "I thought the size of your yacht was greater than the size of your yacht". To return to George IV and *Waverley*, when we say, "George IV wished to know whether Scott was the author of *Waverley*," we normally mean "George IV wished to know whether one and only one man wrote *Waverley* and Scott was that man"; but we *may* also mean: "One and only one man wrote *Waverley*, and George IV wished to know whether Scott was that man". In the latter, "the author of *Waverley*" has a *primary* occurrence; in the former, a *secondary*. The latter might be expressed by "George IV wished to know, concerning the man who in fact wrote *Waverley*, whether he was Scott". This would be true, for example, if George IV had seen Scott at a distance, and had asked "Is that Scott?" A *secondary* occurrence of a denoting phrase may be defined as one in which the phrase occurs in a proposition *p* which is a mere constituent of the proposition we are considering, and the substitution for the denoting phrase is to be effected in *p*, not in the whole proposition concerned. The ambiguity as between primary and secondary occurrences is hard to avoid in language; but it does no harm if we are on our guard against it. In symbolic logic it is of course easily avoided.

The distinction of primary and secondary occurrences also enables us to deal with the question whether the present King of France is bald or not bald, and generally with the logical status of denoting phrases that denote nothing. If "C" is a denoting phrase, say "the term having the property F," then

"C has the property ϕ" means "one and only one term has the property F, and that one has the property ϕ".[12]

If now the property F belongs to no terms, or to several, it follows that "C has the property ϕ" is false for *all* values of ϕ. Thus "the present King of France is bald" is certainly false; and "the present King of France is not bald" is false if it means

"There is an entity which is now King of France and is not bald,"

but is true if it means

"It is false that there is an entity which is now King of France and is bald".

That is, "the King of France is not bald" is false if the occurrence of "the King of France" is *primary*, and true if it is *secondary*. Thus all propositions in which "the King of France" has a primary occurrence are false; the denials of such propositions are true, but in them "the King of France" has a secondary occurrence. Thus we escape the conclusion that the King of France has a wig.

We can now see also how to deny that there is such an object as the difference between A and B in the case when A and B do not differ. If A and B do differ, there is one and only one entity *x* such that "*x* is the difference between A and B" is a

---

[12] This is the abbreviated, not the stricter, interpretation.

true proposition; if A and B do not differ, there is no such entity *x*. Thus according to the meaning of denotation lately explained, "the difference between A and B" has a denotation when A and B differ, but not otherwise. This difference applies to true and false propositions generally. If "*a* R *b*" stands for "*a* has the relation R to *b*", then when *a* R *b* is true, there is such an entity as the relation R between *a* and *b*; when *a* R *b* is false, there is no such entity. Thus out of any proposition we can make a denoting phrase, which denotes an entity if the proposition is true, but does not denote an entity if the proposition is false. E.g., it is true (at least we will suppose so) that the earth revolves round the sun, and false that the sun revolves round the earth; hence "the revolution of the earth round the sun" denotes an entity, while "the revolution of the sun round the earth" does not denote an entity.[13]

The whole realm of nonentities, such as "the round square," "the even prime other than 2," "Apollo," "Hamlet," etc., can now be satisfactorily dealt with. All these are denoting phrases which do not denote anything. A proposition about Apollo means what we get by substituting what the classical dictionary tells us is meant by Apollo, say "the sun-god". All propositions in which Apollo occurs are to be interpreted by the above rules for denoting phrases. If "Apollo" has a primary occurrence, the proposition containing the occurrence is false; if the occurrence is secondary, the proposition may be true. So again "the round square is round" means "there is one and only one entity *x* which is round and square, and that entity is round," which is a false proposition, not, as Meinong maintains, a true one. "The most perfect Being has all perfections; existence is a perfection; therefore the most perfect Being exists" becomes:

> "There is one and only one entity *x* which is most perfect; that one has all perfections; existence is a perfection; therefore that one exists". As a proof, this fails for want of a proof of the

[13] The propositions from which such entities are derived are not identical either with these entities or with the propositions that these entities have being.

premise "there is one and only one entity x which is most perfect".[14]

Mr. MacColl (*Mind*, N.S., No. 54, and again No. 55, p. 401) regards individuals as of two sorts, real and unreal; hence he defines the null-class as the class consisting of all unreal individuals. This assumes that such phrases as "the present King of France," which do not denote a real individual, do, nevertheless, denote an individual, but an unreal one. This is essentially Meinong's theory, which we have seen reason to reject because it conflicts with the law of contradiction. With our theory of denoting, we are able to hold that there are no unreal individuals; so that the null-class is the class containing no members, not the class containing as members all unreal individuals.

It is important to observe the effect of our theory on the interpretation of definitions which proceed by means of denoting phrases. Most mathematical definitions are of this sort: for example, "$m-n$ means the number which, added to n, gives m". Thus $m-n$ is defined as meaning the same as a certain denoting phrase; but we agreed that denoting phrases have no meaning in isolation. Thus what the definition really ought to be is: "Any proposition containing $m-n$ is to mean the proposition which results from substituting for '$m-n$' 'the number which, added to n, gives m' ". The resulting proposition is interpreted according to the rules already given for interpreting propositions whose verbal expression contains a denoting phrase. In the case where m and n are such that there is one and only one number x which, added to n, gives m, there is a number x which can be substituted for $m-n$ in any proposition containing $m-n$ without altering the truth or falsehood of the proposition. But in other cases, all propositions in which "$m-n$" has a primary occurrence are false.

[14] The argument can be made to prove validly that all members of the class of most perfect Beings exist; it can also be proved formally that this class cannot have *more* than one member; but, taking the definition of perfection as possession of all positive predicates, it can be proved almost equally formally that the class does not have even one member.

The usefulness of *identity* is explained by the above theory. No one outside a logic-book ever wishes to say "*x* is *x*," and yet assertions of identity are often made in such forms as "Scott was the author of *Waverley*" or "thou art the man". The meaning of such propositions cannot be stated without the notion of identity, although they are not simply statements that Scott is identical with another term, the author of *Waverley*, or that thou art identical with another term, the man. The shortest statement of "Scott is the author of *Waverley*" seems to be: "Scott wrote *Waverley*; and it is always true of *y* that if *y* wrote *Waverley*, *y* is identical with Scott". It is in this way that identity enters into "Scott is the author of *Waverley*"; and it is owing to such uses that identity is worth affirming.

One interesting result of the above theory of denoting is this: when there is anything with which we do not have immediate acquaintance, but only definition by denoting phrases, then the propositions in which this thing is introduced by means of a denoting phrase do not really contain this thing as a constituent, but contain instead the constituents expressed by the several words of the denoting phrase. Thus in every proposition that we can apprehend (i.e., not only in those whose truth or falsehood we can judge of, but in all that we can think about), all the constituents are really entities with which we have immediate acquaintance. Now such things as matter (in the sense in which matter occurs in physics) and the minds of other people are known to us only by denoting phrases, i.e., we are not *acquainted* with them, but we know them as what has such and such properties. Hence, although we can form propositional functions C (x) which must hold of such and such a material particle, or of So-and-so's mind, yet we are not acquainted with the propositions which affirm these things that we know must be true, because we cannot apprehend the actual entities concerned. What we know is "So-and-so has a mind which has such and such properties" but we do not know "A has such and such properties," where A *is* the mind in question. In such a case, we know the properties of a thing without having acquaintance with the thing itself, and without, consequently, knowing any single proposition of which the thing itself is a constituent.

Of the many other consequences of the view I have been advocating, I will say nothing. I will only beg the reader not to make up his mind against the view—as he might be tempted to do, on account of its apparently excessive complication—until he has attempted to construct a theory of his own on the subject of denotation. This attempt, I believe, will convince him that, whatever the true theory may be, it cannot have such a simplicity as one might have expected beforehand.

18

The Analysis of Belief Sentences

RUDOLF CARNAP

§1. SENTENCES ABOUT BELIEFS

We study sentences of the form 'John believes that . . .' .
If here the subsentence '. . .' is replaced by another sentence
L-equivalent to it, then it may be that the whole sentence
changes its truth-value. Therefore, the whole belief-sentence is
neither extensional nor intensional with respect to the subsentence '. . .' . Consequently, an interpretation of belief-sentences as referring either to sentences or to propositions is not
quite satisfactory. For a more adequate interpretation we need
a relation between sentences which is still stronger than L-
equivalence. Such a relation will be defined in the next section.

We found that '. . . ∨ ---' is extensional with respect to the
subsentence indicated by dots, and that 'N(. . .)' is intensional.
Can there be a context which is neither extensional nor intensional? This would be the case if (but not only if) the
replacement of a subsentence by an L-equivalent one changed
the truth-value and hence also the intension of the whole sentence. In our systems this cannot occur; every sentence in S_1
(and likewise in S_3, to be explained later) is extensional, and
every sentence in S_2 is either extensional or intensional. However, it is the case for a very important kind of sentence with
psychological terms, like 'I believe that it will rain'. Although

From *Meaning and Necessity*, by Rudolf Carnap, pp. 53–59 and 61–62. Copyright 1947 and 1956 by the University of Chicago. Reprinted by permission of
the University of Chicago Press. For a brief explanation of some of Carnap's
basic semantical terms, see the Introduction, pp. 22–24, as well as notes 1 and 2.

sentences of this kind seem to be quite clear and unproblematic at first glance and are, indeed, used and understood in everyday life without any difficulty, they have proved very puzzling to logicians who have tried to analyze them. Let us see whether we can throw some light upon them with the help of our semantical concepts.

In order to formulate examples, we take here, as our object language S, not a symbolic system but a part of the English language. We assume that S is similar in structure to S_1 except for containing the predicator '. . . believes that - - -' and some mathematical terms. We do not specify here the rules of S; we assume that the semantical rules of S are such that the predicator mentioned has its ordinary meaning; and, further, that our semantical concepts, especially 'true', 'L-true', 'equivalent', and 'L-equivalent', are defined for S in accord with our earlier conventions. Now we consider the following two belief-sentences; 'D' and 'D'' are here written as abbreviations for two sentences in S to be explained presently:

(i) 'John believes that D'.
(ii) 'John believes that D''.

Suppose we examine John with the help of a comprehensive list of sentences which are L-true in S; among them, for instance, are translations into English of theorems in the system of [P.M.] and of even more complicated mathematical theorems which can be proved in that system and therefore are L-true on the basis of the accepted interpretation. We ask John, for every sentence or for its negation, whether he believes what it says or not. Since we know him to be truthful, we take his affirmative or negative answer as evidence for his belief or nonbelief. Among the simple L-true sentences, there will certainly be some for which John professes belief. We take as 'D' any one of them, say 'Scott is either human or not human'. Thus the sentence (i) is true. On the other hand, since John is a creature with limited abilities, we shall find some L-true sentences in S for which John cannot profess belief. This does not necessarily mean that he commits the error of believing their negations; it may be that he cannot give an answer either way. We take as 'D″' some sentence of this kind; that is to say, 'D″' is L-true but (ii) is false. Thus the two belief-sentences (i) and (ii) have different truth-values; they are neither equivalent nor L-equivalent. Therefore, the definitions of interchangeability and L-interchangeability (11–1a) lead to the following two results:[1]

1–1. The occurrence of 'D' within (i) is not interchangeable with 'D″'.

1–2. The occurrence of 'D' within (i) is not L-interchangeable with 'D″'.

'D' and 'D″' are both L-true; therefore:

1.3. 'D' and 'D″' are equivalent and L-equivalent.

Examining the first belief-sentence (i) with respect to its subsentence 'D', we see from 1–1 and 1–3 that the condition of extensionality is not fulfilled; and we see from 1–2 and 1–3 that the condition of intensionality is not fulfilled either:[2]

1–4. The belief-sentence (i) is *neither extensional nor intensional* with respect to its subsentence 'D'.

Although 'D' and 'D″' have the same intension, namely, the L-true or necessary proposition, and hence the same extension, namely, the truth-value truth, their interchange transforms the first belief-sentence (i) into the second (ii), which does not have the same extension, let alone the same intension, as the first.

---

[1] [Carnap's definition (11–1a) may be paraphrased roughly as follows: An expression a_1 occurring within a sentence S_1 is (1) *interchangeable*, (2) *L-interchangeable* with another expression a_2 if and only if the result of replacing any occurrence of a_1 in S_1 by a_2 is a sentence S_2 which is (1) equivalent, (2) L-equivalent to S_1. Ed.]

[2] [These conditions (11–2a and 11–3a) may be paraphrased roughly as follows: (11–2a) A sentence is *extensional* with respect to an expression occurring in it if and only if the expression is interchangeable with any expression equivalent to it. (11–3a) A sentence is intensional with respect to an expression occurring in it if and only if the sentence is not extensional and the expression is L-interchangeable with any expression L-equivalent to it. Ed.]

The same result as 1–4 holds also if any other sentence is taken instead of 'D', in particular, any factual sentence.

Let us now try to answer the much-discussed question as to how a sentence reporting a belief is to be analyzed and, in particular, whether such a sentence is about a proposition or a sentence or something else. It seems to me that we may say, in a certain sense, that (i) is about the sentence 'D', but also, in a certain other sense, that (i) is about the proposition that D. In interpreting (i) with respect to the sentence 'D', it would, of course, not do to transform it into 'John is disposed to an affirmative response to the sentence 'D'', because this might be false, although (i) was assumed to be true; it might, for instance, be that John does not understand English but expresses his belief in another language. Therefore, we may try the following more cautious formulation:

> (iii) 'John is disposed to an affirmative response to some sentence in some language, which is L-equivalent to 'D'.'

Analogously, in interpreting (i) with respect to the proposition that D, the formulation 'John is disposed to an affirmative response to any sentence expressing the proposition that D' would be wrong because it implies that John understands all languages. Even if the statement is restricted to sentences of the language or languages which John understands, it would still be wrong, because 'D'', for example, or any translation of it, likewise expresses the proposition that D, but John does not give an affirmative response to it. Thus we see that here again we have to use a more cautious formulation similar to (iii):

> (iv) 'John is disposed to an affirmative response to some sentence in some language which expresses the proposition that D'.

However, it seems to me that even the formulations (iii) and (iv), which are L-equivalent, should not be regarded as anything more than a first approximation to a correct interpretation of the belief-sentence (i). It is true that each of them follows from (i), at least if we take 'belief' here in the sense of 'expressible belief', leaving aside the problem of belief in a wider sense, interesting though it may be. However, (i) does not follow from either of them. This is easily seen if we replace 'D' by 'D″'. Then (iii) remains true because of 1–3; on the other hand, (i) becomes (ii), which is false. It is clear that we must interpret (i) as saying as much as (iii) but something more; and this additional content seems difficult to formulate. If (i) is correctly interpreted in accord with its customary meaning, then it follows from (i) that there is a sentence to which John would respond affirmatively and which is not only L-equivalent to 'D', as (iii) says, but has a still stronger relation to 'D'—in other words, a sentence which has something more in common with 'D' than the intension. The two sentences must, so to speak, be understood in the same way; they must not only be L-equivalent in the whole but consist of L-equivalent parts, and both must be built up out of these parts in the same way. If this is the case, we shall say that the two sentences have the same intensional structure. This concept will be explicated in the next section and applied in the analysis of belief sentences in §3.

§2. INTENSIONAL STRUCTURE

If two sentences are built in the same way out of designators (or designator matrices) such that any two corresponding designators are L-equivalent, then we say that the two sentences are *intensionally isomorphic* or that they have the same *intensional structure*. The concept of L-equivalence can also be used in a wider sense for designators in different language systems; and the concept of intensional isomorphism can then be similarly extended.

We shall discuss here what we call the analysis of the intensional structures of designators, especially sentences. This is meant as a semantical analysis, made on the basis of the semantical rules and aimed at showing, say for a given sentence, in which way it is built up out of designators and what are the intensions of these designators. If two sentences are built in the same way out of corresponding designators with the same intensions, then we shall say that they have the same intensional structure. We might perhaps also use for this relation the term 'synonymous', because it is used in a similar sense by other authors (e.g., Langford, Quine, and Lewis), as we shall see in the next section. We shall now try to explicate this concept.

Let us consider, as an example, the expressions '2 + 5' and "II sum V' in a language *S* containing numerical expressions and arithmetical functors. Let us suppose that we see from the semantical rules of *S* that both '+' and 'sum' are functors for the function Sum and hence are L-equivalent; and, further, that the numerical signs occurring have their ordinary meanings and hence '2' and 'II' are L-equivalent to one another, and likewise '5' and 'V'. Then we shall say that the two expressions are *intensionally ismorphic* or that they have *the same intensional structure*, because they not only are L-equivalent as a whole, both being L-equivalent to '7', but consist of three parts in such a way that corresponding parts are L-equivalent to one another and hence have the same intension. Now it seems advisable to apply the concept of intensional ismorphism in a somewhat wider sense so that it also holds between expressions like '2 + 5' and 'sum(II,V)', because the use in the second expression of a functor preceding the two argument signs instead of one standing between them or of parentheses and a comma may be regarded as an inessential syntactical device. Analogously, if '>' and 'Gr' are L-equivalent, and likewise '3' and 'III', then we regard '5 > 3' as intensionally isomorphic to 'Gr(V,III)'. Here again we regard the two predicators '>' and 'Gr' as corresponding to each other, irrespective of their places in the sentences; further, we correlate the first argument expression of '>' with the first of 'Gr', and the second with the second. Further, '2 + 5 > 3' is isomorphic to 'Gr[sum(II,V),III]', because the corresponding expressions '2 + 5' and 'sum(II,V)' are not only L-equivalent but isomorphic. On the other hand, '7 > 3' and 'Gr[sum(II,V),III]' are not isomorphic; it is true that here again the two predicators '>' and 'Gr' are L-equivalent and that corresponding argument expressions of them are likewise L-equivalent, but the corresponding expressions '7' and 'sum (II,V)' are not isomorphic. We require for isomorphism of two expressions that the analysis of both down to the smallest subdesignators lead to analogous results.

We have said earlier that it seems convenient to take as designators in a system S at least all those expressions in S, but not necessarily only those, for which there are corresponding variables in the metalanguage M. For the present purpose, the comparison of intensional structures, it seems advisable to go as far as possible and take as designators all those expressions which serve as sentences, predicators, functors, or individual expressions of any type, irrespective of the question of whether or not M contains corresponding variables. Thus, for example, we certainly want to regard as isomorphic '$p \lor q$' and 'Apq', where 'A' is the sign of disjunction (or alternation) as used by the Polish logicians in their parenthesis-free notation, even if M, as is usual, does not contain variables of the type of connectives. We shall then regard '$\lor$' and 'A' as L-equivalent connectives because any two full sentences of them with the same argument expressions are L-equivalent.

Frequently, we want to compare the intensional structures of two expressions which belong to different language systems. This is easily possible if the concept of L-equivalence is defined for the expressions of both languages in such a way that the following requirement is fulfilled, in analogy to our earlier conventions: an expression in S is L-equivalent to an expression in S' if and only if the semantical rules of S and S' together, without the use of any knowledge about (extralinguistic) facts, suffice to show that the two expressions have the same extension. Thus, L-equivalence holds, for example, between 'a' in S and 'a″' in S' if we see from the rules of designation for these two individual constants that both stand for the same individual; likewise between 'P' and 'P″', if we see from the rules alone that these predicators apply to the same individuals; between two functors '$+$' and 'sum', if we see from the rules alone that they assign to the same arguments the same values —in other words, if their full expressions with L-equivalent argument expressions (e.g., '$2 + 5$' and 'sum(II,V)') are L-equivalent; for two sentences, if we see from the rules alone that they have the same truth-value (e.g., 'Rom ist gross' in German, and 'Rome is large' in English). Thus, even if the sentences '$2 + 5 > 3$' and 'Gr[sum(II,V),III]' belong to two different systems, we find that they are intensionally isomorphic by establishing the L-equivalence of corresponding signs.

If variables occur, the analysis becomes somewhat more complicated, but the concept of isomorphism can still be defined. We shall not give here exact definitions but merely indicate, with the help of some simple examples, the method to be applied in the definitions of L-equivalence and isomorphism of matrices. Let 'x' be a variable in S which can occur in a universal quantifier '(x)' and also in an abstraction operator '(λx)', and 'u' be a variable in S' which can occur in a universal quantifier 'Πu' and also in an abstraction operator '$\hat{u}$'. If 'x' and 'u' have the same range of values, for example, if both are natural number variables (have natural number concepts as value intensions), we shall say that 'x' and 'u' are L-equivalent, and also that '(x)' and 'Πu' are L-equivalent, and that '(λx)' and '$\hat{u}$' are L-equivalent. If two matrices (sentential or other) of degree n are given, one in S and the other in S', we say that they are L-equivalent with respect to a certain correlation between the variables, if corresponding abstraction expressions are L-equivalent predicators. Thus, for example, '$x > y$' in S and 'Gr(u,v)' in S' are L-equivalent matrices (with respect to the correlation of 'x' with 'u' and 'y' with 'v') because '$(\lambda xy)[x > y]$' and '$\hat{u}\hat{v}[Gr(u,v)]$' are L-equivalent predicators. Intensional isomorphism of (sentential or other) matrices can then be defined in analogy to that of closed designators, so that it holds if the two matrices are built up in the same way out of corresponding expressions which are either L-equivalent designators or L-equivalent matrices. Thus, for example, the matrices '$x + 5 > y$' and 'Gr[sum$(u,\mathrm{V}),v$]' are not only L-equivalent but also intensionally isomorphic; and so are the (L-false) sentences '$(x)(y)[x + 5 > y]$' and 'ΠuΠv [Gr[sum$(u,\mathrm{V}),v$]]'.

These considerations suggest the following definition, which is recursive with respect to the construction of compound designator matrices out of simpler ones. It is formulated in general terms with respect to designator matrices; these include closed designators and variables as special cases. The definition presupposes an extended use of the term 'L-equivalent' with respect to variables, matrices, and operators, which has been indicated in the previous examples but not formally defined. The present definition makes no claim to exactness; an exact definition would have to refer to one or two semantical systems whose rules are stated completely.

2–1. *Definition of intensional isomorphism*

    a. Let two designator matrices be given, either in the same or in two different semantical systems, such that neither of them contains another designator matrix as proper part. They are intensionally isomorphic $=_{\mathrm{Df}}$ they are L-equivalent.

    b. Let two compound designator matrices be given, each of them consisting of one main submatrix (of the type of a predicator, functor, or connective) and n argument expressions (and possibly auxiliary signs like parentheses, commas, etc.). The two matrices are intensionally isomorphic $=_{\mathrm{Df}}$ (1) the two main submatrices are intensionally isomorphic, and (2) for any m from 1 to n, the mth argument expression within the first matrix is intensionally isomorphic to the mth in the second matrix ('the mth' refers to the order in which the argument expressions occur in the matrix).

    c. Let two compound designator matrices be given, each of them consisting of an operator (universal or existential quantifier, abstraction operator, or description operator) and its scope, which is a designator matrix. The two matrices are intensionally isomorphic $=_{\mathrm{Df}}$ (1) the two scopes are intensionally isomorphic with respect to a certain correlation of the variables occurring in them, (2) the two operators are L-equivalent and contain correlated variables.

In accord with our previous discussion of the explicandum, rule b in this definition takes into consideration the order in which argument expressions occur but disregards the place of the main subdesignator. For the intensional structure, in contrast to the merely syntactical structure, only the order of application is essential, not the order and manner of spelling.

§3. APPLICATIONS OF THE CONCEPT OF INTENSIONAL
    STRUCTURE

. . . .

Now let us go back to the problem of the analysis of belief-sentences, and let us see how the concept of intensional structure can be utilized there. It seems that the sentence 'John believes that D' in S can be interpreted by the following semantical sentence:

3–1. 'There is a sentence $\mathfrak{S}_i$ in a semantical system S' such that (a) $\mathfrak{S}_i$ in S' is intensionally isomorphic to 'D' in S and (b) John is disposed to an affirmative response to $\mathfrak{S}_i$ as a sentence of S'.'

This interpretation may not yet be final, but it represents a better approximation than the interpretations discussed earlier (in §1). As an example, suppose that John understands only German and that he responds affirmatively to the German sentence 'Die Anzahl der Einwohner von Chicago ist grösser als 3,000,000' but neither to the sentence 'Die Anzahl der Einwohner von Chicago ist grösser als $2^6 \times 3 \times 5^6$' nor to any intensionally isomorphic sentence, because he is not quick enough to realize that the second sentence is L-equivalent to the first. Then our interpretation of belief-sentences, as formulated in 3–1, allows us to assert the sentence 'John believes that the

number of inhabitants of Chicago is greater than three million'
and to deny the sentence 'John believes that the number of inhabitants of Chicago is greater than $2^6 \times 3 \times 5^6$'. We can do so
without contradiction because the two German sentences,
and likewise their English translations just used, have different
intensional structures. . . . On the other hand, the interpretation of belief-sentences in terms of propositions as objects of
beliefs (like (iv) in §1) would not be adequate in this case,
since the two German sentences and the two English sentences
all express the same proposition.

An analogous interpretation holds for other sentences containing psychological terms about knowledge, doubt, hope, fear,
astonishment, etc., with 'that'-clauses, hence generally about
what Russell calls propositional attitudes and Ducasse epistemic attitudes. The problem of the logical analysis of sentences of this kind has been much discussed,[3] but a satisfactory
solution has not been found so far. The analysis here proposed
is not yet a complete solution, but it may perhaps be regarded
as a first step. What remains to be done is, first, a refinement
of the analysis in terms of linguistic reactions here given and,
further, an analysis in terms of dispositions to nonlinguistic
behavior.

[3] Russell (*An Inquiry into Meaning and Truth* [London: Allen and Unwin,
1940]), gives a detailed discussion of the problem in a wider sense, including
beliefs not expressed in language; he investigates the problem under both an
epistemological and a logical aspect (in our terminology, both a pragmatical and
a semantical aspect), not always distinguishing the two clearly. For C.J. Ducasse's
conception see his paper "Propositions, Opinions, Sentences, and Facts," *The
Journal of Philosophy*, 37 (1940), 701–711.

19

Synonymity, and the Analysis of Belief Sentences

In the paper "Carnap's Analysis of Statements of Assertion and Belief,"[1] Church has advanced some criticisms of the theory of belief-sentences and indirect discourse proposed by Carnap in *Meaning and Necessity*.[2] It appears that these criticisms can be met without a modification of the theory. But certain criticisms by Benson Mates[3] would seem to be more serious; these criticisms are equally forceful against the widely held view that expressions with the same sense are interchangeable in all contexts. In this paper, a revision of this principle will be suggested; and a new definition of intensional isomorphism[4] will be put forward as a basis for the reconstruction of the theory of meaning-analysis in accordance with the suggested principle.

From *Analysis*, 14 (1954), 114–119. Reprinted by permission of the author and the editors.

[1] A. Church, "Carnap's Analysis of Statements of Assertion and Belief," *Analysis*, 10 (1950), 97–99.

[2] R. Carnap, *Meaning and Necessity* (Chicago: Univ. of Chicago Press, 1947). (The sections relevant to the present discussion are included in this volume; see pp. 380–390.)

[3] B. Mates, "Synonymity," in *University of California Publications in Philosophy*, 25 (1950), 201–226; reprinted in *Semantics and the Philosophy of Language*, ed. Leonard Linsky (Urbana, Ill.: Univ. of Illinois Press, 1952), pp. 111–136.

[4] This refers to the explicans proposed for the concept of synonymy in *Meaning and Necessity*. It was Prof. Carnap who pointed out to me the significance of Mates' criticism; this paper owes its existence to his stimulus and help.

I

Church's paper is divided, conveniently for our purposes, into two parts. The criticisms in part I do not apply to Carnap's conception, since Carnap intends that, for the purposes of reconstruction, a "language" is to be thought of as having precise formation rules, designation rules, and truth-rules, or as having had these made precise; and it is to be defined by reference to its rules rather than as "the language spoken in the British Isles in 1941," or something of that kind. The criticism in part II of Church's paper, however, is supposed to apply to Carnap's analysis even when "English" and "German" are construed as semantical systems. Let us therefore turn to this part of Church's criticism.

Church begins by considering the sentence in the system E^5:

(1) Seneca said that man is a rational animal.

and its counterpart in the system G:

(1′) Seneca hat gesagt dass der Mensch ein vernuenftiges Tier sei.

The analysis of (1) in the system E by the technique proposed in Carnap (see footnote 2) leads to the following sentence (we follow Church's enumeration):

(7) There is a sentence S_i in a semantical system S such that (a) S_i as sentence of S is intensionally isomorphic to 'Man is a rational animal' as sentence of E, and (b) Seneca wrote S_i as sentence of S.

and similarly, the analysis of (1′) in the system G leads to:

(7′) Es gibt einen Satz S_i in einem semantischen System S, so dass (a) S_i als Satz von S intensional isomorph zu 'Der Mensch ist ein vernuenftiges Tier' als Satz in G ist, und (b) Seneca S_i als Satz von S geschrieben hat.

[5] The letters 'E' and 'G' will denote the semantical systems corresponding to English and German.

But (7) and (7′), as Church remarks, are "not intensionally isomorphic".[6]

But why should they be? Suppose someone proposes the following as an analysis of 'one' in the simplified theory of types (with "systematic ambiguity"):

(*a*) $\hat{x} \, (\exists y) \, (z) \, (z \epsilon x \equiv z = y)$

and suppose further that someone else proposes instead:

(*b*) $\hat{x} \, (\exists y) \, (y \epsilon x. \, (z) \, (z \epsilon x \supset x = y)$

These are, of course, "not intensionally isomorphic." Yet there would be no contradiction involved in regarding both analyses as correct, for they are logically equivalent. And this is the only requirement that I believe can be imposed on two correct analyses of the same concept.

It is indeed an interesting fact that the analysis of (1) in the system E leads to a result which is not intensionally isomorphic to the result of the analysis of (1′) in the system G even when the analyses are constructed on the same pattern. This is easily seen to be the result of the fact that (7) quotes a sentence of E, while (7′) quotes its translation in G, and the *names* of different intensionally isomorphic expressions are not intensionally isomorphic; in fact they are not synonymous in *any* sense. We could of course construct a sentence in the system E which would be intensionally isomorphic to (7′) (let us call it '(c)'), and a sentence in G which would be intensionally isomorphic to (7) (let us call it '(d)'); but I do not think we should consider it as a theoretical question: 'which is the correct analysis of (1),—(7) or (7′) or (c) or (d)?' If one is correct, then *all* are, and it does not matter that some are intensionally isomorphic and some are not.

In closing this part of my discussion, I should like to remark that if one does wish to emend Carnap's theory so that the analysis of (1) (or of (1) and (1)′) will lead to intensionally isomorphic results in the systems E and G one has only to specify that the quoted sentence should not be "Man is a

---

[6] Cf. "Carnap's Analysis of Statements of Assertion and Belief," p. 99.

rational animal", or "Der Mensch ist ein vernuenftiges Tier", but the translation of this sentence into an arbitrarily selected neutral system, say the system L, corresponding to Latin. Then in (7) the words ' "Man is a rational animal," as sentence of E' become replaced by the words ' "Homo est animal rationale," as sentence of L', and in (7′) the words ' "Der Mensch ist ein vernuenftiges Tier," als Satz von G' are replaced by ' "Homo est animal rationale," als Satz von L'; and (7) and (7′) are then intensionally isomorphic. But I do not think the advantages are sufficiently great to warrant this revision.

We are now in a position to consider Church's final criticism. Church points out that the result of prefixing 'John believes that' to (7) may have a different truth value from the result of prefixing 'John glaubt dass' to (7′). But this, like the remark that (7) and (7′) are "not intensionally isomorphic" is a crushing blow only if one has somehow been led to agree that (7) and (7′) *ought* to be synonymous.

II

Mates remarks:[7] "Carnap has proposed the concept of intensional isomorphism as an approximate explicatum for synonymity. It seems to me that this is the best proposal that has been made by anyone to date. However, it has, along with its merits, some rather odd consequences. For instance, let "D" and "D′" be abbreviations for two intensionally isomorphic sentences. Then the following sentences are also intensionally isomorphic:

(14)   Whoever believes that D, believes that D.
(15)   Whoever believes that D, believes that D′.

But nobody doubts that whoever believes that D believes that D. Therefore, nobody doubts that whoever believes that D believes that D′. This seems to suggest that, for any pair of intensionally isomorphic sentences—let them be abbreviated

[7] "Synonymity," p. 215.

by "D" and "D'"—if anybody even doubts that whoever believes that D believes that D', then Carnap's explication is
incorrect.

This argument seems extremely powerful. Suppose, for the
sake of illustration, that we use 'Hellene' as some newspapers
do, as a synonym for 'Greek'. Then 'All Greeks are Greeks', and
'All Greeks are Hellenes', are intensionally isomorphic. Hence
'Whoever believes that all Greeks are Greeks believes that all
Greeks are Greeks' and 'Whoever believes that all Greeks are
Greeks believes that all Greeks are Hellenes' are intensionally
isomorphic, and so (supposedly) synonymous. Now I do not
myself doubt that 'Whoever believes that all Greeks are Greeks
believes that all Greeks are Hellenes' is true; but it is easy to
suppose that someone *does* doubt this, whereas it is quite likely
that nobody doubts that whoever believes that all Greeks are
Greeks believes that all Greeks are Greeks. Accordingly,

(e) Nobody doubts that whoever believes that all Greeks are
Greeks believes that all Greeks are Greeks.

and

(f) Nobody doubts that whoever believes that all Greeks are
Greeks believes that all Greeks are Hellenes.

may quite conceivably have opposite truth value, and so *cannot*
be synonymous.

Mates only suggests that this may invalidate Carnap's original
proposal; Carnap, however, takes a harsher attitude toward his
own theory; he believes that his theory in its present form
cannot refute this criticism.

Mates goes on: "What is more, *any* adequate explication of
synonymity will have this result, for the validity of the argument
is not affected if we replace the words "intensionally isomorphic" by the word "synonymous" throughout.[8] In short,
on any theory of synonymity, the synonymity of (f) and (e)
must follow if the synonymity of 'Greek' and 'Hellene' is assumed. If we take this seriously, there is but one conclusion to

[8] "Synonymity," p. 215.

which we can come: 'Greek' and 'Hellene' are not synonyms, and by the same argument, neither are any two different terms. This is a conclusion which some authors would be prepared to accept, even on other grounds.[9]

expressions have the same sense, for Mates' argument is

I believe, however, that the felt synonymity of such different expressions as 'snow is white' and 'Schnee ist weiss', or (in the use described above) of 'Greek' and 'Hellene', is undeniable. To maintain this synonymity involves our denying that in fact the synonymity of (f) and (e) follows from the synonymity of 'Greek' and 'Hellene'. Can this be denied?

At first blush, it would seem that it cannot. If two expressions have the same meaning, they can be interchanged in any context. We state this formally:

(g) The sense of a sentence is a function of the sense of its parts.

In a moment we shall criticize the apparent "self-evidence" of this principle. But first let us pause to make one point quite clear: whoever accepts (g) must conclude that no two different expressions have the same sense, for Mates' argument is formally sound if the interchange of expressions with the same sense is permitted in every context; conversely, whoever believes that two different expressions *ever* have the same sense must, by the same token, reject the principle. This would seem to be entirely destructive of some present theories of meaning-analysis, which appear to involve simultaneous acceptance of (g) and the synonymity of some distinct expressions.

Let us now return to (g). 'The sense of a sentence is a function of the sense of its parts'. Let us ask why this seems to be self-evident. There is of course the formal similarity to the 'equals may be substituted for equals' which we learn in high-school mathematics; but let us put this aside as irrelevant. We may suppose that if we were to ask someone who accepts this principle why it is true, he might well reply with the rhetorical

[9] Cf. N. Goodman, "On Likeness of Meaning," *Analysis*, 10 (1949).

question 'Of what else could it be a function?' And just this is the heart of the matter.

Consider, for the moment, a simpler example (a variant of the famous "paradox of analysis"): 'Greek' and 'Hellene' are synonymous. But 'All Greeks are Greeks' and 'All Greeks are Hellenes' do not *feel* quite like synonyms. But what has changed? Did we not obtain the second sentence from the first by "putting equals for equals"? The answer is that the *logical structure* has changed. The first sentence has the form 'All F are F', while the second has the form 'All F are G'—and these are wholly distinct (the first, in fact is L-true, while the second schema is not even L-determinate). This suggests the following revision of the principle:

(h) The sense of a sentence is a function of the sense of its parts *and of its logical structure*.[10]

I believe that a large part of the "self-evidence" of (g) arises from the fact that we do not consider any alternatives: when, in particular, we contrast (g) with (h), I think that we find the latter principle even more plausible than the former. It is easy to illustrate the pervasive importance of logical structure as a factor in meaning: if it is through the names occurring in it that a sentence speaks about the world, it is through its logical structure that a sentence has implication relations to other sentences, and it is upon logical structure, or syntax, that the correctness of all our logical transformations depends.

The foregoing considerations lead us, therefore, to the following modification in the definition of intensional isomorphism:

(i) Two expressions are intensionally isomorphic if they have the same logical structure, and if corresponding parts are L-equivalent.

This amounts to saying that two expressions are intensionally isomorphic if (a) they are intensionally isomorphic in Carnap's sense, and (b) they have the same logical structure. It is pro-

[10] Two sentences are said to have the same *logical structure*, when occurrences of the same sign in one correspond to occurrences of the same sign in the other.

posed that "intensional isomorphism" so defined should be the explicans for synonymity in the strongest sense (interchangeability in belief contexts and indirect discourse).[11]

III

We must now consider another possible solution to the problem posed by Mates: that of Frege. According to Frege, a sentence in an "oblique" context (i.e., a belief context, or indirect discourse) does not have its ordinary nominatum and sense; rather it names the proposition that it normally expresses, i.e., that is normally its sense, and it expresses a new sense called its "indirect" or "oblique" sense:

"There is also . . . indirect discourse, of which we have seen that in it the words have their indirect (oblique) nominata which coincide with what are ordinarily their senses. In this case then the clause has as its nominatum a proposition, not a truth-value; its sense is not a proposition but it is the sense of the words 'the proposition that . . .' ."[12]

This view leads to the following answer to Mates' problem: even if D and D' ordinarily have the same sense, in (14) and (15) they have their indirect senses, and they have their normal sense as nominatum. Hence (14) and (15) do not have the same sense but only the same nominatum (truth-value); and 'nobody doubts that (14)[13] need not even have the same truth-value as 'nobody doubts that (15)', because in 'nobody doubts that (14)' the whole of (14) occurs in its indirect sense.

Mates' paradox is reinstated in a milder but still extremely damaging form for this theory, however, by considering the case in which D and D' are two expressions with the same oblique sense. In such a case (e) and (f) would necessarily have to

---

[11] For a distinction of stronger and weaker concepts of synonymy, see Carnap, "Reply to Leonard Linsky," *Philosophy of Science*, 16 (1949), 347–350.

[12] Cf. G. Frege, "On Sense and Nominatum" (included in this volume), p. 348.

[13] (14), (15), etc., are used as abbreviations here and in similar positions later.

have the same truth-value; and we conclude by the same argument as before that no distinct expressions can ever have the same indirect sense.[14]

This appears to be a serious defect in Frege's theory. Frege himself certainly holds that different expressions may have the same sense in belief contexts. Thus he asserts in the passage quoted that the indirect sense of '. . .' is the same as the sense of the words 'the proposition that . . .', e.g.,

(j) John believes the earth is round.

and

(k) John believes the proposition that the earth is round,

have the same sense. But some philosophers doubt that there are propositions, and hence that (strictly speaking) anyone ever *believes a proposition*. Such a philosopher would doubt that:

(l) If anyone believes the earth is round, he believes the proposition that the earth is round.

But he certainly would not doubt that:

(m) If anyone believes the earth is round, he belives the earth is round.

Therefore 'someone doubts (l)' does not have the same nominatum (truth-value) as 'someone doubts (m)', and accordingly the "indirect" sense of 'the earth is round' is *not* the sense of the words 'the proposition that the earth is round'— contrary to Frege's assertion. But then it becomes difficult to say just what is the indirect sense of 'the earth is round'.

Thus we see that we cannot make the slightest change in the wording of a belief sentence without altering its sense. And it can now be shown that we cannot interchange different expressions in a *reiterated* oblique context (e.g., 'George believes that John says . . .') and hope to maintain even logical equivalence. For consider:

(n) Betty said that John is a Hellene.

[14] This defect in Frege's theory was pointed to me by Carnap.

and

(o) Betty said that John is a Greek.

Since these have different senses, the result of prefixing 'John believes that' to (n) may even have a different truth-value from the result of prefixing 'John believes that' to (o).[15]

Let us imagine a case in which John says: 'Betty said that I am a Greek.' I believe that we should regard:

(p) John believes that Betty said that he is a Greek.

and

(q) John believes that Betty said that he is a Hellene.

as both constituting correct descriptions of this situation. Otherwise, it would appear, we are construing (p) as meaning "John believes that Betty said 'John is Greek' "; and this would amount to taking the quotation as a *direct* quotation, not an indirect one.

In any event, the case for regarding (p) and (q) as equivalent seems exactly as good as the case for regarding (n) and (o) as equivalent. To give up the equivalence of (n) and (o) would of course be to give up indirect quotation altogether; but to maintain it, while denying the equivalence of (p) and (q) is arbitrary.

In concluding, I should like to point out some applications of the concept of synonymy presented in this paper (in (i) above) to some classical problems of meaning analysis. In the first place, let us consider:

(r) George asked whether the property Greek is identical with
     the property Hellene.

and

(s) George asked whether the property Greek is identical with
     the property Greek.

[15] The difference in sense could again be directly established by using our pattern 'nobody doubts that if (n) then (o)'.

This is clearly similar to Russell's "author of Waverley". But certain differences are relevant. The theory of descriptions will not take care of the problem posed by (r) and (s), for 'the property Greek' and 'the property Hellene' may well be designated by constants and not descriptions, even in *Principia Mathematica*. But in the theory presented above, no difficulty arises: 'The property Greek is identical with the property Greek' and 'The property Greek is identical with the property Hellene' are simply not synonymous (this is another instance of the "paradox of analysis"); hence (r) and (s) are not synonymous. The case presented by Carnap in his reply to Linsky,[16] of the sentences '5 is identical with 5' and '5 is identical with V', is disposed of in the same fashion.

Finally, let us look at Mates' sentences (14) and (15) above. If D and D' are different expressions, (14) and (15) are *never* synonymous, on our analysis. Thus we are not disturbed by the fact that 'Nobody doubts that (14)' may have a different truth value from 'Nobody doubts that (15)'.

---

[16] Cf. "Reply to Leonard Linsky."

20

Quantifiers and Propositional Attitudes

W. V. O. QUINE

I

The incorrectness of rendering 'Ctesias is hunting unicorns' in the fashion:

$$(\exists x)(x \text{ is a unicorn . Ctesias is hunting } x)$$

is conveniently attested by the nonexistence of unicorns, but is not due simply to that zoological lacuna. It would be equally incorrect to render 'Ernest is hunting lions' as:

(1) $\qquad (\exists x)(x \text{ is a lion . Ernest is hunting } x),$

where Ernest is a sportsman in Africa. The force of (1) is rather that there is some individual lion (or several) which Ernest is hunting, e.g., stray circus property.

The contrast recurs in 'I want a sloop.' The version:

(2) $\qquad (\exists x)(x \text{ is a sloop . I want } x)$

is suitable insofar only as there may be said to be a certain sloop that I want. If what I seek is mere relief from sloopless-ness, then (2) conveys the wrong idea.

The contrast is that between what may be called the *relational* sense of lion-hunting or sloop-wanting, viz., (1)–(2), and the likelier or *notional* sense. Appreciation of the difference

From *The Journal of Philosophy*, 53 (1956), 177–187. Reprinted by permission of the author and the editors. This paper sums up some points which I have set forth in various lectures at Harvard and Oxford from 1952 onward.

is evinced in Latin and Romance languages by a distinction of mood in subordinate clauses; thus 'Procuro un perro que habla' has the relational sense:

$$(\exists x)(x \text{ is a dog} . x \text{ talks} . \text{I seek } x)$$

as against the notional 'Procuro un perro que hable':

$$\text{I strive that } (\exists x)(x \text{ is a dog} . x \text{ talks} . \text{I find } x).$$

Pending considerations to the contrary in later pages, we may represent the contrast strikingly in terms of permutations of components. Thus (1) and (2) may be expanded (with some premeditated violence to both logic and grammar) as follows:

(3) $(\exists x)(x \text{ is a lion} . \text{Ernest strives that Ernest finds } x)$,

(4) $(\exists x)(x \text{ is a sloop} . \text{I wish that I have } x)$,

whereas 'Ernest is hunting lions' and 'I want a sloop' in their notional senses may be rendered rather thus:

(5) $\text{Ernest strives that } (\exists x)(x \text{ is a lion} . \text{Ernest finds } x)$,

(6) $\text{I wish that } (\exists x)(x \text{ is a sloop} . \text{I have } x)$.

The contrasting versions (3)–(6) have been wrought by so paraphrasing 'hunt' and 'want' as to uncover the locutions 'strive that' and 'wish that,' expressive of what Russell has called *propositional attitudes*. Now of all examples of propositional attitudes, the first and foremost is *belief*; and, true to form, this example can be used to point up the contrast between relational and notional senses still better than (3)–(6) do. Consider the relational and notional senses of believing in spies:

(7) $(\exists x)(\text{Ralph believes that } x \text{ is a spy})$,

(8) $\text{Ralph believes that } (\exists x)(x \text{ is a spy}).$

Both may perhaps be ambiguously phrased as 'Ralph believes that someone is a spy,' but they may be unambiguously phrased respectively as 'There is someone whom Ralph believes to be a spy' and 'Ralph believes there are spies.' The difference is vast; indeed, if Ralph is like most of us, (8) is true and (7) false.

In moving over to propositional attitudes, as we did in (3)–(6), we gain not only the graphic structural contrast between (3)–(4) and (5)–(6) but also a certain generality. For we can now multiply examples of striving and wishing, unrelated to hunting and wanting. Thus we get the relational and notional senses of wishing for a president:

(9)   (∃x)(Witold wishes that x is president),

(10)   Witold wishes that (∃x)(x is president).

According to (9), Witold has his candidate; according to (10) he merely wishes the appropriate form of government were in force. Also we open other propositional attitudes to similar consideration—as witness (7)–(8).

However, the suggested formulations of the relational senses —viz., (3), (4), (7), and (9)—all involve quantifying into a propositional-attitude idiom from outside. This is a dubious business, as may be seen from the following example.

There is a certain man in a brown hat whom Ralph has glimpsed several times under questionable circumstances on which we need not enter here; suffice it to say that Ralph suspects he is a spy. Also there is a gray-haired man, vaguely known to Ralph as rather a pillar of the community, whom Ralph is not aware of having seen except once at the beach. Now Ralph does not know it, but the men are one and the same. Can we say of this *man* (Bernard J. Ortcutt, to give him a name) that Ralph believes him to be a spy? If so, we find ourselves accepting a conjunction of the type:

(11)   w sincerely denies ' . . . ' . w believes that . . .

as true, with one and the same sentence in both blanks. For, Ralph is ready enough to say, in all sincerity, 'Bernard J. Ortcutt is no spy.' If, on the other hand, with a view to disallowing situations of the type (11), we rule simultaneously that

(12)   Ralph believes that the man in the brown hat is a spy,

(13)   Ralph does not believe that the man seen at the beach
        is a spy,

then we cease to affirm any relationship between Ralph and any man at all. Both of the component 'that'-clauses are indeed about the man Ortcutt; but the 'that' must be viewed in (12) and (13) as sealing those clauses off, thereby rendering (12) and (13) compatible because not, as wholes, about Ortcutt at all. It then becomes improper to quantify as in (7); 'believes that' becomes, in a word, referentially opaque.[1]

No question arises over (8); it exhibits only a quantification *within* the 'believes that' context, not a quantification *into* it. What goes by the board, when we rule (12) and (13) both true, is just (7). Yet we are scarcely prepared to sacrifice the relational construction 'There is someone whom Ralph believes to be a spy,' which (7) as against (8) was supposed to reproduce.

The obvious next move is to try to make the best of our dilemma by distinguishing two senses of belief: *belief*$_1$, which disallows (11), and *belief*$_2$, which tolerates (11) but makes sense of (7). For belief$_1$, accordingly, we sustain (12)–(13) and ban (7) as nonsense. For belief$_2$, on the other hand, we sustain (7); and for *this* sense of belief we must reject (13) and acquiesce in the conclusion that Ralph believes$_2$ that the man at the beach is a spy even though he *also* believes$_2$ (and believes$_1$) that the man at the beach is not a spy.

II

But there is a more suggestive treatment. Beginning with a single sense of belief, viz., belief$_1$ above, let us think of this at first as a relation between the believer and a certain *intension*, named by the 'that'-clause. Intensions are creatures of darkness, and I shall rejoice with the reader when they are exorcised, but first I want to make certain points with help of them. Now

[1] See *From a Logical Point of View* (Cambridge, Mass.: Harvard Univ. Press, 1953), pp. 142–159; also "Three Grades of Modal Involvement," in W.V. Quine, ed., *The Ways of Paradox and Other Essays* (New York: Random House, 1966).

intensions named thus by 'that'-clauses, without free variables, I shall speak of more specifically as intensions of degree 0, or propositions. In addition I shall (for the moment) recognize intensions of degree 1, or attributes. These are to be named by prefixing a variable to a sentence in which it occurs free; thus $z(z$ is a spy$)$ is spyhood. Similarly we may specify intensions of higher degrees by prefixing multiple variables.

Now just as we have recognized a dyadic relation of belief between a believer and a proposition, thus:

(14)   Ralph believes that Ortcutt is a spy,

so we may recognize also a triadic relation of belief among a believer, an object, and an attribute, thus:

(15)   Ralph believes $z(z$ is a spy$)$ of Ortcutt.

For reasons which will appear, this is to be viewed not as dyadic belief between Ralph and the proposition *that* Ortcutt has $z(z$ is a spy$)$, but rather as an irreducibly triadic relation among the three things Ralph, $z(z$ is a spy$)$, and Ortcutt. Similarly there is tetradic belief:

(16) Tom believes $yz(y$ denounced $z)$ of Cicero and Catiline,

and so on.

Now we can clap on a hard and fast rule against quantifying into propositional-attitude idioms; but we give it the form now of a rule against quantifying into names of intensions. Thus, though (7) as it stands becomes unallowable, we can meet the needs which prompted (7) by quantifying rather into the triadic belief construction, thus:

(17) $(\exists x)$[Ralph believes $z(z$ is a spy$)$ of x].

Here then, in place of (7), is our new way of saying that there is someone whom Ralph believes to be a spy.

Belief$_1$ was belief so construed that a proposition might be believed when an object was specified in it in one way, and yet not believed when the same object was specified in another way; witness (12)–(13). Hereafter we can adhere uniformly to this narrow sense of belief, both for the dyadic case and for triadic and higher; in each case the term which names the intension (whether proposition or attribute or intension of higher degree) is to be looked on as referentially opaque.

The situation (11) is thus excluded. At the same time the effect of belief$_2$ can be gained, simply by ascending from dyadic to triadic belief as in (15). For (15) does relate the men Ralph and Ortcutt precisely as belief$_2$ was intended to do. (15) does remain true of Ortcutt under any designation; and hence the legitimacy of (17).

Similarly, whereas from:

> Tom believes that Cicero denounced Catiline

we cannot conclude:

> Tom believes that Tully denounced Catiline,

on the other hand we can conclude from:

> Tom believes $y(y$ denounced Catiline) of Cicero

that

> Tom believes $y(y$ denounced Catiline) of Tully,

and also that

(18) $(\exists x)$[Tom believes $y(y$ denounced Catiline) of x].

From (16), similarly, we may infer that

(19) $(\exists w)(\exists x)$[Tom believes $yz(y$ denounced $z)$ of w and x].

Such quantifications as:

> $(\exists x)$(Tom believes that x denounced Catiline),

> $(\exists x)$[Tom believes $y(y$ denounced $x)$ of Cicero]

still count as nonsense, along with (7); but such legitimate purposes as these are served by (17)–(19) and the like. Our names of intensions, and these only, are what count as referentially opaque.

Let us sum up our findings concerning the seven numbered statements about Ralph. (7) is now counted as nonsense, (8) as true, (12)–(13) as true, (14) as false, and (15) and (17) as true. Another that is true is:

(20)   Ralph believes that the man seen at the beach is not a spy,

which of course must not be confused with (13).

The kind of exportation which leads from (14) to (15) should doubtless be viewed in general as implicative. Under the terms of our illustrative story, (14) happens to be false; but (20) is true, and it leads by exportation to:

(21)   Ralph believes z (z is not a spy) of the man seen at the beach.

The man at the beach, hence Ortcutt, does not receive reference in (20), because of referential opacity; but he does in (21), so we may conclude from (21) that

(22)   Ralph believes $z(z$ is not a spy) of Ortcutt.

Thus (15) and (22) both count as true. This is not, however, to charge Ralph with contradictory beliefs. Such a charge might reasonably be read into:

(23)   Ralph believes $z(z$ is a spy . z is not a spy) of Ortcutt,

but this merely goes to show that it is undesirable to look upon (15) and (22) as implying (23).

It hardly needs be said that the barbarous usage illustrated in (15)–(19) and (21)–(23) is not urged as a practical reform. It is put forward by way of straightening out a theoretical difficulty, which, summed up, was as follows: Belief contexts are referentially opaque; therefore it is *prima facie* meaningless to quantify into them (at least with respect to persons or other extensional objects[2]); how then to provide for those indispensable relational statements of belief, like 'There is someone whom Ralph believes to be a spy'?

Let it not be supposed that the theory which we have been examining is just a matter of allowing unbridled quantification into belief contexts after all, with a legalistic change of notation.

[2] See *From a Logical Point of View*, pp. 150–154.

On the contrary, the crucial choice recurs at each point: quantify if you will, but pay the price of accepting situations of the type (11) with respect to each point at which you choose to quantify. In other words: distinguish as you please between referential and nonreferential positions, but keep track, so as to treat each kind appropriately. The notation of intensions, of degree one and higher, is in effect a device for inking in a boundary between referential and nonreferential occurrences of terms.

III

Striving and wishing, like believing, are propositional attitudes and referentially opaque. (3) and (4) are objectionable in the same way as (7), and our recent treatment of belief can be repeated for these propositional attitudes. Thus, just as (7) gave way to (17), so (3) and (4) give way to:

(24) $(\exists x)[x$ is a lion . Ernest strives z(Ernest finds z) of $x]$,

(25) $(\exists x)[x$ is a sloop . I wish z(I have z) of $x]$,

a certain breach of idiom being allowed for the sake of analogy in the case of 'strives.'

   These examples came from a study of hunting and wanting. Observing in (3)–(4) the quantification into opaque contexts, then, we might have retreated to (1)–(2) and foreborne to paraphrase them into terms of striving and wishing. For (1)–(2) were quite straightforward renderings of lion-hunting and sloop-wanting in their relational senses; it was only the notional senses that really needed the breakdown into terms of striving and wishing, (5)–(6).

   Actually, though, it would be myopic to leave the relational senses of lion-hunting and sloop-wanting at the unanalyzed stage (1)–(2). For, whether or not we choose to put these over into terms of wishing and striving, there are other relational cases of wishing and striving which require our consideration anyway— as witness (9). The untenable formulations (3)–(4) may indeed be either corrected as (24)–(25) or condensed back into (1)–(2); on the other hand we have no choice but to correct the untenable (9) on the pattern of (24)–(25), viz., as:

$$(\exists x)[\text{Witold wishes } y(y \text{ is president}) \text{ of } x].$$

The untenable versions (3)–(4) and (9) all had to do with wishing and striving in the relational sense. We see in contrast that (5)–(6) and (10), on the notional side of wishing and striving, are innocent of any illicit quantification into opaque contexts from outside. But now notice that exactly the same trouble begins also on the notional side, as soon as we try to say not just that Ernest hunts lions and I want a sloop, but that *someone* hunts lions or wants a sloop. This move carries us, ostensibly, from (5)–(6) to:

(26) $(\exists w)[w \text{ strives that } (\exists x)(x \text{ is a lion } . w \text{ finds } x)]$,

(27) $(\exists w)[w \text{ wishes that } (\exists x)(x \text{ is a sloop } . w \text{ has } x)]$,

and these do quantify unallowably into opaque contexts.

We know how, with help of the attribute apparatus, to put (26)–(27) in order; the pattern, indeed, is substantially before us in (24)–(25). Admissible versions are:

$$(\exists w)[w \text{ strives } y(\exists x)(x \text{ is a lion } . y \text{ finds } x) \text{ of } w],$$

$$(\exists w)[w \text{ wishes } y(\exists x)(x \text{ is a sloop } . y \text{ has } x) \text{ of } w],$$

or briefly:

(28) $(\exists w)[w \text{ strives } y(y \text{ finds a lion}) \text{ of } w]$,

(29) $(\exists w)[w \text{ wishes } y(y \text{ has a sloop}) \text{ of } w]$.

Such quantification of the subject of the propositional attitude can of course occur in belief as well; and, if the subject is mentioned in the belief itself, the above pattern is the one to use. Thus 'Someone believes he is Napoleon' must be rendered:

$$(\exists w)[w \text{ believes } y(y = \text{Napoleon}) \text{ of } w].$$

For concreteness I have been discussing belief primarily, and two other propositional attitudes secondarily: striving and wishing. The treatment is, we see, closely parallel for the three;

and it will pretty evidently carry over to other propositional attitudes as well—e.g., hope, fear, surprise. In all cases my concern is, of course, with a special technical aspect of the propositional attitudes: the problem of quantifying in.

IV

There are good reasons for being discontent with an analysis that leaves us with propositions, attributes, and the rest of the intensions. Intensions are less economical than extensions (truth values, classes, relations), in that they are more narrowly individuated. The principle of their individuation, moreover, is obscure.

Commonly logical equivalence is adopted as the principle of individuation of intensions. More explicitly: if S and S' are any two sentences with $n(\geq 0)$ free variables, the same in each, then the respective intensions which we name by putting the n variables (or 'that,' if $n = 0$) before S and S' shall be one and the same intension if and only if S and S' are logically equivalent. But the relevant concept of logical equivalence raises serious questions in turn.[3]

Worse, granted certain usual logical machinery (such as is available in *Principia Mathematica*), this principle of individuation can be shown to contradict itself. For I have proved elsewhere,[4] using machinery solely of *Principia*, that if logical equivalence is taken as a sufficient condition of identity of attributes, then mere coextensiveness becomes a sufficient condition as well. But then it follows that logical equivalence is not a necessary condition; so the described principle of individuation contradicts itself.

The champion of intensions can be trusted, in the face of this result, to abandon either that principle of individuation of intensions or some one of the principles from *Principia* which was

---

[3] See "Two Dogmas of Empiricism," in *From a Logical Point of View;* also "Carnap and Logical Truth," in Paul Arthur Schilpp, ed., *The Philosophy of Rudolf Carnap*, Library of Living Philosophers (New York: Tudor, 1963).

[4] At the end of "On Frege's Way Out," *Mind*, 46 (1955), 145–159.

used in the proof. The fact remains that the intensions are at best a pretty obscure lot.

Yet it is evident enough that we cannot, in the foregoing treatment of propositional attitudes, drop the intensions in favor of the corresponding extensions. Thus, to take a trivial example, consider 'w is hunting unicorns.' On the analogy of (28), it becomes:

$$w \text{ strives } y(y \text{ finds a unicorn}) \text{ of } w.$$

Correspondingly for the hunting of griffins. Hence, if anyone w is to hunt unicorns without hunting griffins, the attributes

$$y(y \text{ finds a unicorn}),$$

$$y(y \text{ finds a griffin})$$

must be distinct. But the corresponding classes are identical, being empty. So it is indeed the attributes, and not the classes, that were needed in our formulation. The same moral could be drawn, though less briefly, without appeal to empty cases.

But there is a way of dodging the intensions which merits serious consideration. Instead of speaking of intensions we can speak of sentences, naming these by quotation. Instead of:

$$w \text{ believes that } \ldots$$

we may say:

$$w \text{ believes-true ' } \ldots \text{ ' }.$$

Instead of:

$$(30) \qquad w \text{ believes } y(\ldots y \ldots) \text{ of } x$$

we may say:

$$(31) \qquad w \text{ believes ' } \ldots y \ldots \text{ ' satisfied by } x.$$

The words 'believes satisfied by' here, like 'believes of' before, would be viewed as an irreducibly triadic predicate. A similar shift can be made in the case of the other propositional attitudes, of course, and in the tetradic and higher cases.

This semantical reformulation is not, of course, intended to suggest that the subject of the propositional attitude speaks the language of the quotation, or any language. We may treat a mouses's fear of a cat as his fearing true a certain English sentence. This is unnatural without being therefore wrong. It is a little like describing a prehistoric ocean current as clockwise.

How, where, and on what grounds to draw a boundary between those who believe or wish or strive that p, and those who do not quite believe or wish or strive that p, is undeniably a vague and obscure affair. However, if anyone does approve of speaking of belief of a proposition at all and of speaking of a proposition in turn as meant by a sentence, then certainly he cannot object to our semantical reformulation 'w believes-true S' on any special grounds of obscurity; for, 'w believes-true S' is explicitly definable in *his* terms as 'w believes the proposition meant by S.' Similarly for the semantical reformulation (31) of (30); similarly for the tetradic and higher cases; and similarly for wishing, striving, and other propositional attitudes.

Our semantical versions do involve a relativity to language, however, which must be made explicit. When we say that w believes-true S, we need to be able to say what language the sentence S is thought of as belonging to; not because w needs to understand S, but because S might by coincidence exist (as a linguistic form) with very different meanings in two languages.[5] Strictly, therefore, we should think of the dyadic 'believes-true S' as expanded to a triadic 'w believes-true S in L'; and correspondingly for (31) and its suite.

As noted two paragraphs back, the semantical form of expression:

(32) w believes-true ' . . . ' in L

can be explained in intensional terms, for persons who favor them, as:

(33) w believes the proposition meant by ' . . . ' in L,

thus leaving no cause for protest on the score of relative clarity. Protest may still be heard, however, on a different score: (32)

[5] This point is made by Alonzo Church, "On Carnap's Analysis of Statements of Assertion and Belief," *Analysis*, 10 (1950), 97–99.

and (33), though equivalent to each other, are not strictly equivalent to the 'w believes that . . .' which is our real concern. For, it is argued, in order to infer (33) we need not only the information about *w* which 'w believes that . . .' provides, but also some extraneous information about the language L. Church[6] brings the point out by appeal to translations, substantially as follows. The respective statements:

> *w* believes that there are unicorns,

(35)   *w* glaubt diejenige Aussage, die,,There are unicorns"
         in English

go into German as:

(34)   *w* glaubt, dass en Einhörne gibt,

(35)   *w* glaubt diejenige Aussage, die ,,There are unicorns"
         auf Englisch bedeutet,

and clearly (34) does not provide enough information to enable a German ignorant of English to infer (35).

The same reasoning can be used to show that 'There are unicorns' is not strictly or analytically equivalent to:

> 'There are unicorns' is true in English.

Nor, indeed, was Tarski's truth paradigm intended to assert analytic equivalence. Similarly, then, for (32) in relation to 'w believes that . . .'; a systematic agreement in truth value can be claimed, and no more. This limitation will prove of little moment to persons who share my skepticism about analyticity.

What I find more disturbing about the semantical versions, such as (32), is the need of dragging in the language concept at all. What is a language? What degree of fixity is supposed? When do we have one language and not two? The propositional attitudes are dim affairs to begin with, and it is a pity to have to add obscurity to obscurity by bringing in language variables too. Only let it not be supposed that any clarity is gained by restituting the intensions.

[6] *Ibid.*

21

Substitutivity and Descriptions

LEONARD LINSKY

I

(1) Scott is the author of *Waverley*. (2) George IV wished to know whether Scott was the author of *Waverley*. Therefore, (3) George IV wished to know whether Scott was Scott. Why does this conclusion not follow from the premises? Russell introduced this puzzle in "On Denoting." It was one of three puzzles which he thought a theory of denoting ought to be able to solve. He said of these three puzzles, "I shall show that my theory solves them."[1] Russell accepts the principle of substitutivity to which we appeal in passing to the conclusion from the premises, but he holds the argument to be only apparently sanctioned by it. This principle is formulated as follows, "If *a* is identical with *b*, whatever is true of the one is true of the other, and either may be substituted for the other in any proposition without altering the truth or falsehood of that proposition" (47). There is use-mention confusion in this formulation. Correctly formulated, the principle of substitutivity mentions singular terms of true statements of identity, not the objects denoted by these terms.

When we rewrite the premises (1) and (2) in accordance with Russell's theory, the definite description "disappears on

From *The Journal of Philosophy*, 63 (1966), 673–683. Reprinted by permission of the author and the editors. This article also appeared as a chapter of the author's book, *Referring* (London: Routledge and Kegan Paul, 1967).

[1] "On Denoting," included in this volume; parenthetical page references are also to this volume.

analysis." Thus there is no definite description in (the re-written version of) (2) to be replaced by 'Scott'. The puzzle is caused by a logical mirage. This is Russell's solution. "The puzzle about George IV's curiosity is now seen to have a very simple solution. The proposition 'Scott was the author of *Waverley*', which was written out in its unabbreviated form in the preceding paragraph, does not contain any constituent 'the author of *Waverley*' for which we could substitute 'Scott'" (373 f.). There is a mistake here. (3) cannot be obtained by substituting 'Scott' for 'the author of *Waverley*' in (1). We must make this substitution in (2). I assume that Russell meant to say this and that it is this substitution to which he objects.

II

The "solution" is inadequate. The premises contain no definite description after it has been eliminated in accordance with Russell's analysis. But how does this show that the propositions (1) (2) (3), when thus analyzed, do not constitute a valid argument? After all, we may sometimes substitute proper names for descriptions in propositions (apparently) containing them. Such substitutions are sanctioned by a theorem[2] of *Principia*:

14.15 $\{ (\imath x)(\phi x) = b \} \supset (\{ \psi(\imath x)(\phi x) \} \equiv \{ \psi b \})$

Russell must have been thinking of principles such as this when, after offering the above "solution," he said "This does not interfere with the truth of inferences resulting from making what is *verbally* the substitution of 'Scott' for 'the author of *Waverley*' so long as 'the author of *Waverley*' has what I call *primary* occurrence in the proposition considered" (374). Now consider any argument such that its first premise is of the form of the protasis of 14.15, its second premise of the form of the left-hand side of the apodosis, and its conclusion of the form

[2] 14.15. Here we have replaced the dot notation with a sequence of parentheses and braces.

of the right-hand side. Rewritten in accordance with Russell's theory, this argument contains no descriptions, and it is valid. 14.15 itself contains no descriptions when expanded in accordance with the contextual definitions that introduce descriptions into *Principia*. What, then, is wrong with the view that the argument (1) (2) (3) is sanctioned by 14.15? The answer cannot be that given by the logical-mirage account. It is true that (1) (2) (3) when analyzed contains no descriptions, but the same is true of 14.15.

What is wrong with the view under consideration is that the descriptive phrase is required to have primary occurrence in both of the propositions in which it appears in 14.15, but it has primary occurrence only in the first premise of (1) (2) (3). What Russell tells us in the above quotation is that, in any proposition in which 'the author of *Waverley*' has a primary occurrence, we may validly replace it by 'Scott' (assuming (1), of course). But then it appears that Russell has abandoned the logical-mirage theory in the very paragraph in which he presents it. Now we are told that what is wrong with replacing 'the author of *Waverley*' by 'Scott' in (2) on the basis of (1) is that the description does not have a primary occurrence in (2).

III

Russell does not tell us why the interpretation of premise (2) that sees in it a primary occurrence of the description (hereafter I call this "the primary interpretation") is wrong. But it is wrong; for (2) on this interpretation entails that *Waverley* was not coauthored. If *Waverley* had been coauthored, it would not, on the primary interpretation, be logically possible that George IV should have wished to know whether Scott was the author of *Waverley*. But no plausible analysis of our proposition can have this as a consequence. A sufficient condition for the truth of (2) is that George IV should have asked, in all seriousness, "Is Scott the author of *Waverley*?" Now surely he might seriously have asked this question though *Waverley* had been coauthored. What is the proof that the unwanted consequence follows on the primary interpretation of (2)? On this interpretation, (2) is of the form:

$$(4) \qquad (\exists c)(\{(\phi x) \equiv_x (x = c)\} \& \{\psi c\})$$

Now (4) is an existentially generalized conjunction; so we may distribute the existential quantifier to each of its conjuncts. Simplifying by eliminating the right conjunct, we get:

$$(5) \qquad (\exists c)\{(\phi x \equiv_x (x = c)\}$$

This, by the definition 14.02 of *Principia*, is the definitional expansion of:

$$(6) \qquad E!(\imath x)(\phi x)$$

Consistently with the interpretation we have supplied for the variables above, this says,

$$(7) \qquad \text{One and only one person wrote } \textit{Waverley}.$$

And (7) entails that *Waverley* was not coauthored.[3]

This result brings out in a particularly revealing way what it is that the primary interpretation misses. It misses the feature of (2) that makes it "intentional" in the technical sense. "Intentional" verbs behave in a characteristic way as concerns the existence of their objects. It is not possible to take a bath in a nonexistent tub. But it is possible for someone to want to take a bath in my tub even though I do not, in fact, possess one. Just so, George IV could not have assaulted the author of *Waverley* if *Waverley* had been coauthored, but he could have wanted to know whether Scott was the author of *Waverley* if *Waverley* had been coauthored and even if no such book as *Waverley* had ever been written. Thus the primary interpretation of (2) is not correct, and we escape the conclusion that (1) (2) (3) is a valid argument.

We turn then to the alternative interpretation that Russell's

---

[3] This argument is taken from my paper, "Reference and Referents," in C.E. Caton, ed., *Philosophy and Ordinary Language* (Urbana. Ill.: Univ. of Illinois Press, 1963).

theory provides for (2), the "secondary interpretation." (2) now is taken to be of the form:

$$(8) \qquad \chi\{[(\imath x)(\phi x)]\psi(\imath x)(\phi x)\}$$

The sign '$[(\imath x)(\phi x)]$' in (8) is the "scope operator," and in this formula it indicates that the descriptive phrase is to be eliminated from the subordinate proposition '$\psi(\imath x)(\phi x)$' and not from the whole of (8). That is, the result of eliminating the description from (8) is

$$(9) \qquad \chi\{(\exists c)(\{(\phi x) \equiv_x (x = c)\} \,\&\, \{\psi c\})\}$$

An English sentence that might be translated into (9) would be:

(10)   George IV wished to know whether one and only one person wrote *Waverley* and that person was Scott.

What happens to (2) on the secondary analysis is just that the "intentional" expression 'George IV wished to know whether' is isolated by the scope operator from the subordinate clause from which the descriptive phrase is to be eliminated. The description is then eliminated from this subordinate proposition. There is really just one analysis of propositions of the form $\psi(\imath x)(\phi x)$, and this is what is given by the definition 14.01. The difference between the primary and the secondary interpretations, as I have been calling them, is determined by the part of (2) (proper or improper) taken as $\psi(\imath x)(\phi x)$. On the secondary interpretation, $\psi(\imath x)(\phi x)$ is taken to be not the whole of (2) but the part

(11)   Scott is the author of *Waverley*.

Though we have but one analysis of '$\psi(\imath x)(\phi x)$', we have two (nonequivalent) analyses of (2), according as we take the whole of (2) or the part (11) as $\psi(\imath x)(\phi x)$.

IV

Let us grant (though it has been disputed) that (11) is correctly analyzed by Russell's theory. According to this analysis,

(11) is logically equivalent to

(12)   One and only one person wrote *Waverley*
                and that person is Scott.

We assume that a necessary condition for the correctness of an analysis is that the analysans and the analysandum be logically equivalent. But from the premise that (11) is logically equivalent to (12), it does not follow that (2) is logically equivalent to some proposition of the form (9), e.g., (10). To argue that, since (11) and (12) are logically equivalent, so also are (2) and (10) is to argue fallaciously. There is a fallacy in reasoning that, since *p* and *q* are logically equivalent, so are *f*(*p*) and *f*(*q*) for *any* function of propositions *f*. And we cannot argue thus in this case because (2) does not express an extensional function of the contained proposition (11). Though it would be fallacious to argue in this way, it does not follow that (2) and (10) are not logically equivalent. I want to show that they are not, but I wish first to point out that Russell offers no argument to show that (2) is logically equivalent to (10)—although, of course, he does maintain that it is. He argues that (11) and (12) are logically equivalent and does not pursue the matter further. It is for this reason that I suspect that in "On Denoting" he actually commits the fallacy I have just warned against.

v

Is (2) logically equivalent to (10)? It seems to me that it is not. It seems to me that (10) might be false although (2) is true.[4] Asked whether he wants to know whether one and only one person wrote *Waverley* and that person was Scott, George IV might answer that this is not what he wishes to know since he *already* knows that one and only one individual wrote *Waverley*: what he does not know is whether the author of *Waverley* is Scott. George IV answers thus because he takes it

----

[4] Numbers within parentheses are used sometimes as designations of sentences and sometimes as abbreviations. The context resolves this ambiguity.

that his interlocutor would not put his question as he does unless assuming that he (George IV) does not know that one and only one individual wrote *Waverley*. In this case (2) is true and (10) is not; thus they are not logically equivalent. If this argument is sound, it follows that neither the primary nor the secondary interpretation of (2) is correct and that the theory of descriptions is incapable of providing an analysis of propositions of the type we are considering. But even if the argument is mistaken the conclusion is correct, for it is true that:

(13)   Linsky argued that it might have been the case that George IV wanted to know whether Scott was the author of *Waverley* although George IV did not want to know whether one and only one person wrote *Waverley* and that person was Scott.

Now (13) is a proposition containing a descriptive phrase, and for reasons already given that phrase cannot be regarded as having a primary occurrence there. But neither can it be regarded as having a secondary occurrence, for on this analysis (13) is logically equivalent to:

(14)   Linsky argued that it might have been the case that George IV wanted to know whether one and only one person wrote *Waverley* and that person was Scott although George IV did not want to know whether one and only one person wrote *Waverley* and that person was Scott.

Although this is not a logical contradiction it is certainly false and, thus, not equivalent to (13), which is true. It must be observed that, while (14) seems to be the most natural secondary analysis of (13), it is not the only possible one. A secondary analysis of a proposition containing a descriptive phrase is an analysis that eliminates the description from a subordinate propositional part of the whole proposition in which the description occurs. But such a subordinate propositional part of (13) can be selected in seven possible ways. None of these secondary analyses provides an analysans that is logically equivalent to the analysandum (13). The proof of this contention is left as an exercise for the reader.

VI

It has been established that there are propositions containing descriptive phrases for which any interpretation offered by the theory of descriptions is incorrect—incorrect in the sense that the analysandum and any analysans offered by the theory fail to meet the minimum requirement of L-equivalence. It follows that there are puzzles of essentially the same kind as (1) (2) (3) for which Russell's theory cannot provide a solution, even if (what is not the case) it does supply an adequate treatment of (1) (2) (3). The following is such a "puzzle." (1), (13), therefore:

> Linsky argued that it might have been the case that George IV wanted to know whether Scott was Scott although George IV did not want to know whether one and only one person wrote *Waverley* and that person was Scott.

It is clear that this does not follow from (1) and (13), but the Russellian analysis of (13) into (14) (or into any of the possible alternatives allowed by the theory) is incorrect.

VII

Under the secondary interpretation, (2) has been represented as having the form (8). There is something misleading about thus representing it, and when this has been shown we will be in a position to see more clearly the exact nature of the "solution" that Russell's theory offers for our puzzle. When (2) is given the secondary interpretation, (1) (2) (3) constitutes a valid argument if and only if the following conditional is valid:

$$(15) \quad \{a = (\imath x)(\phi x)\} \supset \{(\chi\{[(\imath x)(\phi x)] \, \psi(\imath x)(\phi x)\}) \supset (\chi\{\psi a\})\}$$

In the final paragraph of Chapter 14 of *Principia* Russell says "It should be observed that the proposition in which $(\imath x)(\phi x)$ has the larger scope always implies the corresponding one in which it has the smaller scope, but the converse implication holds only if either (a) we have $E!(\imath x)(\phi x)$ or (b) the proposition in which $(\imath x)(\phi x)$ has the smaller scope implies $E!(\imath x)(\phi x)$." Part of what Russell is here telling us is put formally thus:

(16) $\{E!(\imath x)(\phi x)\}$
$\supset \{(\chi\{[(\imath x)(\phi x)]\psi(\imath x)(\phi x)\}) \equiv ([(\imath x)(\phi x)]\chi\{\psi(\imath x)(\phi x)\})\}$

Now, if this is true, the interpretation that accords a secondary occurrence to 'the author of *Waverley*' in (2) will yield an argument (call it "1′ 2′ 3′") that is valid. This can be shown as follows. (1) (2) (3) is valid when the descriptive phrase is accorded a primary occurrence in (2); i.e., the following is valid by 14.15:

$$1 \qquad a = (\imath x)(\phi x)$$
$$2 \qquad \chi\{\psi(\imath x)(\phi x)\}$$
$$3 \quad\therefore\ \chi(\psi a)$$

What we want to show is that the argument that accords a secondary occurrence to the description is also valid.

$$1' \qquad a = (\imath x)(\phi x)$$
$$2' \qquad \chi\{[(\imath x)(\phi x)]\psi(\imath x)(\phi x)\}$$
$$\overline{}$$
$$3' \quad\therefore\ \chi(\psi a)$$

Now it is valid; for them 1′ follows 4′: $E!(\imath x)(\phi x)$ (by 14.21). By assumption we have 5′, i.e., (16). From 4′ and 5′ we get

6′ $(\chi\{[(\imath x)(\phi x)]\psi(\imath x)(\phi x)\}) \equiv ([(\imath x)(\phi x)]\chi\{\psi(\imath x)(\phi x)\})$

Now, by 2′ and 6′, we get

7′ $[(\imath x)(\phi x)]\chi\{\psi(\imath x)(\phi x)\}$

But 7′ = 2, and 1′ = 1; hence, since 1 2 3 is valid and 3′ = 3, it follows that 1′ 2′ 3′ is valid.

Russell thought the argument (1) (2) (3), under the interpretation according secondary occurrence to the description in (2), to be invalid; and so it is. The source of the difficulty is that (16) is not true. We have just shown that. Russell says that (16) is true in at least three places in *Principia*. In the Introduction he says, "It will be seen further that when $E!(\imath x)(\phi x)$, we may enlarge or diminish the scope of $(\imath x)(\phi x)$ as much as we please without altering the truth-value of any proposition in which it occurs."[5] Again in Chapter 14 he says "The purpose of the following propositions is to show that, when $E!(\imath x)(\phi x)$, the scope of $(\imath x)(\phi x)$ does not matter to the truth-value of any proposition in which $(\imath x)(\phi x)$ occurs. This proposition cannot be proved generally, but it can be proved in each particular case."[6] We are then given a series of theorems (14.31 to 14.34) in which it is proved that, when $(\imath x)(\phi x)$ occurs in the form $\chi(\imath x)(\phi x)$ and $\chi(\imath x)(\phi x)$ occurs in turn as part of a larger proposition, the scope of $(\imath x)(\phi x)$ does not affect the truth-value of the larger proposition, provided $E!(\imath x)(\phi x)$.

Though principle (16) is not true, there is a theorem corresponding to it which says that, when $\chi(\imath x)(\phi x)$ occurs in a larger proposition built up *truth-functionally* out of it, the scope of $(\imath x)(\phi x)$ does not affect the truth-value of the larger proposition, provided $E!(\imath x)(\phi x)$:

14.3 $\quad (\{\{(p \equiv q) \supset_{p,q}(f(p) \equiv f(g))\} \& \{E!(\imath x)(\phi x)\})$
$\qquad \supset \{(f\{[(\imath x)(\phi x)]\chi(\imath x)(\phi x)\}) \equiv ([(\imath x)(\phi x)]f\{\chi(\imath x)(\phi x)\})\})$

But *this* theorem cannot be used, as (16) was used to establish the validity of 1′ 2′ 3′, because 2′ does not express a truth-function of $\psi(\imath x)(\phi x)$.

VIII

What (16) says is that, provided $E!(\imath x)(\phi x)$, the proposition giving the larger scope to $(\imath x)(\phi x)$, i.e., giving it primary oc-

---

[5] Page 73 of the first edition; p. 70 of the second.
[6] Page 193 of the first edition; p. 184 of the second.

currence, is equivalent to the proposition giving it the smaller scope, i.e., secondary occurrence. This is true provided extensional functions only are involved. Thus 14.3. I do not wish to suggest that Russell believed (16) also to hold for nonextensional functions. For this reason the reduction of 1′ 2′ 3′ to 1 2 3 carried out above is mistaken, and once again we escape the conclusion that our argument is valid. But we also see what is misleading about the view that sees in (2) a proposition of the form (8). If the logic of the theory of descriptions does not apply to propositions of the form (8), for nonextensional values of x, how can the logical form of (2) be exhibited by this formula of the theory? (8) is no more than mere shorthand for (2); it does not display its logical form; for the logic of the theory of descriptions licenses no inferences that turn upon the internal structure of (8) so long as the values of x are nonextensional. What, then, do we finally learn is the mistake in (1) (2) (3)? The primary analysis that Russell's theory gives us of (2) is wrong, and the secondary analysis does not exhibit its form. There is no problem in (1) (2) (3) of logical mirage, of confusing proper names with descriptive phrases. The fault lies with the so-called "principle of the substitutivity of identicals." This principle does not hold for nonextensional contexts.

IX

It is not difficult, I believe, to see why this false principle was accepted. It was accepted because it was not distinguished from another principle, i.e., the principle of the indiscernibility of identicals. This principle states that, if $x = y$, then any property of x is a property of y, and conversely. But this does not at all say what the principle of the substitutivity of identicals says, viz., that the two terms of a true identity statement may be substituted for each other in any true statement, *salve veritate*.

We can see the confusion of those two principles in Russell's formulation of the principle of substitutivity. "If a is identical with *b*, whatever is true of the one is true of the other, and either may be substituted for the other in any proposition without altering the truth or falsehood of that proposition." It is true that if *a* is identical with *b*, whatever is true of the one is true of the other, if this means that every property of *a* is a property of *b* and conversely. Russell then goes on to say, "and either may be substituted for the other in any proposition," *salve veritate*. But what we substitute in a proposition is not *a* for *b* (or conversely) but names (or other designations) for *a* and *b*. So, if we correct the use-mention confusion in this formulation we obtain the following principle: "If *a* is identical with *b*, whatever is true of the one is true of the other, and names (or other designations) for *a* and *b* may be substituted for each other in any proposition *salve veritate*." But this principle is false, as is evidenced by (1) (2) (3). The assumption that makes it seem that we must accept this false principle is the assumption that every open sentence expresses a property. On this assumption the (corrected) principle of substitutivity is entailed by the principle of the indiscernibility of identicals. But we have only to state this assumption to see that it is wrong.

To leave the matter here is clearly unsatisfactory. Some open sentences do express properties. Which do and which do not? I do not know how to draw this line, nor so far as I know does anybody else. To that extent it still remains a problem what is wrong with (1) (2) (3).

x

Statements of propositional attitudes, nonextensional propositions generally, are in a certain sense Janus-faced. One of their faces is extensional (or transparent); the other face is intentional (or "opaque," as Quine uses this term). These propositions are ambiguous. Consider the proposition, 'Oedipus wanted to marry his mother'. Understood opaquely, what I have said is wildly false. But it can be understood transparently, and, so understood, it is true. Oedipus wanted to marry Jocasta; and who was she? His mother! Thus, plainly, Oedipus did want to marry his mother. Understood transparently, 'Oedipus wanted to marry his mother' may be paraphrased as saying "Oedipus wanted to marry a person who, in fact, was his mother." This would naturally be understood not to imply that Oedipus knew that the person he wanted to marry was, in fact, his mother. Certainly in the case of the proposition we are considering, at least out of any special context, the opaque interpretation is the more natural. It strikes us, straight off, as wildly false. But 'Oedipus wanted to marry his mother' can be understood transparently; thus it exemplifies my thesis, viz., that a proposition whose principal verb is a verb of propositional attitude is, in the way indicated, ambiguous.

How is all of this related to Russell's distinction between the primary and the secondary interpretations of propositions containing descriptive phrases? It seems to me that Russell was aware of the ambiguity I have attempted to bring out. There is reason to suppose that he believed that the distinction between the primary and secondary interpretation of propositions which, like (2), have verbs of propositional attitude as their main verbs, corresponds to the distinction between their transparent and their opaque interpretations. There is reason to suppose that Russell believed that his primary interpretation gave a correct analysis of such propositions transparently understood and that his secondary analysis gave a correct analysis of these propositions opaquely understood.

If Russell did believe these things, then I think that he was partly right and partly wrong. He was right if he thought that his primary analysis was correct for propositions expressing propositional attitudes when these latter are transparently understood. Let us see how this works out for (2). Suppose that George IV sees Scott dimly in the distance through a thick English fog. He wonders who it is and makes a guess. He asks, "Is that man possibly Scott?" It seems to me that it would be perfectly correct to report this incident by saying "George IV wished to know whether Scott was Scott." These words *might* mislead an audience, but, if the situation were entirely clear to that audience, there would be no reason, as far as I can see, for saying that the report was false. The incident might equally correctly be reported in these words: "George IV wished to know, concerning the man who had in fact written *Waverley*, whether he was Scott," or again "One and only one individual both wrote *Waverley* and was such that George IV wanted to know whether he was Scott." Now both of these last two say what, on Russell's analysis, is said by (2) on the primary analysis. I believe that they give possible interpretations of (2), though certainly these interpretations are not the natural ones. If this interpretation is taken, (1) (2) (3) is a valid argument. This alone brings out how unnatural the primary interpretation is.

But I have also argued that Russell was partly wrong. He was wrong if he thought that his secondary analysis was correct for propositions expressing propositional attitudes opaquely understood. This is what I have been trying to demonstrate in sections v and vi of this paper. But it should be clear that this failure concerns only the opaque interpretations of the propositions in question [(2), (13)]. It is also clear that Russell understands (2) opaquely; for Russell assumes from the beginning, as we all naturally do, that the argument is invalid.

The claim of section iii of this paper is that the primary interpretation of (2) is wrong. In view of the distinction brought out here, this claim must be modified to say that the primary interpretation of (2), opaquely understood, is wrong; for the primary interpretation of (2), transparently understood, seems to me to be unobjectionable. The two terms of a true identity statement may replace each other, *salve veritate*, even in positions governed by verbs of propositional attitude, so long as the propositions containing these verbs are transparently understood. Substitutivity of terms of a true identity statement in a given context is often cited as a criterion for the extensionality of that context. Thus we can see how, in another way, Russell's program of analysis is tied to the extensional point of view. His analysis of propositions expressing propositional attitudes is correct when these propositions are interpreted extensionally.

22

Semantics for Propositional Attitudes

JAAKKO HINTIKKA

I. THE CONTRAST BETWEEN THE THEORY OF REFERENCE AND THE THEORY OF MEANING IS SPURIOUS

In the philosophy of logic a distinction is often made between the *theory of reference* and the *theory of meaning*. In this paper I shall suggest (*inter alia*) that this distinction, though not without substance, is profoundly misleading. The theory of reference is, I shall argue, the theory of meaning for certain simple types of language. The only entities needed in the so-called theory of meaning are, in many interesting cases and perhaps even in all cases, merely what is required in order for the expressions of our language to be able to refer in certain more complicated situations. Instead of the theory of reference and the theory of meaning we perhaps ought to speak in some cases of the theory of simple and of multiple reference, respectively. Quine has regretted that the term 'semantics', which etymologically ought to refer to the theory of meaning, has come to mean the theory of reference.[1] I submit that this usage is happier than Quine thinks, and that large parts of the theory of meaning in reality are—or ought to be—but semantical theories for notions transcending the range of certain elementary types of concepts.

From *Philosophical Logic*, eds. J.W. Davis, D.J. Hockney, and W.K. Wilson. Copyright 1969 by D. Reidel Publishing Co., Dordrecht, Holland. Reprinted by permission of the publisher.

[1] See, e.g., W.V.O. Quine, *From a Logical Point of View*, 2nd ed. (Cambridge, Mass.: Harvard Univ. Press, 1961), pp. 130–132.

It seems to me in fact that the usual reasons for distinguishing between meaning and reference are seriously mistaken. Frequently, they are formulated in terms of a first-order (i.e., quantificational) language. In such a language, it is said, knowing the mere references of individual constants, or knowing the extensions of predicates, cannot suffice to specify their meanings because the references of two individual constants or the extensions of two predicate constants 'obviously' can coincide without there being any identity of meaning.[2] Hence, it is often concluded, the theory of reference for first-order languages will have to be supplemented by a theory of the 'meanings' of the expressions of these languages.

The line of argument is not without solid intuitive foundation, but its implications are different from what they are usually taken to be. This whole concept of meaning (as distinguished from reference) is very unclear and usually hard to fathom. However it is understood, it seems to me in any case completely hopeless to try to divorce the idea of the meaning of a sentence from the idea of the *information* that the sentence can convey to a hearer or reader, should someone truthfully address it to him.[3] Now what is this information? Clearly it is just information to the effect that the sentence is true, that the world is such as to meet the truth-conditions of the sentence.

Now in the case of a first-order language these truth-conditions cannot be divested from the references of singular

---

[2] For a simple recent argument of this sort (without a specific reference to first-order theories), see, e.g., William P. Alston, *Philosophy of Language* (Englewood Cliffs, N.J.: Prentice-Hall, 1964), p. 13. Cf. also Quine, *From a Logical Point of View*, pp. 21–22.

[3] In more general terms, it seems to me hopeless to try to develop a theory of sentential meaning which is not connected very closely with the idea of the information which the sentence can convey to us, or a theory of meaning for individual words which would not show how understanding them contributes to appreciating the information of the sentences in which they occur. There are of course many nuances in the actual use of words and sentences which are not directly explained by connecting meaning and information in this way, assuming that this can be done. However, there do not seem to be any obstacles in principle to explaining these nuances in terms of pragmatic, contextual, and other contingent pressures operating on a language-user. For remarks on this methodological situation, see my paper "Epistemic Logic and the Methods of Philosophical Analysis," *Australasian Journal of Philosophy*, 46 (1968), 37–51.

terms and from the extensions of its predicates. In fact, these references and extensions are precisely what the truth-conditions of quantified sentences turn on. The truth-value of a sentence is a function of the references (extensions) of the terms it contains, not of their 'meanings'. Thus it follows from the above principles that a theory of reference is for genuine first-order languages the basis of a theory of meaning. Recently, a similar conclusion has in effect been persuasively argued for (from entirely different premises and in an entirely different way) by Donald Davidson.[4] The references, not the alleged meanings, of our primitive terms are thus what determine the meanings (in the sense explained) of first-order sentences. Hence the introduction of the 'meanings' of singular terms and predicates is strictly useless: In any theory of meaning which serves to explain the information which first-order sentences convey, these 'meanings' are bound to be completely idle.

What happens, then, to our intuitions concerning the allegedly obvious difference between reference and meaning in first-order languages? If these intuitions are sound, and if the above remarks are to the point, then the only reasonably conclusion is that our intuitions do not really pertain to first-order discourse. The 'ordinary language' which we think of when we assert the obviousness of the distinction cannot be reduced to the canonical form of an applied first-order language without violating these intuitions. How these other languages enable us to appreciate the real (but frequently misunderstood) force of the apparently obvious difference between reference and meaning I shall indicate later (see Section VI below).

II. FIRST-ORDER LANGUAGES

I conclude that the traditional theory of reference, suitably extended and developed, is all we need for a full-scale theory of meaning in the case of an applied first-order language. All that is

---

[4] Donald Davidson, "Truth and Meaning," *Synthese*, 17 (1967), 304–323.

needed to grasp the information that a sentence of such a language yields is given by the rules that determine the references of its terms, in the usual sense of the word. For the purposes of first-order languages, to specify the meaning of a singular term is therefore nearly tantamount to specifying its reference, and to specify the meaning of a predicate is for all practical purposes to specify its extension. As long as we can restrict ourselves to first-order discourse, the theory of truth and satisfaction will therefore be the central part of the theory of meaning.

A partial exception to this statement seems to be the theory of so-called 'meaning postulates' or 'semantical rules' which are supposed to catch nonlogical synonymies.[5] However, I would argue that whatever nonlogical identities of meaning there might be in our discourse ought to be spelled out, not in terms of definitions of terms, but by developing a satisfactory semantical theory for the terms which create these synonymies. In those cases in which meaning postulates are needed, this enterprise no longer belongs to the theory of first-order logic.

In more precise terms, one may thus say that to understand a sentence of first-order logic is to know its interpretation in the actual world. To know this is to know the interpretation function ϕ. This can be characterized as a function which does the following things:

(1.1)   For each individual constant a of our first-order language, $\phi(a)$ is a member of the domain of individuals l.

The domain of individuals l is of course to be thought of as the totality of objects which our language speaks of.

(1.2)   For each constant predicate Q (say of n terms), $\phi(Q)$ is a set of n-tuples of the members of l.

If we know ϕ and if we know the usual rules holding the satisfaction (truth), we can in principle determine the truth-values of all the sentences of our first-order language. This is the cash value of the statement made above that the extensions of our

---

[5] See Quine, *From a Logical Point of View*, pp. 32–37.

individual constants and constant predicates are virtually all that we need in the theory of meaning in an applied first-order language.[6]

These conditions may be looked upon in slightly different ways. If ϕ is considered as an arbitrary function in $(1.1)-(1.2)$, instead of that particular function which is involved in one's understanding of a language, and if 1 is likewise allowed to vary, we obtain a characterization of the concept of interpretation in the general model-theoretic sense.

III. PROPOSITIONAL ATTITUDES

We have to keep in mind the possibility that ϕ might be only a partial function (as applied to free singular terms), i.e., that some of our singular terms are in fact empty. This problem is not particularly prominent in the present paper, however.[7] If what I have said so far is correct, then the emphasis philosophers have put on the distinction between reference and meaning (e.g., between *Bedeutung* and *Sinn*) is motivated only insofar as they have implicitly or explicitly considered concepts which go beyond the expressive power of first-order languages.[8] Probably the most important type of such concept is a propositional attitude.[9] One purpose of this paper is to sketch some salient features of a semantical theory of such concepts. An interesting problem will be the question as to what extent we have to assume entities other than the usual individuals (the members of 1) in order to give a satisfactory account of the meaning of prop-

[6] The main reason why the truth of these observations is not appreciated more widely seems to be the failure to consider realistically what the actual use of a first-order language (say for the purpose of conveying information to another person) would look like.

[7] The basic problems as to what happens when this possibility is taken seriously are discussed in my paper, "Studies in the Logic of Existence and Necessity I," *The Monist*, 50 (1966), 55–76.

[8] This is certainly true of Frege. His very interest in oblique contexts seems to have been kindled by the realization that they cannot be handled by means of the ideas he had successfully applied to first-order logic.

[9] The term seems to go back to Bertrand Russell, *An Inquiry into Meaning and Truth* (London: George Allen and Unwin, 1940).

ositional attitudes. As will be seen, what I take to be the true answer to this question is surprisingly subtle, and cannot be formulated by a simple 'yes' or 'no'.

What I take to be the distinctive feature of all use of propositional attitudes is the fact that in using them we are considering more than one possibility concerning the world.[10] (This consideration of different possibilities is precisely what makes propositional attitudes propositional, it seems to me.) It would be more natural to speak of different possibilities concerning our 'actual' world than to speak of several possible worlds. For the purpose of logical and semantical analysis, the second locution is much more appropriate than the first, however, although I admit that it sounds somewhat weird and perhaps also suggests that we are dealing with something much more

---

[10] An important qualification here is that for deep logical reasons one cannot usually distinguish effectively between what is 'really' a logically possible world and what merely 'appears' on the face of one's language (or thinking) to be a possibility. This, in a sufficiently sharp analysis, is what destroys the pleasant invariance of propositional attitudes with respect to logical equivalence. Even though p and q are equivalent, i.e., even though the 'real' possibilities concerning the world that they admit and exclude are the same,

$$a \quad \begin{matrix} \text{knows} \\ \text{believes} \\ \text{remembers} \\ \text{hopes} \\ \text{strives} \end{matrix} \quad \text{that } p$$

and

$$a \quad \begin{matrix} \text{knows} \\ \text{believes} \\ \text{remembers} \\ \text{hopes} \\ \text{strives} \end{matrix} \quad \text{that } q$$

need not be equivalent, for the apparent (to a) possibilities admitted by p and q need not be identical.

I have studied this concept of an 'apparent' possibility and its consequences at some length elsewhere, especially in the second and third paper printed in *Deskription, Analytizität und Existenz*, ed. Paul Weingartner (Pustet, Salzburg, and Munich, 1966), in "Are Logical Truths Analytic?," *The Philosophical Review*, 74 (1965), 178–203, in "Surface Information and Depth Information," in *Information and Inference*, eds. K.J.J. Hintikka and P. Suppes (Dordrecht: D. Reidel Publishing Co., 1969), and in "Are Mathematical Truths Synthetic a Priori?," *The Journal of Philosophy*, 65 (1968), 640–651.

It is an extremely interesting concept to study and codify. However, it is not directly relevant to the concerns of the present paper, and would in any case break its confines. Hence it will not be taken up here, except by way of this *caveat*.

unfamiliar and unrealistic than we are actually doing. In our sense, whoever has made preparations for more than one course of events has dealt with several 'possible courses of events' or 'possible worlds'. Of course, the possible courses of events he considered were from his point of view so many alternative courses that the actual events might take. However, only one such course of events (at most) became actual. Hence there is a sense in which the others were merely 'possible courses of events', and this is the sense on which we shall try to capitalize.

Let us assume for simplicity that we are dealing with only one propositional attitude and that we are considering a situation in which it is attributed to one person only. Once we can handle this case, a generalization to the others is fairly straightforward. Since the person in question remains constant throughout the first part of our discussion, we need not always indicate him explicitly.

IV. PROPOSITIONAL ATTITUDES AND 'POSSIBLE WORLDS'

My basic assumption (slightly oversimplified) is that an attribution of any propositional attitude to the person in question involves a division of all the possible worlds (more precisely, all the possible worlds which we can distinguish in the part of language we use in making the attribution) into two classes: into those possible worlds which are in accordance with the attitude in question and into those which are incompatible with it. The meaning of the division in the case of such attitudes as knowledge, belief, memory, perception, hope, wish, striving, desire, etc., is clear enough. For instance, if what we are speaking of are (say) a's memories, then, these possible worlds are all the possible worlds compatible with everything he remembers.

There are propositional attitudes for which this division is not possible. Some such attitudes can be defined in terms of attitudes for which the assumptions do hold, and thus in a sense can be 'reduced' to them. Others may fail to respond to this kind of attempted reduction to those 'normal' attitudes which we shall be discussing here. If there really are such recalcitrant propositional attitudes, I shall be glad to restrict the scope of my treatment so as to exclude them. Enough extremely important notions will still remain within the purview of my methods.

There is a sense in which in discussing a propositional attitude, attributed to a person, we can even restrict our attention to those possible worlds which are in accordance with this attitude.[11] This may be brought out, e.g., by paraphrasing statements about propositional attitudes in terms of this restricted class of all possible worlds. The following examples will illustrate these approximate paraphrases:

a believes that $p =$ in all the possible worlds compatible with what *a* believes, it is the case that p;

a does not believe that p (in the sense 'it is not the case that *a* believes that p') $=$ in at least one possible world compatible with what *a* believes it is not the case that p.

V. SEMANTICS FOR PROPOSITIONAL ATTITUDES

What kind of semantics is appropriate for this mode of treating propositional attitudes? Clearly what is involved is a set Ω of possible worlds or of models in the usual sense of the word. Each of them, say $\mu\epsilon\Omega$, is characterized by a set of individuals $1(\mu)$ existing in that 'possible world'. An interpretation of individual constants and predicates will now be a two-argument

[11] There is a distinction here which is not particularly relevant to my concerns in the present paper but important enough to be noted in passing, especially as I have not made it clear in my earlier work. What precisely are the worlds 'alternative to' a given one, say μ? A moment's reflection on the principles underlying my discussion will show, I trust, that they must be taken to be worlds compatible with a certain person's having a definite propositional attitude in μ, and not just compatible with the content of his attitude, for instance, compatible with someone's knowing something in μ and not just compatible with what he knows. I failed to spell this out in my *Knowledge and Belief* (Ithaca, N.Y.: Cornell Univ. Press, 1962), as R. Chisholm in effect pointed out in his review article, "The Logic of Knowing," *The Journal of Philosophy*, 60 (1963), 773–795.

function $\phi(a, \mu)$ or $\phi(Q, \mu)$ which depends also on the possible world μ in question. Otherwise an interpretation works in the same way as in the pure first-order case, and the same rules hold for propositional connectives as in this old case.

Simple though this extension of the earlier semantical theory is, it is in many ways illuminating. For instance, it is readily seen that in many cases earlier semantical rules are applicable without changes. *Inter alia,* insofar as no words for propositional attitudes occur inside the scope of a quantifier, this quantifier is subject to the same semantical rules (satisfaction conditions) as before.

VI. MEANING AND THE DEPENDENCE OF REFERENCE ON 'POSSIBLE WORLDS'

A new aspect of the situation is the fact that the reference $\phi(a, \mu)$ of a singular term now depends on μ—on what course the events will take, one might say. This enables us to appreciate an objection which you probably felt like making earlier when it was said that in a first-order language the theory of meaning is the theory of reference. What really determines the meaning of a singular term, you felt like saying, is not whatever reference it happens to have, but rather the way in which this reference is determined. But in order for this to make any difference, we must consider more than one possibility as to what the reference is, depending on the circumstances (i.e., depending on the course events will take). This dependence is just what is expressed by $\phi(a, \mu)$ when it is considered as a function of μ. (This function *is* the meaning of a, one is tempted to say.) Your objection thus has a point. However, it does not show that more is involved in the theory of meaning for first-order languages than the references of its terms. Rather, what is shown is that in order to spell out the idea that the meaning of a term is the way in which its reference is determined we have to consider how the reference varies in different possible worlds, and therefore go beyond first-order languages, just as I suggested above. Analogous remarks apply of course to the extensions of predicates.

Another novelty here is the need of picking out one distinguished possible world from among them all, viz., the world that happens to be actualized ('the actual world').

VII. DEVELOPING AN EXPLICIT SEMANTICAL THEORY: ALTERNATIVENESS RELATIONS

How are these informal observations to be incorporated into a more explicit semantical theory? According to what I have said, understanding attributions of the propositional attitude in question (let us assume that this is expressed by 'B') means being able to make a distinction between two kinds of possible worlds, according to whether they are compatible with the relevant attitudes of the person in question. The semantical counterpart to this is of course a function which to a given individual person assigns a set of possible worlds.

However, a minor complication is in order here. Of course, the person in question may himself have different attitudes in the different worlds we are considering. Hence this function in effect becomes a relation which to a given individual *and to a given possible world* μ associates a number of possible worlds which we shall call the *alternatives* to μ. The relation will be called the alternativeness relation. (For different propositional attitudes, we have to consider different alternativeness relations.) Our basic apparatus does not impose many restrictions on it. The obvious requirement that ensues from what has been said is the following:

(S.B.) $B_a p$ is true in a possible world μ if and only if p is true in all the alternatives to μ.

$B_a p$ may here be thought of as a shorthand for 'a believes that p'. We can write this condition in terms of an interpretation function ϕ. What understanding B means is to have a function ϕ_B which to a given possible world μ and to a given individual a

associates a set of possible worlds ϕ_B (a, μ), namely, the set of all alternatives to μ.[12] Intuitively, they are the possible worlds compatible with the presence of the attitude expressed by B in the person a in the possible world μ.

In terms of this extended interpretation function, (S.B) can be written as follows:

$B_a p$ is true in μ if and only if p is true in every member of $\phi_B(a, \mu)$.

VIII. RELATION TO QUINE'S CRITERION OF COMMITMENT

The interesting and important feature of this truth-condition is that it involves quantification over a certain set of possible worlds. By Quine's famous criterion, we allegedly are ontologically committed to whatever we quantify over.[13] Thus my semantical theory of propositional attitudes seems to imply that we are committed to the existence of possible worlds as a part of our ontology.

This conclusion seems to me false, and I think that it in fact constitutes a counterexample to Quine's criterion of commitment *qua* a criterion of ontological commitment. Surely we must in some sense be *committed* to whatever we quantify over. To this extent Quine seems to be entirely right. But why call this a criterion of *ontological* commitment? One's ontology is what one assumes to exist in one's world, it seems to me. It is, as it were, one's census of one's universe. Now such a census is meaningful only in some particular possible world. Hence Quine's criterion can work as a criterion of *ontological* commit-

---

[12] As the reader will notice, I am misusing (in the interest of simplicity) my terminology systematically by speaking elliptically of 'the person a', etc., when 'the person referred to by a' or some such thing is meant. I do not foresee any danger of confusion resulting from this, however.

[13] Quine, *From a Logical Point of View*, pp. 1–14, and *Word and Object* (Cambridge, Mass.: MIT Press; New York and London: John Wiley, 1960), pp. 241–243. It is not quite clear from Quine's exposition, however, precisely how much emphasis is to be put on the word 'ontology' in his criterion of ontological commitment. My discussion which focuses on this word may thus have to be taken as a qualification to Quine's criterion rather than as outright criticism.

ment only if the quantification it speaks of is a quantification over entities belonging to some one particular world. To be is perhaps to be a value of a bound variable. But to exist in an ontologically relevant sense, to be a part of the furniture of the world, is to be a value of a special kind of a bound variable, namely one whose values all belong to the same possible world. Thus the notion of a possible world serves to clarify considerably the idea of ontological commitment so as to limit the scope of Quine's dictum.

Clearly, our quantification over possible worlds does not satisfy this extra requirement. Hence there is a perfectly good sense in which we are not ontologically committed to possible worlds, however important their role in our semantical theory may be.

Quine's distinction between *ontology* and *ideology*, somewhat modified and put to a new use, is handy here.[14] We have to distinguish between what we are committed to in the sense that we believe it to exist in the actual world or in some other possible world, and what we are committed to as a part of our ways of dealing with the world conceptually, committed to as a part of our conceptual system. The former constitute our ontology, the latter our 'ideology.' What I am suggesting is that the possible worlds we have to quantify over are a part of our ideology but not of our ontology.

The general criterion of commitment is a generalization of this. Quantification over the members of one particular world is a measure of ontology, quantification that crosses possible worlds is often a measure of ideology. Quine's distinction thus ceases to mark a difference between two different types of studies or two different kinds of entities within one's universe. It now marks, rather, a distinction between the object of reference and certain aspects of our own referential apparatus. Here we can perhaps see what the so-called distinction between theory of reference and theory of meaning really amounts to.

It follows, incidentally, that if we could restrict our attention

[14] Quine, *From a Logical Point of View*, pp. 130–132.

to *one* possible world only, Quine's restriction would be true without qualifications. Of course, the restriction is one which Quine apparently would very much like to make; hence he has a legitimate reason for disregarding the qualifications for his own purposes.

Our 'ideological' commitment to possible worlds other than the actual one is neither surprising nor disconcerting. If what we are dealing with are the things people do—more specifically, the concepts they use—in order to be prepared for more than one eventuality, it is not at all remarkable that in order to describe these concepts fully we have to speak of courses of events other than the actual one.

IX. SINGULAR TERMS AND QUANTIFICATION IN THE CONTEXT OF PROPOSITIONAL ATTITUDES

Let us return to the role of individual constants (and other singular terms). Summing up what was said before, we can say that what the understanding of an individual constant amounts to in a first-order language is knowing what individual it stands for. Now it is seen that in the presence of propositional attitudes this statement has to be expanded to say that one has to know what the singular term stands for in the different possible worlds we are considering.

Furthermore, in the same way as these individuals (or perhaps rather the method of specifying them) may be said to be what is 'objectively given' to us when we understand the constant, in the same way what is involved in the understanding of a propositional attitude is precisely that distinction which in our semantical apparatus is expressed by the function which serves to define the alternativeness relation. This function is what is 'objectively given' to us with the understanding of a word for a propositional attitude.

These observations enable us to solve almost all the problems that relate to the use of identity in the context of propositional attitudes. For instance, we can at once see why the

familiar principle of the substitutivity of identity is bound to
fail in the presence of propositional attitudes when applied to
arbitrary singular terms.[15] Two such terms, say a and b, may
refer to one and the same individual in the actual world
($\phi(a, \mu_0) = \phi(b, \mu_0)$ for the world μ_0 that happens to be actualized), thus making the identity '$a = b$' true, and yet fail to
refer to the same individual in some other (alternative) possible world. (I.e., we may have $\phi(a, \mu_1) \neq \phi(b, \mu_1)$ for some
$\mu_1 \epsilon \phi_B(c, \mu_0)$ where c is the individual whose attitudes are being
discussed and B the relevant attitude.) Since the presence of
propositional attitudes means (if I am right) that these other
possible worlds have to be discussed as well, in their presence
the truth of the identity '$a = b$' does not guarantee that the
same things can be said of the references of a and b without
qualification, i.e., does not guarantee the intersubstitutivity of
the terms a and b.

Our observations also enable us to deal with quantification
in contexts governed by words for propositional attitudes as
long as we do not quantify *into* them. However, as soon as we
try to do so, all the familiar difficulties which have been so
carefully and persuasively presented by Quine and others will
apply with full force.[16] An individual constant occurring within
the scope of an operator like B which expresses a propositional
attitude does not specify a unique individual. Rather, what it
does is to specify an individual in each of the possible worlds we
have to consider. Replace it by an individual variable, and you
do not get anything that you could describe by speaking of
the individuals over which this variable ranges. There are (it
seems) simply no uniquely defined individuals here at all.

It is perhaps thought that the way out is simply to deny that
one can ever quantify into a nonextensional context. However,

---

[15] For a discussion of the problems connected with the substitutivity principle,
see my exchange with Føllesdal: Dagfinn Føllesdal, "Knowledge, Identity, and
Existence," *Theoria*, 33 (1967), 1–27; Jaakko Hintikka, "Existence and Identity in Epistemic Contexts," *ibid.*, 138–147.
[16] See Quine, *From a Logical Point of View*, ch. 8; *Word and Object*, ch. 6;
The Ways of Paradox and Other Essays (New York: Random House, 1966),
chs. 13–15.

this way out does not work.[17] As a matter of fact, in our ordinary language we often quantify into a grammatical construction governed by an expression for a propositional attitude. Locutions like 'knows who', 'sees what', 'has an opinion concerning the identity of' are cases in point, and so is almost any (other) construction in which pronouns are allowed to mix with words for propositional attitudes. Beliefs about 'oneself' and 'himself' yield further examples, and an account of their peculiarities leads to an interesting reconstruction of the traditional distinction between so-called modalities *de dicto* and *de re*.[18]

Another general fact is that we obviously have beliefs about definite individuals and not just about whoever happens to meet a certain description. I want to suggest that such beliefs (and the corresponding attitudes in the case of other propositional attitudes) are precisely what one half of the *de dicto – de re* distinction amounts to.[19]

Furthermore, it does not do to try to maintain that in these constructions the propositional attitude itself has to be taken in an unusual extensional or 'referentially transparent' sense. Such senses can in fact be defined in terms of the normal senses of propositional attitudes. However, these definitions already involve the objectionable quantification into opaque contexts, and if one tries to postulate the defined senses as irreducible primitive senses, they do not have the properties which they ought to

[17] Some arguments to this effect were given in *Knowledge and Belief* (note 11 above), pp. 142–146. The only informed criticism of this criticism that I have seen has been presented by R. L. Sleigh, in a paper entitled "A Note on an Argument of Hintikka's," *Philosophical Studies*, 18 (1967), 12–14. As I point out in my reply, "Partially Transparent Senses of Knowing," *Philosophical Studies*, 20 (1969), 5–8, Sleigh's argument turns on an ambiguity in my original formulation which is easily repaired. Neither the ambiguity nor its elimination provides any solace to the adherents of the view I have criticized, however.

[18] One thing at which this old distinction aims is obviously the distinction (which I am about to explain) between statements about whoever or whatever meets a description and statements about the individual who in fact does so. For the distinction, cf. Jaakko Hintikka, "Individuals, Possible Worlds, and Epistemic Logic," *Noûs*, 1 (1967), 32–62, especially 46–49, as well as "'Knowing Oneself' and Other Problems in Epistemic Logic," *Theoria*, 32 (1966), 1–13.

[19] Cf. below (sec. XII).

have in order to provide the resulting quantified statements with the logical powers they in fact have in ordinary language. For instance, Quine's attempt to postulate a sense of (say) knowledge in which one is allowed to quantify into a context governed by a transparently construed construction 'knows that' has the paradoxical result that

$$(\exists x) \text{ Jones knows that } (x = a)$$

is implied by any (transparently interpreted) statement of the form

$$\text{Jones knows that } (b = a)$$

and even by a similarly interpreted sentence

$$\text{Jones knows that } (a = a).$$

This I take to show that the first of these three sentences can scarcely serve as a formulation of 'Jones knows who (or what) a is' in our canonical idiom. Yet no other paraphrase of this ubiquitous locution has been proposed, and none is likely to be forthcoming. (For what else can there be to Jones' knowing who a is than his knowing of some well-defined individual that Jones is that very individual?) And it is Quine who always insists as strongly as anyone else that the values of bound variables have to be well-defined individuals. It is not much more helpful to try to maintain that no true sentences of the form

$$\text{Jones knows that } (b = a)$$

(with the transparent sense of 'knows') are forthcoming whenever Jones fails to know who a is. The transparent sense in which this would be the case has never been explained in a satisfactory way, and I do not see how it can be done in a reasonable way without falling back to my own analysis. (What can it conceivably mean, e.g., for Jones not to know in the transparent sense that an a, whom he knows to exist, is not self-identical? Can this self-identity fail to be true in a possible world compatible with everything Jones knows?)

Hence we have to countenance quantification into a context governed by an expression for an ('opaquely construed) propositional attitude. Our semantical theory at once suggests a way of handling these problems. For instance, in order for existential generalization to be applicable to a singular term *b* occurring, say, in a context where *a*'s beliefs are being discussed, it has to be required that *b* refers to the *same* individual in the different possible worlds compatible with what *a* believes (plus, possibly, in the actual world). This, naturally, will be expressed by a statement of the form

(*) $$(\exists x)[B_a(x=b)\&(x=b)]$$

or, if we do not have to consider the actual world, of the form $(\exists x)B_a(x=b)$.

X. METHODS OF CROSS-IDENTIFICATION

This solution is simple, straightforward, and workable. It generalizes easily to other propositional attitudes. However, it hides certain interesting conceptual presuppositions. With what right do we speak of individuals in the different possible worlds as being *identical?* This is the problem to which we have to address ourselves.

It is not difficult to see what more there is given to us with our ordinary understanding of propositional attitudes that we have not yet dealt with. For instance, consider a man who has a number of beliefs as to what will happen tomorrow to himself and to his friends. Consider, on his behalf, a number of possible courses of events tomorrow. If I know what our man believes, I can sort these into those which are compatible with his beliefs as distinguished from those which are incompatible with them. But this is not all that is involved. Surely the same or largely the same individuals must figure in these different sequences of events. Under different courses of events a given individual may undergo different experiences, entertain different beliefs and hopes and fears; he may behave rather differently and perhaps even look somewhat different. Nevertheless our man can be (although he need not be) and usually is completely confident that, whatever may happen, he is going to be able to recognize (re-identify) his friends under these various courses of events, at least in principle. He may admit that courses of events are perhaps logically possible under which he would fail to do so; but these would not be compatible with his beliefs as to what will happen. Given full descriptions of two different courses of events tomorrow, both compatible with what our man believes ('believes possible', we sometimes say with more logical than grammatical justification), he will be able to recognize what individuals figuring in one of these descriptions are identical with which individual in the other, even if their names are being withheld. (Of course our man need not believe all this but my point is merely that he *can* and very often *does* believe it.)

The logical moral of this story is that together with the rest of our beliefs we are often given something more than we have so far incorporated into our semantical theory. We are given ways of *cross-identifying* individuals, that is to say, ways of understanding questions as to whether an individual figuring in one possible world is or is not identical with an individual figuring in another world.[20]

This is one point at which the obviousness of my claim may be partially obscured by my terminology. Let us recall what these 'possible worlds' are in the case of a propositional attitude. They are normally possible states of affairs or courses of events compatible with the attitude in question in some specified person. Now normally these attitudes may be attitudes toward definite persons or definite physical objects. But how is it that we may be sure, sight unseen, that the attitudes are directed toward the right persons or objects? Only if in all the possible worlds compatible with the attitude in question we can pick out the recipient of this attitude, i.e., the individual at its receiving end. Although in many concrete situations the possibility of doing so is obvious, it has not been built into our semantical

[20] Cf. here my paper, "On the Logic of Perception," in *Perception and Personal Identity*, eds. N. Care and R. Grimm (Cleveland: Case Western Reserve Univ. Press, 1969).

apparatus so far. There is so far nothing in our semantical theory which enables us to relate to each other the members of the different domains of individuals $1(\mu)$. In many, though not necessarily all, applications of such relations are given to us as a part of our understanding of the concepts involved. For such cases, we have to build a richer semantical theory.

The way to do so is to postulate a method of making cross-identifications. One possible way to do so is to postulate a set of functions F each member f of which picks out at most one individual $f(\mu)$ from the domain of individuals $1(\mu)$ of each model μ. We must allow that there is no such value for some models μ. In other words, $f \epsilon F$ may be a partial function. Furthermore, we must often require that, given $f_1, f_2 \epsilon F$, if $f_1(\mu) = f_2(\mu)$ then $f_1(\lambda) = f_2(\lambda)$ for all alternatives λ to μ. In other words, an individual cannot 'split' when we move from a world to its alternatives. This question may seem to be a mere matter of detail, but it is easily seen that the question whether an individual can split in the sense just explained is tantamount to the question whether the substitutivity of identity can fail for bound (individual) variables, i.e., to the question whether a sentence

$$(x)(y)(x = y \supset B_a(x = y))$$

can fail to be logically true. This, again, is tantamount to the question whether a sentence of the form

$$(x)(y)(x = y \supset (Q(x) \supset Q(y)))$$

(with just one layer of operators for propositional attitudes in Q) can fail to be logically true.

In terms of the set F, the question whether $a\epsilon l(\mu)$ is identical with $b\epsilon l(\lambda)$ amounts to the question whether there is a function of $f \epsilon F$ such that $f(\mu) = a$, $f(\lambda) = b$.

XI. THE ROLE OF INDIVIDUATING FUNCTIONS

Instead of speaking of a set of functions correlating to each other the individuals existing in the different possible worlds, it is often more appropriate to speak of these domains of individuals as being partly identical (overlapping). Then there would be no need to speak of correlations at all. This point of view is useful in that it illustrates the fact that the apparently different individuals which are correlated by one of the functions $f \epsilon \mathbf{F}$ is just what we ordinarily mean by one and the same individual. It is the concrete individual which we speak about, which we give a name to, etc. In fact, the members of $\mathbf{F}$ might in fact be thought of as names or individual constants of a certain special kind, namely those having a unique reference in all the different worlds we are speaking of and hence satisfying formulas of the form (*). Indeed, I shall assume in the sequel that a constant of this kind can be associated with each function $f \epsilon \mathbf{F}$.

However, emphasizing the role of the functions $f \epsilon \mathbf{F}$ is useful for several purposes. First and foremost, it highlights an extremely important nontrivial part of our native conceptual skills, namely, our capacity to recognize one and the same individual under different circumstances and under different courses of events. What the set $\mathbf{F}$ of functions embodies is just the totality of ways of doing this. The nontrivial character of the possibility of this recognition would be lost if we should simply speak of the members of the different possible worlds as being partly identical.

For another thing, the structure formed by the relations of cross-world identity (David Kaplan calls them "trans-world heir lines") may be so complex as to be indescribable by speaking simply of partial identities between the domains of individuals of the different possible worlds. Above, it was said that in the case of many propositional attitudes an individual cannot 'split' when we move from a world to its alternatives. Although this seems to me to be the case with all the propositional attitudes I have studied in any detail, it is not quite clear to me precisely why this should always be the case. At any rate, there seem to be reasons for suspecting that the opposite 'irregularity' can occasionally take place with some modalities: individuals can 'merge together' when we move from a world to its alternatives. An analogy with temporal modalities may be instructive here.[21] If we presuppose some suitable system of cross-identifications between individuals existing at different times which turn on continuity, it seems possible in principle that a singular term should refer to the same physical system at all the different moments of time we are considering although this system 'merges' with others at times and occasionally 'splits up' into several. Some of these complications seem to be impossible to rule out completely in the case of some propositional attitudes, and because of them the idea of partly overlapping domains seems to me seriously oversimplified.

An extremely important further reason why we cannot reify the members of **F** into ordinary individuals is the possibility of having two different methods of cross-identification between the members of the same possible worlds, i.e., two different sets of 'individuating functions' although we are dealing with precisely the same sets of possible worlds. I have argued elsewhere that this kind of situation is not only possible to envisage but is actually present in our own ways with perceptual contexts.[22] It would take us too far to show precisely what is involved in such cases. Suffice it to point out that this claim, if true, would strikingly demonstrate the dependence of our methods of cross-identification on our own conceptual schemes and hence on things of our own creation. The apparent simplicity of our idea of an 'ordinary' individual, safe as it may seem in its solid commonplace reality, is thus seen to be merely a reflection of the familiarity and relatively deep customary entrenchment of one particular method of cross-identification, which *sub specie aeternitatis* (i.e., *sub specie logicae*) nevertheless enjoys but a relative privilege as against a host of others.

The methods of cross-identification represented by the set **F**

---

[21] For temporal modalities, see, e.g., A. N. Prior, *Past, Present and Future* (Oxford: Clarendon Press, 1967). I am not saying that our actual methods of cross-identification in the case of temporal modalities (i.e., on ordinary methods of re-identification) turn on continuity quite as exclusively as I am about to suggest. It suffices for my purposes to present an example of methods of cross-identification that allows both 'branching' and 'merging', and it seems to me at least conceivably that temporal modalities might under suitable circumstances create such a situation.

[22] This is argued in "On the Logic of Perception" (note 20 above).

of 'individuating functions', as we might call them, also call for several further comments.

The main function of this part of our semantical apparatus is to make sense of quantification into contexts of propositional attitudes. The truth-conditions of statements in which this happens can be spelled out in terms of membership in F. As an approximation we can say the following: A sentence of the form $(\exists x)\, Q(x)$ is true in μ if and only if there is an individual constant (say b) associated with some $f \epsilon F$ such that $Q(b)$ is true in μ. This approximation shows, incidentally, how close we can stick to the simple-minded idea that an existentially quantified sentence is true if and only if it has a true substitution instance. The only additional requirement we need is that the substitution-value of the bound variable has to be of the right sort, to wit, has to specify the same individual in all the possible worlds we are speaking of in the existential sentence in question. This is what is meant by the requirement that b has to be associated with one of the functions $f \epsilon F$.

This approximation, although not unrepresentative of the general situation, requires certain modifications in order to work in all cases. The set F has to be relativized somewhat in the same way the unrestricted notion of a possible world was replaced by the notion of an alternative in the truth-criterion (S.B) above. (Not everyone is in all situations 'familiar with' all the relevant methods of individuation, it might be said.) I shall not discuss the ensuing complications here, however, for they do not change the overall picture in those respects which are relevant in the rest of this paper.

XII. STATEMENTS ABOUT DEFINITE INDIVIDUALS *VS.* STATEMENTS ABOUT WHOEVER OR WHATEVER IS REFERRED TO BY A TERM

The possibility of quantifying across an operator which expresses a propositional attitude enables us to explicate the logic of the locutions in which we need this possibility in the first place. Perhaps the most important thing we can do here is to make a distinction between propositional attitudes directed to whoever (whatever) happens to be referred to by a term and attitudes directed toward a certain individual, independently of how he happens to be referred to. This distinction was hinted at above. Now it is time to explain it more fully. For instance, someone may have a belief concerning the next Governor of California, whoever he is or may be, say that he will be a Democrat. This is different from believing something about the individual who, so far unbeknownst to all of us, in fact is the next Governor of California.

In formal terms, the distinction is illustrated by the pair of statements

$$B_a(g \text{ is a Democrat})$$

$$(\exists x)\,((x = g)\,\&\,B_a(x \text{ is a Democrat})).$$

Notice, incidentally, that my way of drawing this distinction implies that one can have (say) a belief concerning the individual who in fact is a only if such an individual actually exists, whereas one can in principle have a belief concerning a, 'whoever he is', even though there is no such person. This, of course, is just as it ought to be.

The naturalness of our semantical conditions, and their close relation to the realities of actual usage, can be illustrated by applying them to what I have called a statement about a definite individual. As an example, we can use

'a believes of the man who in fact is Mr. Smith that he is a thief',

in brief,

$$(\exists x)\,(x = \text{Smith} \,\&\, B_a(x \text{ is a thief})).$$

In order for this to be true, there has to be some $f \epsilon \mathbf{F}$ such that the value of f in the actual world (call it μ_0) exists and is Smith and that $f(\mu)$ has the property of being a thief whenever $\mu \epsilon \phi_B(a, \mu_0)$, i.e., in all the alternatives to the actual world.

What the requirement of the existence of f amounts to is clear enough. If it is true to say that a has a belief about *the particular individual* who in fact is Smith, then a clearly must believe that he can characterize this individual uniquely. In other words, he must have some way of referring to or characterizing this individual in such a way that one and the same individual is in fact so characterized in all the worlds compatible with what he believes. This is precisely what the existence of f amounts to. If no such function existed, a would not be able to pick out the individual who in fact is Smith under all the courses of events he believes possible, and there would not be any sense in saying that a's belief is *about* the particular individual in question.

XIII. INDIVIDUATING FUNCTIONS *vs.* INDIVIDUAL CONCEPTS

One important consequence of my approach is that not every function which from each μ picks out an individual can be said to specify a unique individual. In fact, many perfectly good free singular terms fail to do so in the context of many propositional attitudes. Even proper names fail to do so in epistemic contexts, for one may fail to know who the bearer of a given proper name is.

Such arbitrary functions may be important for many purposes. They are excellent approximations in our theory to the 'individual concepts' which many philosophers have postulated.[23] (In Section VI above we already met a number of such 'individual concepts' in the form of the functions $\phi(a, \mu)$ with a fixed a.) Each such individual concept specifies or 'contains', as Frege would say, not just a reference (in the actual world) but also the way in which this reference is given to us. Each of them would thus qualify for a sense (*Sinn*) of a singular term *à la* Frege.[24] However, we do not need the totality of such arbitrary functions in the semantics which I am building up and which

[23] Cf., e.g., R. Carnap, *Meaning and Necessity*, 2nd ed. (Chicago: Univ. of Chicago Press, 1956), pp. 41, 180–181, and sec. VI above.

[24] Cf. Gottlob Frege, "Ueber Sinn und Bedeutung," *Zeitschrift für Philosophie und philosophische Kritik*, 100 (1892), 25–50, especially p. 26, last few lines (English trans. by H. Feigl, "On Sense and Nominatum," included in this volume).

(I want to argue) is largely implicit in our native conceptual apparatus. Quine's criterion, however misleading it may be as a criterion of ontological commitment, still works as a criterion of commitment. If it is applied here, it shows that we are not committed (ontologically or 'ideologically') to these arbitrary functions, since we do not have to quantify over them, only over the members of the much narrower class **F**.

The other side of the coin is that in our semantical apparatus we do have to quantify over the members of **F**. Does it follow that they 'exist' or 'are part of our ontology'? An answer to this question can be given along the same lines as to the corresponding question concerning 'possible world'. The members of **F** are not members of any possible world; they are not part of anybody's count of 'what there is'. They may 'subsist' or perhaps 'exist', and they are certainly 'objective', but they do not have any ontological role to play. The need to distinguish between ontology and 'ideology' is especially patent here.

The functions that belong to **F** may of course be considered special cases of the 'individual concepts' postulated by some philosophers of logic or as special cases of Frege's 'senses' (*Sinne*). No identification is possible between the two classes, however, for we saw earlier that not every arbitrary singular term (say b) which picks out an individual from each $l(\mu)$ we are considering goes together with an $f \epsilon \mathbf{F}$, although every such term is certainly meaningful and hence has a Fregean 'sense' and perhaps even gives us an 'individual concept'. As I have put it elsewhere, members of **F** do not only involve a 'way of being given' as Frege's senses do, but also *a way of being individuated.*[25] The primary care is in our approach devoted to ordinary concrete individuals. Singular terms merit a special honorary mention only if they succeed in picking out a unique individual of this sort.

Let us say that an $f \epsilon \mathbf{F}$ is (gives us) an individuating concept, and let us say that a term b does *individuate* (in the context of discussing a's beliefs) in so far as

[25] Cf. "On the Logic of Perception" (note 20 above).

$$(\exists x)\ B_a(x = b)$$

is true. Then we could have individuation without reference and reference without individuation: Both

$$(\exists x)\ B_a(x = b)\ \&\sim (\exists x)\,(x = b)$$

and

$$(\exists x)\,(x = b)\ \&\sim (\exists x)\ B_a(x = b)$$

can be true. We could even have both, but without matching:

$$(\exists x)\,(x = b)\,\&(\exists x)\ B_a(x = b)\,\&\sim (\exists x)\,((x = b)\,\&B_a(x = b))$$

is satisfiable. Only if

$$(\exists x)\,((x = b)\,\&B_a(x = b))$$

is true does the successful individuation give us the individual which the term b actually refers to.

XIV. THE THEORY OF REFERENCE AS REPLACING THE THEORY OF MEANING

Here we are perhaps beginning to see what I meant when I said in the beginning of this paper that what is often called the theory of meaning is better thought of as the theory of reference for certain more complicated conceptual situations. Some of the most typical concepts used in the theory of meaning, such as Frege's *Sinne* and the 'individual concepts' of certain other philosophers of logic, were in the first place introduced to account for such puzzles as the failure of the substitutivity of identity and the difficulty of quantifying into opaque contexts (e.g., into a context governed by a word for a propositional attitude). I have argued, however, that a satisfactory semantical theory which clears up these puzzles can be built up without using Frege's *Sinne* and without any commitment to individual concepts in any ordinary sense of the word. Instead, what we need are the individuating functions, i.e., the members of **F**. And what these functions do is not connected with the ideas of the traditional theory of meaning. What they do is precisely to give us the individuals which we naively think our singular statements to be about and which we think our singular terms as referring to. This naive point of view is essentially correct, it seems to me. The functions of $f\epsilon F$ are the prime vehicles of our references to individuals when we discuss propositional attitudes. What is not always realized, however, is how much goes into our ordinary concepts of an individual and a reference. These are not specified in a way which works only under one particular course of events. They are in fact specified in a way which works under a wide variety of possible courses of events. But, in order to spell out this idea, we are led to consider several possible worlds, with all the problems with which we have dealt in this paper, including very prominently the problem of cross-identifying individuals.

The function of our 'individuating functions', i.e., the members of the set **F**, is to bring out these hidden—or perhaps merely overlooked—aspects of our concept of an individual (definite individual). This close connection between the set **F** and the concept of an individual appears in a variety of ways. One may for instance think of the role which the membership in **F** plays in the truth-conditions which we set up above for quantification into modal contexts. When it is asked in such a context whether there exists an individual of a certain kind, a singular term specifies such an individual only if its references match the values of a unique member of **F** in all the relevant possible worlds. Thus it is these functions that in effect give us the individuals which can serve as values of bound variables. As we saw above, it is mainly the possible subtlety and multiplicity of relations of cross-identity that prevent us from simply making the domains of the different possible worlds partly identical and thus hypostatizing my individuating functions into commonplace individuals.

This connection between individuating functions and the concept of an individual is part of what justifies us in thinking that in the traditional dichotomy their theory would belong primarily to the theory of reference, in spite of the fact that their main function in our semantical theory is to solve some of the very problems which the traditional theory of meaning was calculated to handle. This role is perhaps especially clear in connection with the substitutivity of identity. As we have seen, this principle does not hold for arbitrary singular terms a, b. However, if it is required in addition that both of these terms specify a well-defined individual, i.e., satisfy expressions like ($*$), depending on the context, then the substitutivity of identicals is easily seen to hold, presupposing of course here the prohibition against merging that was mentioned above. What this observation shows is clear enough. The failure of the substitutivity of identity poses one of the most typical problems for the treatment of which meanings, individual concepts, and other paraphernalia of the theory of meaning were introduced in the first place. If the substitutivity of identity fails, clearly we cannot be dealing with ordinary commonplace individuals, it was alleged, for if two such individuals are in fact identical, surely precisely same things can be said of them. This is what prompts the quest for individuals of some nonordinary sort, capable of restoring the substitutivity principle when used as references of our terms. (This is almost precisely Frege's strategy.) We have seen, however, that the (apparent) failure of the substitutivity is due simply to the failure of some free singular terms to specify the same individual in the different 'possible worlds' we have to consider. Moreover, we have seen that this apparent failure is automatically corrected in precisely those cases in which it ought to be corrected, viz. in the cases where the two terms in question really do specify a *unique* individual. (That this depends on certain specific requirements concerning our methods of cross-identification, viz. on a prohibition against 'splitting', does not affect my point.) Substitutivity of identity is restored, in brief, not by requiring that our singular terms refer to the entities postulated by the so-called theory of meaning, but by requiring (in the form of an explicit premise) that they really succeed in specifying uniquely the kind of ordinary individual with which the theory of reference typically deals. One can scarcely hope to find a more striking example of the breakdown of the distinction between a theory of meaning and a theory of reference.[26]

XV. TOWARD A SEMANTIC NEOKANTIANISM

The aspect of my observations most likely to upset many contemporary philosophers is the ensuing implicit dependence of our concept of an individual on our ways of cross-identifying members of different 'possible worlds'. These 'possible worlds' and the supply of individuating functions which serve to interrelate their respective members may enjoy, and in my view do enjoy, some sort of objective reality. However, their existence is not a 'natural' thing. They may be as solidly objective as houses or books, but they are as certainly as these created by men (however unwittingly) for the purpose of facilitating their transactions with the reality they have to face. Hence my reasoning ends on a distinctly Kantian note. Whatever we say of the world is permeated throughout with concepts of our own making. Even such *prima facie* transparently simple notions as that of an individual turn out to depend on conceptual assumptions dealing with different possible states of affairs. As far as our thinking is concerned, reality cannot be in principle wholly disentangled from our concepts. A *Ding an sich*, which could be described or even as much as individuated without relying on some particular conceptual framework, is bound to remain an illusion.

[26] Views closely resembling some of those which I am putting forward here (and in some cases anticipating them) have been expressed by David Kaplan, Richard Montague, Dagfinn Føllesdal, Stig Kanger, Saul Kripke, and others. Here I am not trying to relate my own ideas to theirs. It is only fair, however, to emphasize my direct and indirect debts to these writers.

supplement

The Rosenthal-Sellars Correspondence on Intentionality

DAVID M. ROSENTHAL AND
WILFRID SELLARS

Professor Wilfrid Sellars
Department of Philosophy
University of Pittsburgh
Pittsburgh, Pennsylvania

July 5, 1965

Dear Professor Sellars:

In response to your kind offer to read through portions of the typescript of my thesis pertaining to your views on intentionality, I am sending you a copy of an introductory section to such a chapter.[1] The enclosed typescript represent a first draft, for which I apologize, but I thought it might be useful to get any comments you might have in at the ground floor, so to speak.

The section I am sending leaves off, as you will see, just at the point at which I intend to take up the question of the reducibility (or not) of semantical discourse to psychological discourse. For this reason, I have not yet discussed the problems involved with the 'means'-rubric, nor those examined in

This correspondence is published here for the first time with the kind permission of David M. Rosenthal and Wilfrid Sellars.

[1] Since these letters were written my ideas have crystallized, and I have expanded my earlier views on the disagreement between Sellars and Chisholm. See my Ph.D. dissertation, "Intentionality: A Study of the Views of Chisholm and Sellars" (Princeton University, 1968), ch. IV. [D.M.R.]

"Some Reflections on Language Games."[2] Aside from these omissions, however, I wrote this section by way of attempting to present a general account of your position respecting mental discourse (leaving out, of course, language which exhibits what you call, in "Being and Being Known,"[3] pseudo-intentionality).

A point at which I was conscious of not understanding your views as well as I would like to involves the nature of the so-called reporting role which uses of sentences in the language of Jones' theory come to have. I have gone through, in particular, the letter of yours to Castañeda (p. 4 ff.)[4] in which you discuss this, and the passages you refer to, but I find myself unclear on what, for the language of thoughts, is meant by your statement that "the conditioning is itself caught up in a conceptual framework" (p. 6). I would be immensely grateful for any help you could give me with this.

Sincerely yours,
David Rosenthal

Mr. David Rosenthal
82 Graduate College
Princeton, New Jersey

September 3, 1965

Dear Rosenthal,

I must apologize for the delay in responding to your letter and chapter. I moved from the Princeton frying pan to the Pittsburgh fire. The Spring (!) term is now over and I am caught up and ready to move ahead. I read the Ms through when it first arrived, and have just gone through it again. It

---

[2] *Philosophy of Science*, 21 (1954), 204–228. Reprinted with revisions in W. Sellars, *Science, Perception, and Reality* (London: Routledge and Kegan Paul, 1964).

[3] *Proceedings of The American Catholic Philosophical Association*, 34 (1960), 28–49. Reprinted in *Science, Perception, and Reality*.

[4] This and the following page reference are to an unpublished correspondence between W. Sellars and H-N. Castañeda. See the Bibliography, p. 512, entry [125].

strikes me as remarkably lucid and to the point. I think you give an excellent account of what I was up to in "Empiricism and the Philosophy of Mind."[5] There is nothing which sets my teeth on edge. I shall limit myself to a few comments which relate what you have said to the larger setting of the problem.

My argument requires that the term 'psychological' is ambiguous in a way which, if not taken into account, can muddy the waters.

(a) Pre-Jonesean psychological statements employ Rylean resources enriched by semantical devices and categories. They *neither* coincide with behavior-plus-dispositions-to-behave statements nor reach beyond them to inner episodes proper. Pre-Jonesean psychological statements are, of course, entailed by statements characterizing someone as speaking nonparrotingly. But that is simply because to say of an utterance that it is nonparroting is to say that it is, for example, a tokening of a certain proposition—e.g., the proposition that two plus two equals four. If the utterance is a ·two plus two equals four· or a ·I shall raise my hand· or a ·Lo! this is green· then it is, respectively, a *thinking-out-loud* that two plus two equals four, an *intending-out-loud* that he himself will raise his hand, or a *taking-out-loud* of something to be green. The relation of

Utterance U_1 is a ·two plus two equals four·

to

Utterance U_1, being a 'zwei und zwei gleicht vier', means *two plus two equals four*

and

'zwei und zwei gleicht vier's are ·two plus two equals four·s

was explored in "Abstract Entities" and "Notes on Intentionality."[6]

---

[5] (Hereafter abbreviated as EPM). In H. Feigl and M. Scriven, eds., *Minnesota Studies in the Philosophy of Science*, vol. I (Minneapolis: Univ. of Minnesota Press, 1956). Reprinted in *Science, Perception, and Reality*. The part of this essay to which references are made in this correspondence is included in this volume. Parenthetical references will be to this volume.

[6] "Abstract Entities," *Review of Metaphysics*, 16 (1963), 627–671; "Notes on Intentionality" is included in this volume.

(b) Inner episode psychological statements. When I have denied that

'. . .' means ———

is to be analyzed in terms of psychological statements, I have had primarily in mind the claim that it is to be analyzed in terms of inner episode psychological statements. Notice, however, that even in the case of the pre-Jonesean use of 'means' the above rubric is not to be *analyzed* in (pre-Jonesean) psychological terms. It is, rather, constitutive of the latter. For, roughly, to be a pre-Jonesean psychological event is to *mean something;* where to mean something is to be a ·two plus two equals four·, etc. [where I am taking the liberty of using pre-Jonesean English in employing the dot-quoting device].

The above considerations are, I believe, relevant to your discussion on pp. 25–26.[7] It also indicates my strategy with respect to Chisholm's claim which you summarize on p. 6.

As for the question which you ask in your letter: "What . . . is meant by . . . 'the conditioning is itself caught up in a conceptual framework'?"—the point I had in mind was intended to be the exact counterpart of the additional condition I lay down for "Lo! this is green" to count as observational knowledge, and which you correctly describe on p. 20.

One small point. On p. 21 you suggest that my view requires that pre-Joneseans do not think 'in the full sense.' I do not see this. I would rather say that since they do not have the concept of thought in the full sense, the *scope* of their thinking is substantially restricted.

Finally, it is 'I think . . .' rather than 'I am thinking . . .' which I count as a dispositional cousin of 'I believe . . .'. 'I am thinking . . .' is a variegated locution which implies a sequence of episodes (a process?), and can also be used to refer to processes-cum-dispositions ("I am thinking much, these days, about Vietnam").

I enjoyed reading the pages you sent me, and promise to be

---

[7] This and the following three page references are to the draft of Mr. Rosenthal's paper.

more prompt in responding to any further material you might wish to send.

Cordially,
Wilfrid Sellars

October 2, 1965

Dear Professor Sellars:

Thank you very much for your letter of September 3. I found your comments very helpful and encouraging, and was delighted to learn that you felt I had given an adequate account of your views. I had intended, by this time, to have written the sequel to that paper, dealing with the points of controversy between you and Chisholm which I had mentioned. The comments you made in your letter, however, led me to formulate some new ideas on that subject. Because I am not confident that I have not misunderstood you, I thought I might first write you to ask whether you think that the following rough line of argument does justice to what you had in mind.

You write that your "argument requires that the term 'psychological' is ambiguous" in a way which you then trace out. This fact indicates, you write, your "strategy with respect to Chisholm's claim" about the analysis of 'means'-statements in terms of psychological statements. If I am correct at the end of my paper, then, this would indicate your strategy in dealing with the problem of showing that sentences like ' "Spectre" means ghost' and 'John's utterance was meaningful (in the sense of nonparrotingly uttered)' do not mean the same as certain psychological sentences.

As I understand your views, your construction of psychological discourse from certain linguistic resources revolves around the claim that thoughts (and other psychological items) may be viewed as exhibiting two aspects, viz., their intentionality and their determinate factual character. It is in terms of these two aspects which, as I understand the broad line of your argument, you hope to account for the various properties which we believe psychological items to have, and, as you say in your March 11, 1962, letter to Castañeda, "save the appearances."

In your letter, you contrasted pre-Jonesean psychological statements with what you called inner episode psychological statements. It occurred to me that it might be possible to understand this contrast in terms, roughly, of the contrast between the intentionality and the determinate factual character of psychological items. Let me try to sketch what I have in mind in a bit more detail.

You write that the 'means'-rubric is constitutive of pre-Jonesean psychological terms, and for this reason it is fruitless to attempt to analyze 'means'-statements in terms of pre-Jonesean psychological statements. You also write that "statements characterizing someone as speaking nonparrotingly entail pre-Jonesean psychological statements," and that this amounts to saying no more than that a nonparrotingly uttered utterance tokens a certain proposition (or something of this sort). This suggests to me that you may wish to assert that 'means'-statements are no more nor less than *pre-Jonesean* psychological statements.

This assertion seems on the face of it, however, to be mistaken. For while 'means'-statements are about more than simply the "characteristics which marks and noises can have *as marks and noises*" (Chisholm-Sellers Correspondence, p. 223),[8] they are also about certain marks and noises *as marks and noises.* Pre-Jonesean psychological statements, I would assume, are not.

As we may with psychological items, we may assert of semantic items that they exhibit two aspects, their intentionality and their (linguistic) determinate factual character. We might then distinguish two senses of ' "means"-statement' as follows: (1) 'statement which is about intentional items (no matter what their determinate factual character is)' and (2) 'statement which is about items which exhibit intentionality and a certain (linguistic) determinate factual character'. Would it be possible, then, to re-express what you have in mind by saying

[8] Included in this volume. Hereafter abbreviated as CSC.

that pre-Jonesean psychological statements are no more nor less than 'means'-statements (in sense (1))?

This sort of account suggested to me that we might be able to describe the ambiguity of 'psychological' which you discuss as follows. Pre-Jonesean psychological statements are simply statements about intentional items; we may call them psychological because psychological items are intentional. Post-Jonesean (presentday) psychological statements, on the other hand, are psychological because they are about items which exhibit both intentionality and the sort of determinate factual character (whatever it may be) that psychological items exhibit.

One might summarize these points as follows. 'Psychological statement' can mean the same as (1) 'statement about intentional items (no matter what their determinate factual character)' or the same as (2) 'statement about items which exhibit both intentionality and a certain (psychological) determinate factual character'. It is clear that psychological statements (in sense (2)) "reach beyond," to use your words, "behavior-plus-disposition-to-behave statements . . . to inner episodes proper." We might then, if I understand your views correctly, go on to claim the following. It is a sufficient condition for a statement to be about items with a certain (psychological) determinate factual character that it be about inner episodes proper. (Would it, as I believe, be a necessary condition as well?) So we can characterize what you call inner episode psychological statements as being psychological statements (in sense (2)).

(At this point I am unclear concerning whether you may, by 'inner episode psychological statement', have meant 'statement about inner episodes (no matter whether intentional or not)' rather than 'psychological statement (in sense (2))'. Although I shall assume that you meant the former, I believe that what follows can be suitably altered if, in fact, you did mean the latter.)

Given this account, it is clear that 'means'-statements (in sense (1)) *are* psychological statements (in sense (1)). But it also seems clear that neither 'means'-statements (in sense (1)) nor 'means'-statements (in sense (2)) either are, or are to be analyzed in terms of, psychological statements (in sense (2)). If this is correct so far, then I am, I believe, in a position to compare this account with that advanced by Chisholm.

Let us take Chisholm's claim as being that 'means'-statements may be (or are to be) analyzed in terms of, or entail, psychological statements. But Chisholm does not draw (nor, it seems, does he recognize) a distinction between our two senses of 'psychological statement' nor between our two senses of ' "means"-statement'. So it is necessary to see if Chisholm's claim can be made more precise. (It may be easier if, in what follows, I use 'P1 statement' as an abbreviation for 'psychological statement (in sense (1))', 'M1 statement' for ' "means"-statement (in sense (1))', etc.)

It is clear that by ' "means"-statement' Chisholm means the same as either 'M1 statement' or 'M2 statement.' For the alternative is that he is talking about marks and noises *as marks and noises*. But it is just this that he appears to wish to deny by saying that "linguistic entities (sentences, etc.) are also intentional" (CSC, p. 239, sentence (C-2)). Similarly, by 'psychological statement' Chisholm must mean the same as either 'P1 statement' or 'P2 statement'. For he denies that he is talking about merely inner episodes (no matter whether intentional or not) when he says that "thoughts (i.e., beliefs, desires, etc.) are intentional . . ." (CSC, p. 239, sentence (C-1)). Since, further, we may assume that he is talking about psychological statements and 'means'-statements in the ordinary, everyday (and, therefore, post-Jonesean) sense, it is reasonable to suppose that he is claiming that M2 statements entail P2 statements.

Both M2 statements and P2 statements are about items which exhibit a certain determinate factual character. M2 statements, however, are about marks and noises which exhibit certain characteristic patterns, while P2 statements are about inner episodes of a certain sort. From these considerations, it appears that neither M2 statements entail P2 statements (as we supposed that Chisholm was claiming) nor do P2 statements entail M2 statements.

We might, however, wish to assert that P2 statements do entail P1 statements (conjoined with certain statements about inner episodes), just as we might assert that M2 statements entail M1 statements (conjoined with certain statements about marks and noises *as marks and noises*). We would then be in a position to claim that M2 statements entail P1 statements (for P1 statements *are* M1 statements). But since P1 statements do not, in turn, entail P2 statements, we are still without warrant in supposing that M2 statements entail P2 statements.

If we were to understand Chisholm to be claiming that M2 statements entail P1 statements, then the account sketched above would lend support to his claim. The reason that we had for believing that Chisholm was talking about P2 statements was our supposition that he was talking about ordinary, present-day (and, therefore, post-Jonesean) psychological statements. Such statements could not, on our account, be P1 statements. We might, therefore, diagnose Chisholm's error as being the result of his belief that (at least in certain respects) presentday psychological statements behave as if they were (prc-Jonesean) P1 statements.

This diagnosis seems to gain some plausibility from the following considerations. In CSC (p. 239), Chisholm characterizes thoughts as being intentional items and a "source of intentionality." He does not, so far as I can discover, speak of thoughts as the sort of item which has a certain determinate factual character; nor does he speak of thoughts as inner episodes. It appears that Chisholm may be using 'psychological statement' to mean, in effect, the same as 'statement about intentional items without any determinate factual character'.

I have suggested, roughly, that Chisholm is so set on stressing the intentionality of thoughts that he ignores their determinate factual character as inner episodes. Can Chisholm, in responding to this suggestion, reasonably claim that ordinary psychological statements *are* P1 statements? An argument that he cannot might run as follows. Our ordinary ideas about thoughts involve that they are inner episodes which are (although we may not know how) causally connected to other episodes and to dispositions to behave. (They are, for example, causally connected to overt, nonverbal behavior, to certain forms of sensory stimulation, to dispositions to utter 'I have a thought that-p', etc.) Therefore if a statement is about thoughts *as thoughts*, then it is about thoughts *as inner episodes*.

How might Chisholm respond to this line of argument? If we accept that 'John thinks that-p' is a psychological statement, then it seems that Chisholm would have to deny that it is about an inner episode *as inner episode* (and that it entails a statement which is). This denial might open the door, then, for a view which looked upon the statement as expressing a relation between John and an intentional item (namely, the thought that-p), or as expressing an intentional relation between John and whatever the proposition that-p is about. (Such a view might be suggested by Chisholm's formulation of the problem of intentionality in the abstract of his paper for the symposium at West Virginia last May.)[9]

In any event, it seems that Chisholm would be committed to looking on the statement 'John has a thought that-p' as if it were about "a thought in Frege's sense—an 'abstract entity' rather than a mental episode" ("Notes on Intentionality," p. 322). And if I understand your distinction in that paper, looking upon such a statement as being about a thought in Frege's sense alone would be no more nor less than looking upon it as being a P1 statement. (Similarly, an M1 statement might be "Spectre"'s are the same as 'ghost's', or would this be a mistake?)

I know that this account, as it stands, would at the very least need to be both tidied up and filled out in point of detail. It represents, however, an account which I thought you might have in mind on the basis of reading your remarks on the ambiguity of the term 'psychological'. I would be immensely grateful to learn if you think that I am on the right track, and for any suggestions and comments you might have.

[9] Abstract of "The Problem of Intentionality," presented at a symposium entitled "The Problem of Intentionality" with papers by R. Chisholm and W. Sellars, held at West Virginia University, May 13, 1965.

There are two other points in your letter on which I am not clear and about which I thought I might write you. One is the question that I asked in my earlier letter about what you meant by "the conditioning is itself caught up in a conceptual framework." If I understand your account, it runs as follows. If (a) 'I have a thought that-p' expresses direct self-knowledge, then both (b) the uttering was the result of a conditioned response and (c) "the conditioning is itself caught up in a conceptual framework" (Castañeda Correspondence, p. 6). But if (c), then (d) the person who utters 'I have a thought that-p' must recognize the speech act as a report (EPM, p. 298 f.). Finally if (d), then (e) the speaker recognizes that such reports are reliable symptoms of the person who reports having the thought that-p (EPM, p. 298). If this is correct so far, then my question is whether (e) is a sufficient condition for (c), and whether (c) and (b) are jointly sufficient for (a). Put differently, is any more involved in what you call meta-thinking (Castañeda Correspondence, p. 17) than conditions (b) and (e)?

It was on the basis of this same page in your correspondence with Castañeda that I took your view to require "that pre-Joneseans do not think 'in the full sense'." What I believed that you had in mind might be suggested by saying that 'John thinks that-p, but he is never able to be aware of what he is thinking' is incoherent. You write that pre-Joneseans "do not have the concept of thought in the full sense," and that for this reason "the *scope* of their thinking is substantially restricted." Is what you meant by this (and by the passage at the top of the page 17 of your correspondence with Castañeda) no more than that pre-Joneseans do not think about their own thoughts (or, for that matter, about others' thoughts), and for this reason they "do not have the concept of thought in the full sense"? If so, might one then say that the respect in which they lack the concept of thought in the full sense is that whatever they think of as thoughts, they do not think of them as inner episodes? (That is, that their concept of thought is no more nor less than a concept of intentional items.)

Let me once again thank you for having read my earlier typescript, and for your very helpful comments on it. Let me also apologize for the undue length of this letter; I hope that the ideas I have sketched reflect, to some extent, what you had in mind in your letter, and I shall be immensely grateful for any help you can give me concerning respects in which they do not.

Sincerely yours
David Rosenthal

Dear Mr. Rosenthal,

Thank you very much for your clear and searching letter. You are, indeed, on the right track, though I have some reservations, as you will see, about the exact way the points are to be made.

1. To begin with, all pre-Jonesean statements are (by definition) Rylean in character—at least as far as their descriptive content is concerned. The same will be true *a fortiori* of pre-Jonesean statements. Thus, though we may allow pre-Joneseans to conceive of intentional items in a way which makes abstraction from 'determinate factual character', we may not allow them to conceive of intentional items, as *we* can, in a way which makes abstraction from their Rylean character. Thus, their concept of an explanation of behavior in terms of intentional items coincides with their concept of an explanation in terms of what *we* would call Rylean intentional items, i.e., in terms (roughly) of what people say or are disposed to say. Thus, even though these items are specified by them in a way which, as in indirect discourse, abstracts from the determinate linguistic materials (including gestures, etc.) in which the relevant linguistic roles are embodied, they are not specified by them in a way which makes abstraction from their Rylean character.

2. Thus the contrast between explanation in terms of determinate Rylean role-players (thus, by quoting) and explanation in terms of Rylean items which are specified in terms of the role which they play, but *not* in terms of determinate lin-

guistic materials, is as close as they could come to *our* contrast between explanation in terms of Rylean role-players, whether or not their determinate factual character is specified, and explanation in terms of Ockhamite role-players (inner episodes).

3. Thus, pre-Jonesean psychological explanations are explanations in terms of what I have called 'thoughts-out-loud' and long- or short-term dispositions to have thoughts-out-loud. Of course, if we were to follow ordinary usage and restrict the phrase 'psychological explanation' to explanation in terms of thoughts (inner episodes) and dispositions to have them, then we would not speak of pre-Jonesean explanations of behavior as 'psychological', but rather as 'linguistic' or 'symbol-behavioral'. But if the argument of EPM is sound, there is every reason to extend the term 'psychological' to cover pre-Jonesean explanations—provided the necessary distinctions are drawn.

4. Thus, when you suggest (p. 466, par. 3) that pre-Jonesean psychological statements are not (in addition to being about items *qua* intentional) about marks and noises *as marks and noises*, I am unhappy. For though they do not require their subject matter to be marks or noises[10] (they may be gestures), let alone specific kinds of marks or noises, their subject matter must, in order for them to qualify as pre-Jonesean psychological statements, be conceived in Rylean terms.

5. Now since *we* can distinguish between explanation in terms of Rylean intentional items and explanations in terms of Ockhamite intentional items, it is possible for us to draw a distinction between at least three senses of 'psychological explanation'.

(1) Explanation in terms of Rylean intentional items: PR
(2) Explanation in terms of Ockhamite intentional times: PO
(3) Explanation in terms of intentional items, Rylean or Ockhamite: PRO

Of these three, the third (PRO) corresponds *roughly* to your P1, the second (PO) to your P2. Thus (on p. 467) you charac-

---

[10] I would rather say mark-makings and utterances, for marks and noises are psychological only as actually or potentially related to the linguistic activity of persons.

terize a P1 statement as a "statement about intentional items (no matter what their determinate factual character)" and a P2 statement as a "statement about items which exhibit both intentionality and a certain (psychological) determinate factual character," by which latter expression you mean their inner-episode-ness or Ockhamicity.

6. Thus, when we speak of a Jonesean statement as psychological, we can mean either that it is PR or that it is PRO. As for 'means'-statements, philosophers may stretch this expression by using the 'means'-rubric not only to give the sense of Rylean linguistic expressions (its normal use) but also the sense of Ockhamite linguistic expressions (inner speech). If we do so stretch it, then we have three senses of 'means'-statement.

(1) 'Means'-statement about Rylean items: MR
(2) 'Means'-statement about Ockhamite items: MO
(3) 'Means'-statement about either Rylean or Ockhamite items: MRO

The first of these (MR) is roughly your M2. It is not quite so easy to pick out a counterpart to your M1.

7. Now MR statements are the core of pre-Jonesean psychological statements in the sense that pre-Joneseans explain what people do in terms of the rubrics,

> x thought-out-loud that-p

> x was disposed to think-out-loud that-p

where

> x thought-out-loud that-p

has the sense of

> x (nonparrotingly) uttered a ·p·

which, in turn, has the sense of

> x (nonparrotingly) uttered something which means p

given that

> '___' means ———

has the sense of

'——'s are ·————·s

for to classify an utterance as a ·————· is to classify it in terms of its role in a language-life game, i.e., in terms of its place in a system of language entry transitions, language departure transitions and intralinguistic moves.

8. Thus, there is a perfectly legitimate sense in which MR statements are psychological statements. They are PR statements.

9. Again, MO statements are PO statements. To say of a thought that it means that-p is to classify it as an inner episode which is doing a job analogous in important respects to that which thinkings-out-loud that-p do in behavior built around spontaneous Rylean intentional episodes. Such inner episodes, furthermore, are conceived to be links in the explanation of, among other things, the occurrence of thinkings-out-loud that-p and of short term propensities to think-out-loud that-p.

10. Now it is clear from the above that an MR statement is not a PO statement, nor is it to be analyzed in terms of PO statements. Thus when you write (p. 468, par. 4) "neither 'means'-statements (in sense 1) nor 'means'-statements (in sense 2) either are, or are to be analyzed in terms of, psychological statements in sense 2," I agree with this in so far as it concerns 'means'-statements (in sense 2). As for 'means'-statements in sense 1 the situation is less clear. For the class of 'means'-statements about either Rylean or Ockhamite items includes the class of 'means'-statements about Ockhamite items and these *are* psychological statements in your sense 2. They are "about items which exhibit both intentionality and the sort of determinate factual character (whatever it may be) that psychological items exhibit." Here, I take it, by the expression "the sort of determinate factual character . . ." you have in mind the determinate inner-episode-ness, whatever it may be, by virtue of which it does its job.

11. Thus, I would prefer to say that from the fact that a statement is a 'means'-statement in your sense 1 it *does not*

follow that it either is or is to be analyzed in terms of a psychological statement in your sense 2. For the class of 'means'-statements about either Rylean or Ockhamite items *also* includes the class of 'means'-statements about Rylean items, and these, as we saw, neither are nor are to be analyzed in terms of Ockhamite psychological statements—my PO statements, your P2 statements.

12. You now go on to explore what Chisholm's claim might be, against the background of the sort of distinctions we have been drawing. You give it an initial formulation as

> 'means'-statements may be (or are to be) analyzed in terms of, or entail, psychological statements.

You continue by pointing out that Chisholm does not draw the sort of distinctions we have drawn, and you raise the question whether Chisholm's claim can be made more precise by using these distinctions.

13. You then write:

> It is clear that by ' "means"-statement' Chisholm means the same as either 'M1 statement' or 'M2 statement'. For the alternative is that he is talking about marks and noises *as marks and noises*. But it is just this that he appears to wish to deny by saying that "linguistic entities (sentences, etc.) are also intentional" (CSC, p. 239, (C-2)).

Clearly, for Chisholm a 'means'-statement does not simply characterize marks and noises in terms of mark or noise characteristics (e.g., pitch, shape, color, intensity, etc.). On the other hand he surely thinks that the *subject* of a 'means'-statement is always a linguistic item in the literal sense (as contrasted with the analogical sense in which Ockhamite items are 'linguistic'). Is not, for Chisholm, a subject of a 'means'-statement always a mark or a noise or a gesture (or class of such?) and, hence, Rylean in character? I conclude, provisionally, that a 'means'-statement, for Chisholm, is always about a Rylean item.

15. If so, then according to my account of the 'means'-rubric, the role attributed to a Rylean item by a 'means'-statement must be a Rylean role, thus

'Himmel's (in German) are ·sky·s

uses a Rylean 'sky' in our language to characterize the job Rylean 'Himmel's do in German.

16. Now the Rylean 'means'-statement

'Himmel' (in German) means *sky*

has as its 'relational' counterpart (see "Notes on Intentionality")

'Himmel' (in German) stands for the concept sky.

If this is put as

'Himmel' (in German) expresses the concept sky

there is danger of confusing the sense of 'express' in this context with the relation of *manifesting*, and 'concept of . . .' in this context with 'ability to have thought-episodes about. . . .' The result would be to confuse

'Himmel' (in German) stands for the concept sky

with

'Himmel' (in German) manifests an ability to think about the sky.

This confusion would infect the original 'means'-statement, generating the idea that

'Himmel' (in German) means sky

is to be analyzed as something like

German sentences containing 'Himmel' give expression to thought episodes about the sky.

17. You write that

Similarly, by 'psychological statement' Chisholm must mean the same as either 'P1 statement' or 'P2 statement'. For he denies that he is talking about merely inner episodes (no matter whether intentional or not) when he says that "thoughts (i.e., beliefs, desires, etc.) are intentional" (CSC, p. 239, sentence (C-1)).

My guess is that Chisholm so uses 'psychological' that it includes items which I would deny to be intentional (e.g., sense
impressions, tickles, pains, etc.)—cf. my discussion of the
pseudo-intentionality of sense impressions in EPM. Unless I
am very much mistaken, he thinks that they have intentionality, or confuses a broader sense of 'intentionality' in which non-
conceptual items can have intentionality, with a narrrower
sense in which they cannot. But this is an aside. In any event,
I agree with you that "it is reasonable to suppose that he is
claiming that M2 statements entail P2 statements."

18. In other words, he is claiming that 'means'-statements
about Rylean items entail psychological statements about
Ockhamite inner episodes. In my terminology, he is claiming
that MR statements entail PO statements.

19. The only sense in which I am prepared to grant that MR
statements entail PO statements is that in which statements
about bricks entail statements about molecules.

20. You proceed (p. 469) to raise the possibility that Chisholm is to be understood as "claiming that M2 statements entail P1 statements" (i.e., that statements about linguistic intentional items entail statements about intentional items,
which do not specify that they are linguistic). If this were his
thesis it would, of course, be correct. You find that

> this diagnosis seems to gain some plausibility from the following
> considerations. . . . He does not, so far as I can discover, speak
> of thoughts as the sort of item which has a certain determinate
> factual character, nor does he speak of thoughts as inner epi-
> sodes. It appears that Chisholm may be using 'psychological
> statement' to mean, in effect, the same as 'statement about in-
> tentional items without any determinate factual character.'

I take it that by "without any determinate factual character"
you mean *without specifying their determinate factual character*. If so, then Chisholm would be holding that a statement
which made no reference to inner episodes, or dispositions pertaining to such, might properly be called a psychological statement. You suggest that he might have, at the back of his mind,

the Fregean sense of 'thought' (*Gedanke*) in which this term refers not to inner episodes, but to certain abstract entities, so that to say that a piece of verbal behavior expresses a *thought* would not be to say that it is related to inner episodes, but rather to say that it stands in a direct relation to a nonmental, public, abstract entity, the sort of thing that Platonists call a proposition.

21. But the above can be interpreted either as a tough- or as a tender-minded thesis. The tough thesis would be that there are no thoughts in the sense of non-Rylean inner episodes. This would amount to a philosophical behaviorism which differs from the usual form only by postulating an irreducible relation of 'expressing' to obtain between meaningful verbal behavior and Fregean abstract entities. The tender-minded version would be that both inner episodes and verbal behavior can 'express' Fregean abstract entities.

22. The tough-minded view would hold that 'means'-statements imply psychological statements because statements about what marks and sounds mean imply statements about meaningful verbal behavior, which latter are psychological not because they relate verbal behavior to thoughts as inner episodes, but because they relate verbal behavior to Fregean abstract entities. Clearly, this tough-minded view is simply the pre-Jonesean outlook transformed into a self-conscious philosophy which misconstrues meaning as a relation.

23. Needless to say, I find it difficult to believe that Chisholm is a Fregean pre-Jonesean. But it would be interesting to press him on this point. As for the tender-minded version, his thesis could approach mine if it amounted to the claim that 'means'-statements about marks and sounds are to be analyzed in terms of psychological statements in the sense of Ryle-Frege statements about verbal behavior and dispositions to behave; and also *imply* (by virtue of the connection of verbal behavior and dispositions to behave with thoughts *in the sense of inner episodes*), psychological statements in the more ordinary sense.

24. My own conviction is that if confronted with these dis-

tinctions, Chisholm would claim that verbal behavior gets its
relation to Fregean entities *indirectly* by virtue of being the
manifestation of inner episodes which alone are *directly* related to these entities, and that he would correspondingly claim
that verbal behavior and dispositions to behave are psychological only by virtue of the fact that they manifest such inner
episodes. Thus, as I see it, he is claiming that what I have called
PR statements are to be analyzed in terms of PO statements,
and hence that MR statements are to be analyzed in terms of
PO statements. In other words, he is claiming that M2 statements are to be analyzed in terms of P2 statements, which is
the thesis we have already rejected.

25. It seems to me to be clear that the distinctions we are
developing enable a spelling out of the issues. I look forward
eagerly to your reactions.

26. As for your concluding questions, you are again on the
right track with respect to my views on direct self-knowledge.
The only point I would want to add is that the avowal 'I have
a thought that-p' (1) asserts the occurrence of a thought that-p,
(2) is a reliable symptom of the occurrence of such a thought,
and (3) gives overt expression to a metathought (inner episode) which is an ·I have a thought that-p·, exactly as the
candid assertion 'it is raining' gives overt expression to a non-
metathought which is an ·it is raining·. In other words, we
post-Joneseans explain the connection between thoughts that
it is raining and reports that one has a thought that it is raining,
in terms of a connection between thoughts that it is raining
and apperceptive metathoughts that one has a thought that it is
raining. Thus I cannot agree that nothing more "is . . . involved in what [I] call meta-thinking . . . than conditions
(b) and (e)."

27. As for your next to last paragraph, I would not say that
"their concept of thought is no more nor less than a concept of
intentional item" not because it is false, but because it is misleading. It must be stressed that their concept of thought is a
concept of thought-out-loud. In other words *we* must emphasize that their concept includes only Rylean intentional items

—though, of course, they do not have the contrast Rylean *versus* non-Rylean.

Let me say, once again, how admirably lucid I found your letter. I hope you will find my comments helpful.

Sincerely yours,
Wilfrid Sellars

December 17, 1965

Dear Professor Sellars:

Thank you very much for your detailed letter of November 8. I was delighted to learn your reactions to my October 2 letter, and have found your remarks a very great help in clarifying the points of discussion. I was also grateful to learn of several ways in which my distinctions did not accurately capture what you had in mind. Because of this initial misunderstanding on my part, I think it may be worthwhile to say something more about these distinctions.

1. I must first apologize for an unfortunate phrase which I used throughout my letter. On page 466, I describe 'means'-statements (in sense (1)) as statements which are "about intentional items (no matter what their determinate factual character is)," and on page 467, I describe psychological statements (in sense (1)) in the same way. This phrase does not express what I intended. Again, at the bottom of page 469, I considered statements "about intentional items without any determinate factual character." In paragraph 20, you suggest that "by 'without any determinate factual character' . . . [I] mean *without specifying their determinate factual character.*" This is correct, and also correctly expresses what I had in mind on pages 466 and 467, but to make the point more clearly, I might say that such statements fail to specify anything at all about the determinate factual character of the items in question.

2. As I envisaged the distinction, therefore, the statement 'George said that-p' is not an M_1 statement. For although the statement does not specify the determinate factual character of the intentional item (that is, what words were uttered), it does

specify that the determinate factual character of the item is
Rylean.

3. Although I am no longer altogether happy with the notion
of a P1 (or M1) statement, I believe that instances of such
statements might be statements about Fregean intentional items,
that is, Fregean *Gedanken*. For Fregean intentional items are
abstract entities, and thus have no determinate factual character. Statements about them, therefore, will not specify anything about such factual character. (I no longer believe, as I
went on to suggest on page 470, paragraph 2, that '·Spectre·s
are the same as ·ghost·s' is an M1 statement, for it clearly
specifies that the items under discussion have a Rylean character, whatever it may be.) An instance of a statement about
(only) Fregean intentional items might be 'The Fregean
Gedanke that-p contains (or involves, or entails) the Fregean
Gedanke that-q'.

4. The fact that pre-Jonesean psychological statements are
about certain events is one reason why such statements cannot,
as you point out in paragraph 4, be P1 statements.

5. In paragraph 1, however, you write that "we may not allow
. . . [pre-Joneseans] to conceive of intentional items, as we
can, in a way which makes abstraction from their Rylean
character." (I assume that this amounts to roughly the same
thing as your writing, in paragraph 4, that the "subject matter
[of statements] must, in order for them to qualify as *pre-
Jonesean* psychological statements, be conceived in Rylean
terms" [emphasis mine].) Am I correct in supposing that you
mean to rule out the possibility that pre-Joneseans might use
their enriched Rylean language to talk about abstract entities,
and, *a fortiori*, about Fregean abstract entities?

6. Since a statement which specifies anything about the determinate factual character of the items involved cannot be a
P1 or M1 statement, the classes of PRO and MRO statements
cannot correspond, respectively, to the classes of P1 and M1
statements. For all PR and PO statements are PRO statements,
and all MR and MO statements are MRO statements. But
PR, PO, MR, and MO statements all specify at least the *sort*

of determinate factual character involved. (Although your PR/PO/PRO distinctions are among various sorts of explanations in terms of, I assume they may be recast as distinctions among various sorts of statements about (and in terms of.)[11]

7. Thus I think that your reservations about my statement which you quote in paragraph 10 may have resulted from a confusion which was due to my unclear account of P1 and M1 statements. For although M1 statements neither are, nor are to be analyzed in terms of, P2 statements, I agree, as you suggest in paragraph 11, that "from the fact that a statement is . . . [an MRO statement] it *does not follow* that it either is or is to be analyzed in terms of a psychological statement in . . . [my] sense 2."

8. In any event, I think that the distinctions developed in your letter are far preferable for the purpose of giving an account of the issues to the ones suggested in my October 8 letter. In what follows, therefore, I shall use your terminology, as I understand it. (I agree that P2 and PO statements are the same, just as M2 and MR statements are.) It will be useful for what I would like to say, however, to add a fourth class of psychological statements for consideration:

(4) Statements about Fregean intentional items: PF

(Here I do not wish to exclude statements which, although about Fregean abstract entities, are also about items conceived as having a certain (sort of) determinate factual character.)

9. I would like, then, to try again to characterize Chisholm's position as I understand it. The clearest shortcoming of my earlier characterization was that it failed to account for Chisholm's conviction that thoughts are uniquely a source of intentionality. Thus I entirely agree when you write that

> if confronted with these distinctions, Chisholm would claim that verbal behavior gets its relation to Fregean entities *indirectly* by virtue of being the manifestation of inner episodes

[11] I assume that you talk initially about explanation in terms of certain intentional items in order to make plausible the extension of the term 'psychological' discussed in paragraph 3.

which alone are directly related to these entities, and that he would correspondingly claim that verbal behavior and dispositions to behave are psychological only by virtue of the fact that they manifest such inner episodes (paragraph 24).

10. You continue by writing that "he is claiming that . . . PR statements are to be analyzed in terms of PO statements, and hence that MR statements are to be analyzed in terms of PO statements." Although Chisholm may indeed be claiming this, it does not seem to me that such a claim is embodied in the claims attributed to him above. For it seems to me possible to interpret that latter claims as being about no more than matters of fact, and not about the appropriateness of analyzing certain statements in terms of others. So the task remains of trying to give an account of those of Chisholm's claims described in the previous paragraph.

11. From the fact that George nonparrotingly utters (or writes) 'p', together with Jones' theory, we can infer, I believe, that George had some thoughts. (If he speaks or writes sincerely, for example, then he had the thought that-p.) Let us suppose, now, that there simply are no thoughts. From this fact together with the truth of Jones' theory, we can infer that if there are any utterances or inscriptions, then they are not meaningful (in the sense of nonparrotingly produced). We cannot, however, turn this inference around. From the fact that there are no nonparrotingly produced utterances or inscriptions together with the truth of Jones' theory, we cannot infer that there are no thoughts.

12. It is reasonable to suppose, I think, that if an utterance or an inscription is not meaningful (in the sense of nonparrotingly produced), then it is not intentional. For such an utterance would not be produced *as* tokening a properly formed expression in a language, nor would it properly be understood *as* tokening such an expression (although it might, mistakenly, be so understood). For this reason, it would be properly described as not being about anything, and hence as not intentional. (It does not seem to me that the converse holds, for an utterance may fail to be about something by failing, in fact,

to token a properly formed expression, even though it was non-parrotingly uttered.)

13. If these remarks are correct, then we can infer the following. From the fact that there are no thoughts together with Jones' theory it follows that if there are any utterances or inscriptions, they are not intentional. On the other hand, from the fact that there are no intentional utterances or inscriptions, together with Jones' theory, we cannot infer that thoughts are not intentional.

14. On page 239 of CSC, Chisholm explicates his claim that "thoughts are a 'source of intentionality'" as follows: "Thoughts would be intentional even if there were no linguistic entities. . . . But if there were no thoughts, linguistic entities would not be intentional" ((C-4), (C-5)).

15. Chisholm goes on to caution that he is not talking here about the proper analysis of one sort of statement into another (CSC, p. 239, (C-6)). The alternative seems to be that he is talking instead about causal relationships. Thus I would like to suggest that one way of spelling out what he is claiming is provided by my remarks in paragraphs 11 through 13. That is, Chisholm may be viewed as claiming the following. If we take certain statements about causal relations as axioms, then 'There are no thoughts' entails 'If there are any utterances or inscriptions, they are not intentional', but 'There are no linguistic entities' does not entail 'If there are thoughts, they are not intentional'. One suitable choice of causal axioms would, of course, be Jones' theory. So Chisholm might be viewed as in effect taking for granted the truth of Jones' theory, even though in fact he does not state the causal axioms which he is presupposing. At the least, it seems that Chisholm's claims are certainly, thus far, compatible with the truth of Jones' theory.

16. These remarks may possibly provide an account of the claims Chisholm makes in the seven labeled sentences on page 239 of CSC. It remains, however, to consider his (independent) claim that 'means'-statements are to be analyzed in terms of psychological statements, and not conversely (CSC, page 221).

17. In my October 2 letter I suggested that we might view

Chisholm's position as being partially the result of an implicit belief "that (at least in certain respects) presentday psychological statements behave as if they were . . . P₁ statements" (p. 469). We would thus understand his claim that 'means'-statements are to be analyzed in terms of thoughts as amounting to the claim that MR statements entail statements about Fregean *Gedanken,* that is, that they entail PF statements.

18. You suggest in paragraph 21 that such a claim is compatible with each of two sorts of positions—a tough-minded thesis and a tender-minded thesis. You describe the former as involving the assertion that "there are no thoughts in the sense of non-Rylean inner episodes." Thus whenever Chisholm talks about thoughts, he would, on this reading, be talking either about Rylean items or about Fregean abstract entities.

19. In spite of my having written that Chisholm "does not, so far as I can discover, speak of thoughts as the sort of item which has a certain determinate factual character, nor does he speak of thoughts as inner episodes" (*vide* your paragraph 20), he does speak of people having thoughts, and of the thoughts *of* people. (For example, in the last sentence of CSC.) This seems to rule out the possibility that he is talking about Fregean abstract entities *tout court,* for it is difficult to imagine that he supposes that people *have* abstract entities of any sort whatever. Thus since it seems unlikely that he is using 'thought' to talk about Rylean items, I share your skepticism with regard to Chisholm's position being the tough-minded thesis.

20. What you describe as the tender-minded thesis seems to hold more promise. This thesis involves, if I understand you correctly, that there exist both Ockhamite and Rylean intentional items, and that, although MR statements do not entail PO statements, both MR and PO statements do entail PF statements. Thus, for example, 'He had a thought that-p' would be held to entail 'He had an inner episode which stands in relation to the Fregean *Gedanke* that-p', just as ' ". . ." means - - -' would be held to entail ' ". . ." stands in relation to the Fregean *Gedanke* - - -'.

21. These claims would indeed be independent of the ques-

tion of what sort of items are related directly, and what sort are related indirectly, to Fregean abstract entities. For this question would be settled by appealing to considerations of the sort described in paragraph 15. Thus, a certain sort of intentional item would be said to stand in direct, immediate relation to Fregean items if, and only if, from the nonexistence of such items, together with the truth of Jones' theory, it follows that other sorts of items (which are in fact intentional) would not, if they existed at all, be intentional. Intentional items would then be said to be indirectly, and in a mediated way, related to Fregean items if, and only if, they do not satisfy the above condition. And we could express this by saying that the former (but not the latter) sort of items are a source of intentionality.

22. In paragraph 23 you write that

> As for the tender-minded version, his thesis could approach
> . . . [yours] if it amounted to the claim that 'means'-statements
> about marks and sounds are to be analyzed in terms of psycho-
> logical statements in the sense of Ryle-Frege statements about
> verbal behavior and dispositions to behave; and also *imply* (by
> virtue of the connection of verbal behavior and dispositions to
> behave with thoughts *in the sense of inner episodes*), psycho-
> logical statements in the more ordinary sense.

I am unclear concerning the second part of this claim, because I am not sure how to understand the force of 'imply', but it seems to me that you may have in mind the following. The connection between Rylean and Ockhamite items which you mention is, I assume, a causal connection. If so, then perhaps what you are saying might be expressed as follows: that MR statements together with the truth of Jones' theory, logically *entail* certain PO statements. I shall in any case return to this idea in paragraphs 28–30.

23. The chief feature of this thesis in which it departs from yours, as I understand it, is that it (like the tough-minded thesis) "misconstrues meaning as a relation" (paragraph 22). For whereas you enrich Rylean discourse so as to include semantical categories by means (essentially) of the semantical device of dot-quotes, the tender-minded thesis is able to deal

with intentionality rather because it posits Fregean abstract entities. Thus on the tender-minded view MR statements entail PF statements. And if I am correct in my suggestion at the end of paragraph 5, pre-Joneseans simply do not have the linguistic resources required to talk about abstract entities at all. (Since your account in "Notes on Intentionality" seems to involve that MR statements entail PF statements, it may be more accurate to say that the tender-minded thesis, unlike yours, must involve claiming that such an entailment holds if it is to be able to account for intentionality.)

24. These differences raise the question as to what advantages the use of dot-quotes has as against the introduction of Fregean abstract entities, and the related (and more basic) question as to whether or not meaning is, after all, a relation. If the tender-minded thesis does accurately represent Chisholm's position, then his qualms at this point might be expressed by the question whether it is possible to explicate the term 'role' used (in "Notes on Intentionality," p. 325) in explicating the function of dot-quotes. In the context of the present discussion, I would be more interested in learning why, aside from reasons of parsimony, pre-Joneseans must be disallowed from having the concept of an abstract entity.

25. These questions aside, however, attributing the tender-minded thesis to Chisholm has the following disadvantage. Since Chisholm uses 'thought' to cover both Fregean and Ockhamite items, either he is using the term ambiguously, or he is simply confusing two senses.

26. Your remarks in paragraph 16, if I understand them correctly, are intended to make plausible the idea that Chisholm might have in fact simply confounded two senses of 'thought'. In terms of your distinctions in "Notes on Intentionality," Chisholm would have confused the logical and causal senses of 'express', and the sense of 'thought' as *thinking* with its sense as *proposition* (p. 322). (The way in which I tried to make this confusion plausible was slightly different, I think, for it rested on the idea that "Chisholm is so set on stressing the intentionality of thoughts that he ignores their determinate factual

character as inner episodes" (p. 469). Thus Chisholm would be viewed as having allowed himself, *at certain times*, to treat inner-episode thoughts *as if* they were intentional items which altogether lack a determinate factual character, that is, as if they were Fregean abstract entities.) In any event, if Chisholm held the tender-minded view, then his claim as indicated in paragraph 16 would, properly understood, be about inner episodes and not about Fregean abstract entities.

27. If we must choose between supposing that Chisholm has used 'thought' ambiguously (but without even so much as indicating that he has) and supposing that he has just confounded two senses of the term, the latter does seem more likely. Having backed off somewhat, however, in paragraph 19, I think it may be worthwhile looking for an interpretation which does not so indict him.

28. If my suggestions in paragraph 15 are correct, then it follows that 'There exist Rylean intentional items' (logically) entails 'Either there exist Ockhamite intentional items or Jones' theory is incorrect'. But if we take Jones' theory as axiomatic, then it follows that 'There exist Rylean intentional items' entails 'There exist Ockhamite intentional items' *tout court*. But any MR statement entails that there are some Rylean intentional items, and 'There exist Ockhamite intentional items' might be taken as itself a PO statement. So if we take Jones' theory as axiomatic, then any MR statement entails at least one PO statement.

29. But we can go farther than this, along the lines of my suggestion in paragraph 22. If Jones' theory is true, then if '. . .' means --- then whenever '. . .' is nonparrotingly and sincerely uttered it (causally) expresses a thought episode that ---. Thus if we take Jones' theory as axiomatic, then ' ". . ." means ---' entails 'Whenever '. . .' is nonparrotingly and sincerely uttered, it (causally) expresses a thought episode that ---'. But this last statement is about Ockhamite (as well as Rylean) intentional items. It is thus a PO statement. In general, then, if we take Jones' theory as axiomatic, then MR statements entail PO statements.

30. I assume that this account does not depart very far from your own, for you write that "the only sense in which . . . [you are] prepared to grant that MR statements entail PO statements is that in which statements about bricks entail statements about molecules. Thus one might claim that if we adopt '(x)(x is a brick only if x is a set of molecules)' as an axiom, then statements about bricks *will* entail statements about molecules (*vide* your paragraph 19).

31. Moreover, this account is compatible with one which does not admit Fregean abstract entities. (On the other hand, of course, Chisholm's position, if so understood, would not involve any *account* of the intentionality of mental or linguistic items; it would be purely descriptive.)

32. Against this background, we might diagnose Chisholm's error as follows. Being a post-Jonesean, Chisholm uses a language which presupposes Jones' theoretical innovations. But he so takes for granted the truth of Jones' theory, that he treats the claims made by Jones as if they were logical axioms (or rules of inference). Thus he makes assertions about entailment (or logical derivability) relations which if true, presuppose the use of Jones' claims as axioms (or rules of inference). His error would thus consist in his presupposing certain causal truths without making them explicit, and his treatment of them as if they were logical truths. (The strength of his conviction that such axioms may be taken for granted might be indicated by his insistence that pre-Joneseans, if they "gave just *a little bit* of thought to the semantical statements they make," would see that such statements entail statements about thought-episodes [CSC, p. 248, emphasis mine].)

33. I am not by any means completely persuaded by the account I have sketched in the last five paragraphs, but I think it may capture what Chisholm is claiming more accurately than an account which commits him to having talked about Fregean *Gedanken*. (He writes, for example, in the third footnote of "Sentences About Believing" that he wishes to prevent confusion of "what . . . [he wants] to say with what Frege had to say about meaning." This may make it less plausible to suppose

that he confuses thought-episodes with the abstract entities in terms of which Frege discusses meaning. But this is, of course, not at all conclusive.) In any case, I would be very grateful to learn your reactions to these suggestions, as well as to my earlier ideas.

34. I have one final remark, which is not directly relevant to the foregoing discussion. In paragraph 17 you write that you believe that "Chisholm so uses 'psychological' that it includes items which . . . [you] would deny to be intentional (e.g., sense impressions, tickles, pains, etc.). . . ." Although I agree with this, I do not think that Chisholm wants to claim that all psychological items are intentional, that is, about something. For he is careful, in formulating his thesis of intentionality, to restrict it to "*certain* psychological phenomena—in particular . . . thinking, believing, perceiving, seeing, knowing, wanting, hoping, and the like . . ." ("Sentences About Believing",[12] p. 35 [emphasis mine]). And although he does include perceiving and seeing, I think that he is clear that he is talking about episodes which involve not only sense impressions but a judgmental component as well. Thus he claims, for example, that seeing a cat involves *taking* something to be a cat (*ibid*, p. 39). And in general, it seems to me that he avoids, for the most part, talking at all about the sort of items to which, in "Being and Being Known," you ascribe pseudo-intentionality. But I may well have overlooked features of his position that are relevant here.

I want again to express my appreciation for your very helpful and detailed letter. I have very much enjoyed studying it, and it has clarified a great number of points for me, as well as helping me to see more clearly, I think, the general setting of your position. I shall be immensely grateful to learn any reactions you may have to my suggestions.

Sincerely yours,
David Rosenthal

[12] Inclued in this volume.

January 4, 1966

Dear Rosenthal,

I was delighted to receive your admirably lucid letter of December 17. From my point of view, at least, we are making exceptional progress in clarifying the ambiguities and misunderstandings which prevented the Chisholm correspondence from achieving more decisive results.

Now for some specific reactions.

1. Your paragraph 3 introduces a number of themes which run throughout your letter. The most important of these is the concept of a 'Fregean intentional item'. (I assume that our aim is not Frege exegesis, but to spell out and evaluate the implications of the distinctions drawn in "Notes on Intentionality" for the questions with which this correspondence began.) I am not clear whether by 'Fregean intentional items' you are referring to particulars (e.g., episodes) or to nonparticulars. Your characterization of them as abstract entities strongly suggests the latter; but your use of the phrase 'determinate factual character' which we have been using in connection with psychological episodes and dispositions suggests the former. Since your paradigm seems to be a Fregean *Gedanke,* I suppose you are saying (a) that Fregean *Gedanken* do not have 'determinate factual character'—which is reasonable enough, though I shall comment further below; (b) That a particular (e.g., episode) which gains its intentionality through its relation to a Fregean *Gedanke* is not thereby pinned down to one determinate factual character rather than another—or do you mean that it need have no determinate factual character at all? Again, by 'determinate factual character' you might mean such characters as Ryleanism or Ockhamicity in general, or you might have in mind more appropriately determinate forms of these (e.g., on the Rylean side, specific patterns of meaningful English behavior).

2. My view, of course, is that every particular which has intentionality has, and necessarily has, a determinate factual character. A Rylean intentional item must have a determinate factual character by virtue of which it is a rule to play a certain

Rylean role. An Ockhamite intentional item must have a determinate factual character (though we may not know what it is, save in very generic terms) by virtue of which it is able to play a certain Ockhamite role (conceived by analogy with a certain Rylean role). What, then, in my view, would Fregean intentional items be? The unpolished answer is that they are items which are either Rylean intentional items or Ockhamite intentional items.

3. Let me introduce three different quotation devices: *dot* quotes for Rylean items, *diamond* quotes for Ockhamite items and *dot-diamond* quotes for Fregean items. Consider the following statements

(a) α is a ·not·
(b) α is a ◊not◊
(c) α is a ·◊not·◊

The first, (a), tells us that α is an item playing the Rylean role played in our language by 'not's; (b) tells us that α is an item playing in inner speech an Ockhamite role analogous to that played in our language by 'not's (and in any language by ·not·s); (c) tells us that α is either a ·not· or a ◊not◊. In each case α must have a determinate factual character in order to be a ·not·, a ◊not◊, or a ·◊not·◊.

4. Notice that each of the three statements (a), (b) and (c), has the form '. . . is a K', so that what corresponds to 'K' is a predicative expression rather than an abstract singular term of the form 'K-kind'. Now just as

Fido is a dog

has as its 'platonistic' counterpart

Fido is an instance of dog-kind

so (a), (b), and (c) have as platonistic counterparts

(a′) α is an instance of ·not·-kind
(b′) α is an instance of ◊not◊-kind
(c′) α is an instance of ·◊not·◊-kind

which correspond, on my analysis, to

(a″) α stands for Rylean negation
(b″) α stands for Ockhamite negation
(c″) α stands for Fregean negation

Thus, just as, on my analysis, the move down from the platonistic

> Fido is an instance of dog-kind

to

> Fido is a dog

is to be reconstructed, in first approximation, as

> Fido is an instance of dog-kind
> The ·is a dog· is true of ·Fido·
> The ·Fido is a dog· is true
> Fido is a dog

so the move down from the platonistic

α stands for Rylean negation [= α is an instance of Rylean negation]

to

α is a ·not· [= (Rylese) α means *not*]

is to be reconstructed as

> α stands for Rylean negation
> The ·is a ·not· · is true of ·α·
> The ·α is a ·not· · is true
> α is a ·not·.

5. Notice that, as the above examples bring out, not all abstract entities (in our examples, kinds) are instantiated by intentional items. Negation is, dog-kind is not, at least as they occur in our examples. Thus the distinction between Rylean, Ockhamite, and Fregean abstract entities with which we are concerned applies only to a subset of abstract entities—those which have as their instances psychological items, that is, either Rylean or Ockhamite particulars. If, therefore, we assume that our (post-Jonesean) use of '-kind' (also '-ity', '-hood', '-ness',

'that-', etc.) can be reconstructed as dot-diamond (Fregean) quoting, both

dog-kind

and

·◊not·◊-kind

would be Fregean in the sense that they both *use* Fregean quotes, thus

The ·◊dog·◊
The ·◊(·◊not·◊)·◊

(I throw in the parentheses to make the iterated quotes more perspicuous), whereas the latter but not the former is a Fregean quoting of a Fregean quote. Thus the latter, *but not the former,* can be contrasted with

◊not◊-kind

which, given the above interpretation of '-kind' translates into

The ·◊(◊not◊)·◊

which is a Fregean quoting of an Ockhamite quote.

6. Notice that the above assumption about the role of '-ity', '-hood', '-ness', etc., is appropriate only to the post-Jonesean era. Obviously neither ◊-quoting nor ·◊-quoting can occur in pre-Jonesean times, for they are conceptually bound up with the Jonesean revolution. Pre-Jonesean philosophers, confronted with the problem of abstract entities, might hit upon the device of reconstructing singular-term-making prefixes and suffixes as a use of quotes to make sortal words, together with that use of 'the' which turns sortal words into distributive singular terms, but the quotes would have to be, from our point of view, Rylean, for the sortal words they formed would apply to overt linguistic behavior. Thus a pre-Jonesean Sellars would construe

Dog-kind

as

The ·dog·

and would have no use for

The ◇dog◇

or

The ˙◇dog˙◇.

7. Against this background, let us take another look at the statement I quoted from your paragraph 3: "Fregean intentional items are abstract entities, and *thus* can have no determinate factual character" (I now underline the 'thus'). It seems likely that Frege did not think that his abstract entities had 'determinate factual character', if 'factual' is to have anything like the sense of 'empirical'. On the other hand, he would, I take it, have granted that, though they could not be, e.g., red or noisy, they nevertheless *could* have structural characteristics, indeed that they *must* have structural characteristics. This enables me to bring out one of the strengths of my view. Consider

(d) Triangularity is noisy.

There is clearly *something* odd about this statement. Now on my view (d) is equivalent to

(d′) The ˙◇triangular˙◇ is noisy

and this, in turn, to

(d″) ˙◇triangular˙◇s are, as such, noisy.

In other words, the oddity of this predication can be explained as a clash between the implication of dot-diamond quotes and the predicate '(as such) noisy'. The dot-diamond quotes permit events of any kind to be ˙◇triangle˙◇s in 'outer' or 'inner' speech provided that they can participate in the same structurally characterized moves and transitions as our word 'triangle'. The predicate '(as such) noisy' clashes with this permission. In this respect

The ˙◇triangular˙◇ is noisy

resembles

The pawn is made of ivory.

8. The above remarks on 'intentional items' and 'abstract entities' are designed to show that a reference to 'abstract entities' does not bypass the conceptual progression Rylean-Ockhamite-Fregean, where 'Fregean' stands for the common character of Rylean and Ockhamite roles. Bluntly put, the abstract entities which intentional items 'stand for' (i.e., instantiate) divide into Rylean (R), Ockhamite (O) and Fregean (RO) exactly as do the intentional items which 'stand for' (i.e., instantiate) them. Indeed, on the above analysis, this claim is a truism.

9. Thus, when, in paragraph 5, you ask "Am I correct in supposing that you mean to rule out the possibility that pre-Joneseans might use their enriched Rylean language to talk about abstract entities, and, *a fortiori*, about Fregean abstract entities?" The answer is a ringing "No"! I take it for granted that pre-Joneseans can speak of abstract entities (use abstract singular terms). These abstract entities are, however, from our point of view, Rylean abstract entities. Since, *ex hypothesi*, they cannot speak of Ockhamite abstract entities, it follows—and here is where '*a fortiori*' belongs—that they cannot speak of Fregean abstract entities, *if* the concept of a Fregean abstract entity is the concept of what is common to Rylean intentional items and Ockhamite counterparts. If, therefore, my "Notes on Intentionality" is on the right track, one cannot minimize the revolutionary character of the Jonesean development by arguing that *independently of this development* pre-Joneseans could come to conceive of intentional items which are freed from Rylean limitation, by conceiving of them as standing for Fregean abstract entities.

10. Let me expand a bit on pre-Jonesean discourse about abstract entities. I pointed out above that of those abstract entities which can be said to have instances some are instanced by intentional items (as when an intentional item stands for [instantiates] the *Gedanke* that snow is white), while some are instantiated either by particulars which lack intentionality (e.g., triangularity) or by intentional items in some respect other than their intentionality (e.g., temporal precedence). Let

me call the former *intentional* abstract entities and the latter *nonintentional* abstract entities. The pre-Jonesean conception of the nonintentional abstract entity triangularity, would be the conception of the ·triangular·.

11. It might be thought that since ·triangular·s are Rylean intentional items, the account I am giving is incoherent, for how could triangularity be a nonintentional abstract entity, if triangularity is the ·triangular· and ·triangular·s are intentional. The answer is that an intentional abstract entity was defined as one which is instantiated by intentional items *qua* intentional, and the statements

(e) Triangularity is exemplified by pyramids, which are non-intentional items

and

(f) Triangularity = the ·triangular·

are perfectly compatible with

(g) The ·triangular· is an intentional item (i.e., ·triangular·s are intentional items).

For, according to my analysis, (e) has, roughly, the sense of

(e′) The ·triangular· is true of pyramids, and pyramids are non-intentional items

and this, in turn, the sense of

(e″) The ·pyramids are triangular· is true, and pyramids are nonintentional items.

Thus, whereas in (g) the dichotomy *intentional/nonin-tentional* is considered with respect to that which is expressed by the construction which uses the word 'triangular', in (e) it is considered with respect to that which is expressed by the use of the word 'pyramid'.

12. If I am right, then pre-Joneseans can, by the use of their enriched Rylean language, make such statements as

(h) 'Dreieckig' (in G) means *triangular*
(i) 'Dreieckig's (in G) are ·triangular·s

(j) The ˙pyramids are triangular' is true

(k) The ˙triangular˙ is true of pyramids.

Of these (k) is the reconstruction of

(k′) Triangularity is exemplified by pyramids.

They can also say

(l) 'Dreieckig's (in G) are included in ˙triangular˙-kind

which is the 'platonistic', or abstract, counterpart of (i), and thus belongs at the level of double dot-quoting, thus

(l′) The ˙are ˙triangular˙s˙ is true of ˙'dreieckig's (in G)˙

which is equivalent to

(l″) The 'dreieckig's (in G) are ˙triangular˙s˙ is true

which reduces, by the truth-move, to (i). Since (e) is the proposed reconstruction of

(m) 'Dreieckig's (in G) stand for triangularity

it follows that, according to the account I am giving, the *word* 'triangularity' has different senses in the two contexts

. . . stands for triangularity

and

. . . exemplifies triangularity.

In the former, it is at the level of double dot-quoting, in the latter of single dot-quoting. The fact that words appropriate to the gap in the second context refer, in their *primary* use, to nonintentional items is, as I see it, the root of the mistaken idea that abstract entities are independent of intentionality. Another root is the classical equation of the *psychological* with the *individual-psychological,* which led people to suppose that to equate the domain of the intentional with the domain of the psychological is to make unintelligible the *inter*subjectivity of the intentional. One must recognize the 'community' character of the psychological in order to appreciate the fundamental truth of conceptualism.

13. As for intentional abstract entities, our pre-Joneseans can, of course, conceive only of the Rylean variety. To see this we need only pull together some points made in paragraph 11. Thus, if we specify that an abstract entity is instantiated by intentional items *qua* intentional, we are specifying, in effect, that the expression for that abstract entity is at the double quote level of the available type of quote (R, O, or RO). For, on the above analysis,

(intentional items) exemplify, *qua* intentional, (abstract entity)

has the sense of

(intentional items) stand for (abstract entity)

and is the abstract counterpart of a statement having the form

(intentional items) are ·——·s

which is the schema for the reconstruction of the Rylean 'means'-statement

(intentional items) means ———

which is available, *ex hypothesi*, to our pre-Joneseans. Thus, pre-Jonesean concepts of abstract entities generally are Rylean concepts, and pre-Jonesean concepts of intentional abstract entities are Rylean concepts of Rylean abstract entities. To reiterate a point made in paragraph 9, the ability of pre-Joneseans to conceive of abstract entities does not provide them with an independent route (independent, that is, of the Jonesean revolution) for arriving at a notion of an intentional item which is freed from Rylean restriction.

14. In paragraph 8 you express the belief that it is "useful for what [you] would like to say . . . to add a fourth class of psychological statements for consideration:

(4) statements about Fregean intentional items: PF."

You are quite right, for this addition enables a sharpening of the issues. Thus, if to conceive of PF items were not the same thing as to conceive of PRO items, then the possibility would remain that although pre-Joneseans cannot conceive of PO items, and hence, *a fortiori* of PRO items, they *can* conceive of non-Rylean psychological items, for they can conceive of items which are PF without being overt verbal behavior or dispositions to behave. (The question would remain, of course, as to just what these items are conceived by pre-Joneseans to be, if they are not construed as Ockhamite inner episodes.) The peculiar relevance of the problem of abstract entities to problems pertaining to intentionality and the mental should not be prejudged by a terminology which requires that all intentional items be either Rylean or Ockhamite or Rylean-Ockhamite.

15. In paragraph 11 you write "let us suppose, now, that there simply are no thoughts. From this fact, together with the truth of Jones' theory. . . ." I do not understand this. From the truth of the Jonesean theory, it follows that where there is nonparroting speech, there are thoughts. Notice that pre-Joneseans have a concept of nonparroting speech (thinking-out-loud), and that Jones' theory adds to this concept a theoretical framework which, by providing an explanation of nonparroting speech, generates a richer concept of nonparroting speech.

16. From the truth of Jones' theory one can, however, draw the *weaker* conclusion that if *on a particular occasion* there are no thoughts, then on that occasion there is no meaningful (nonparroting) speech. And you are correct in saying that the reverse implication does not hold. Jones' theory allows, of course, for unexpressed thoughts.

17. In paragraph 13 you make what seems to be an illegitimate step from 'we cannot infer that there are no thoughts' to 'we cannot infer that thoughts are not intentional'. You overlook, perhaps, that it is an analytic feature of the Jonesean concept of a thought that thoughts are intentional.

18. Your point in paragraphs 14ff. begins to probe to the heart of the issue. You will remember that in commenting on Chisholm's (C-1) through (C-7) (CSC, p. 239) I write "it isn't so much that I disagree with your seven sentences, for I can use each of them separately, with varying degrees of discomfort, to say something which needs to be said. . . . It is rather that I am unhappy about the force they acquire in the overall frame-

work in which you put them" (CSC, p. 243). The key sentence
of your paragraph 15 is "So Chisholm might be viewed as in
effect taking for granted the truth of Jones' theory. . . ." Exactly. Taken together with your paragraph 16, the chips are
down.

19. Your definition of mediate and immediate relation to a
Fregean item contains the passage "from the nonexistence of
such items, *together with the truth of Jones' theory*, it follows
that other sorts of items (which are in fact intentional) would
not, if they existed at all, be intentional." My comments in
paragraphs 15–17 apply here as well.

20. Your interpretation in paragraph 22 is essentially correct.

21. You write, in paragraph 23, "on the tenderminded
view MR statements entail PF statements." Only in conjunction
with Jones' theory is this true. You also repeat your suggestion of paragraph 5 that pre-Joneseans do not have the
linguistic resources required to talk about abstract entities at
all. This, of course, I have rejected.

22. Your paragraphs 24–25 raise the whole question of abstract entities and their connection with intentionality. I agree
that the issue is central, and have tried to throw some light on
it in the earlier parts of this letter.

23. It is certainly true, as you remark in paragraph 28, that
"if we take Jones' theory as axiomatic, then any MR statement
entails at least one PO statement." But this doesn't help the
pre-Joneseans.

24. Your remarks in paragraphs 30–32 are pretty much on
target.

25. I agree, on the whole, with what you say in paragraph
34 about Chisholm's use of the term 'psychological'. I quite
agree that he explicitly attributes intentionality only to "episodes which involve not only sense impressions but a judgmental component as well." On the other hand his equation in
Perceiving[13] of *sensing* with *being appeared to,* and his failure
explicitly to distinguish between a sense of 'appear' in which

---

[13] R. M. Chisholm, *Perceiving* (Ithaca, N.Y.: Cornell Univ. Press, 1957),
ch. 8.

being appeared to has a conceptual character from a putative sense in which it does not, lead me to suspect that the intentionality of conceptual episodes and the pseudo-intentionality of sensory states have not adequately been distinguished in Chisholm's philosophical psychology.

On rereading this letter, I am afraid that it contains whole sections of a rather opaque character. But, then, I could scarcely have omitted the paragraphs on abstract entities; while if I had expanded them in the hope of reducing this opacity, this letter would have turned into "Abstract Entities." Every philosophical issue, seriously pursued, takes one everywhere. I can only hope that our other philosophical experiences have had enough in common to make this correspondence a genuine and fruitful dialogue. That, at least, is how it appears to me.

Sincerely yours,
Wilfrid Sellars

Bibliography

This bibliography is not intended to be comprehensive but only to guide the reader through a maze of philosophical literature which, in the opinion of the editor, has immediate or indirect relevance to the issues dealt with in this anthology. I have attempted, as much as possible, to single out basic topics and point out ideal lines of development. Each entry is preceded by a number in square brackets in order to aid cross references and help separate visually the citations from the commentary. Numbers preceded by an asterisk indicate that the work being cited is reprinted (in whole or part) in the present volume. The following abbreviations have been used:

A–*Analysis*
APF–*Acta Philosophica Fennica*
APQ–*American Philosophical Quarterly*
ASP–*Aristotelian Society Proceedings*
ASSV–*Aristotelian Society Supplementary Volume*
JP–*Journal of Philosophy*
JSL–*Journal of Symbolic Logic*
MSPS–*Minnesota Studies in the Philosophy of Science*, vol. I, eds. H. Feigl and M. Scriven; vol. II, eds. H. Feigl, M. Scriven, and G. Maxwell (Minneapolis: Univ. of Minnesota Press, 1956 and 1958).
POS–*Philosophy of Science*
PPR–*Philosophy and Phenomenological Research*
PQ–*Philosophical Quarterly*
PR–*Philosophical Review*
PS–*Philosophical Studies*
RM–*Review of Metaphysics*

I. HISTORICAL BACKGROUND. SCHOLASTICISM, NEO-SCHOLASTICISM, AND
PHENOMENOLOGY

For Scholastic and neo-Scholastic bibliographical sources on the con-
cept of intentionality the reader is referred to P. Mandonnet and J. De-
strez, [1] *Bibliographie thomiste*, rev. ed., M.D. Chenu (Paris, 1960);
V.J. Bourke, [2] *Thomistic Bibliography, 1920–1940* (St. Louis: The
Modern Schoolman, 1945); [3] *Bulletin thomiste*, I–XII (Le Saulchoir,
Belgium), which has reported on practically all books and articles on
Thomistic philosophy from 1921 to 1965.[1] A fully annotated catalogue
of Aquinas' works is included in E. Gilson, [4] *The Christian Philosophy
of St. Thomas Aquinas* (New York: Random House, 1956). Of specific
interest are: A. Hayen, S.J., [5] *L'Intentionnel selon saint Thomas*, 2nd
ed. (Paris: Desclée, 1953); and R.W. Schmidt, [6] *The Domain of
Logic According to Saint Thomas Aquinas* (The Hague: Martinus
Nijhoff, 1966), ch. V and selective bibliography for the same chapter.
See also the article entitled [7] "Intenzionalità" in *Enciclopedia Filoso-
fica*, ed. by Centro Studi Filosofici di Gallarate (Florence: Sansoni
Editore, 1967).

Those interested in the phenomenological development of the concept
of intentionality are referred to H. Spiegelberg, [8] *The Phenomenological Movement: A Historical Introduction*, 2 vols., 2nd ed. (The Hague:
Martinus Nijhoff, 1969), which contains a valuable selective bibliography
at the end of each chapter. See also M. Farber, [9] *The Foundations of
Phenomenology* (Cambridge, Mass.: Harvard Univ. Press, 1943), which
contains a detailed analytical exposition of Husserl's works, and Q. Lauer,
[10] *Phenomenology: Its Genesis and Prospects* (New York: Fordham
Univ. Press, 1958). Selections from the writings of Brentano, Meinong,
and Husserl are included in R. Chisholm, ed., [11] *Realism and the
Background of Phenomenology* (Glencoe, Ill.: Free Press, 1960), which
also contains a valuable introduction and a helpful bibliography.

II. DEFINITIONS OF INTENTIONALITY

In vol. I, bk. II, ch. i of [12] *Psychologie vom empirischen Standpunkt* (2 vols., 1874; reprint, Hamburg: Felix Meiner, 1956–59), Franz
Brentano had suggested that intentionality is a distinguishing charac-

---

[1] Starting in 1966, [3] has been replaced by *Rassegna di Letteratura Tomistica*
(Naples: Edizioni Domenicane Italiane), the first volume of which appeared
in 1969.

teristic of psychological phenomena. An English translation of this chapter may be found in Chisholm [11]. In recent years R.M. Chisholm has attempted to reformulate Brentano's claim by specifying a set of criteria for distinguishing psychological from nonpsychological *sentences* (or other locutions). Chisholm's first systematic attempt to specify such criteria was made in [13] "Intentionality and the Theory of Signs," *PS* 3 (1952), 56–63. This initial attempt, criticized by B.A. Farrell in a paper by the same title ([14], *PPR* 15 (1954–55), 500–511), and further pursued by Chisholm in [15] "On the Uses of Intentional Words," *JP* 51 (1954), 436–441, culminated later in Chisholm's important paper [*16] "Sentences About Believing" and in ch. 11 of [17] *Perceiving: A Philosophical Study* (Ithaca, N.Y.: Cornell Univ. Press, 1957). Discussions of Chisholm's proposal followed in papers by J.W. Cornman, [*18] "Intentionality and Intensionality"; S.C. Brown, [19] "Intentionality Intensified," *PQ* 13 (1963), 357–360; Cornman again, in reply to Brown, [20] "The Extent of Intentionality," *PQ* 14 (1964), 355–357; then again Brown in [21] "Intentionality without Grammar," *ASP* 65 (1964–65), 123–146; M. Clark, [22] "Intentional Objects," *A*, suppl. vol. 25 (1965), 123–128; H. Heidelberger, [23] "On Characterizing the Psychological," *PPR* 26 (1965–66), 529–536; D.J. O'Connor, [24] "Tests for Intentionality," *APQ* 4 (1967), 173–178; and A. Marras, [*25] "Intentionality and Cognitive Sentences." Further discussion of Chisholm's criteria in [17] may also be found in ch. 9 of A. Kenny, [26] *Action, Emotion and Will* (London: Routledge and Kegan Paul, 1963).

A fresh start was made again by Chisholm in a paper read at a symposium on the philosophy of mind held at Wayne State University in December, 1962, where Chisholm attempted to provide a definition of an intentional prefix (such as 'S believes that'). This paper was later published as [27] "On Some Psychological Concepts and the 'Logic' of Intentionality" in H-N. Castañeda, ed., [28] *Intentionality, Minds and Perception* (Detroit: Wayne State Univ. Press, 1967), together with comments by R.C. Sleigh [29] and a reply by Chisholm [30]. A critical discussion of the Chisholm-Sleigh exchange may be found in R.W. Binkley, [31] "Intentionality, Minds and Behaviour," *Noûs* 3 (1969), 49–60. A further attempt along the lines of [27] and [30] was made again by Chisholm in [*32] "Notes on the Logic of Believing," followed by replies by D.R. Luce in [*33] "On the Logic of Belief" and R.C. Sleigh in [*34] "Notes on Chisholm on the Logic of Believing," and by a rejoinder by Chisholm in [*35] "Believing and Intentionality: A Reply to Mr. Luce and Mr. Sleigh." See also Chisholm's article [36] "Intentionality" in P. Edwards, ed., [37] *The Encyclopedia of Philosophy* (New York: Macmillan, 1967), vol. 4. Further discussion of certain claims made by Chisholm in [27] and [*32] is found in A. Marras, [38] "Properties and Beliefs about Existence," *Logique et Analyse* 13 (1970),

438–451. Chisholm's most recent criterion of intentionality is proposed in [39] "Brentano on Descriptive Psychology and the Intentional," in E.N. Lee and M. Mandelbaum, eds., [40] *Phenomenology and Existentialism* (Baltimore: Johns Hopkins Univ. Press, 1967).

An independent attempt to specify criteria of intentionality, in the context of a discussion of perception verbs, may be found in G.E.M. Anscombe, [41] "The Intentionality of Sensation: A Grammatical Feature," in R. J. Butler, ed., [42] *Analytical Philosophy*, 2nd ser. (Oxford: Blackwell, 1968). (By Anscombe see also her analysis of the concept of intention, in the context of a discussion of human action, in [43] *Intention* (Oxford: Blackwell, 1957).) Discussions of Anscombe's paper may be found in G.N.A. Vesey, [44] "Miss Anscombe on the Intentionality of Sensation," A 26 (1965–66), 135–137; C.V. Borst, [45] "Perception and Intentionality," *Mind* 79 (1970), 115–121; and W.G. Lycan, [*46] "On 'Intentionality' and the Psychological." Lycan's paper also contains a very useful critical survey of various definitions of intentionality, as well as the author's own contribution to the subject. Other recent discussions of criteria of intentionality include L.J. Cohen, [47] "Criteria of Intensionality," AASV 42 (1968), and D.H. Sanford, [48] "On Defining Intentionality," *Proceedings of the XIVth International Congress of Philosophy*, vol. 2 (Vienna: Herder, 1970).

III. INTENTIONALITY, PHYSICALISM AND MIND-BODY PROBLEM

The linguistic thesis of intentionality (that all and only psychological sentences are intentional) was defended by Chisholm in [13], [*16], and [17] against various physicalistic attempts to "reduce" psychological sentences to nonintentional sentences about verbal or nonverbal behavior or other physical states. See also, with particular reference to perception sentences, Chisholm's paper [49] "Reichenbach on Observing and Perceiving," *PS* 2 (1951), 45–48, which contains a critique of a physicalistic analysis of perception sentences offered by H. Reichenbach in sec. 49 of his book [50] *Elements of Symbolic Logic* (New York: Macmillan, 1947). See also Reichenbach's reply [51] "On Observing and Perceiving," *PS* 2 (1951), 92–93, and Chisholm's rejoinder [52] "Reichenbach on Perceiving," *PS* 3 (1952), 82–83. Basic to Chisholm's criticism of any attempt to define psychological concepts in terms of linguistic behavior is his claim that linguistic behavior must itself be understood in terms of psychological concepts. On this, see especially Chisholm's [53] "A Note on Carnap's Meaning Analysis," *PS* 6 (1955), 87–89, written in reply to a proposal by R. Carnap in [54] "Meaning and Synonymy in Natural Languages," *PS* 6 (1955), 33–47 (reprinted in Carnap, [200] below), to analyze linguistic meaning in terms of linguistic responses.

See also Carnap's rejoinder in [55] "On Some Concepts of Pragmatics,"
PS 6 (1955), 89–91 (reprinted in Carnap, [200] below). For more on
the question of the relation between linguistic meaning and intentionality see section IV of this bibliography. The relation between the linguistic thesis of intentionality and various forms of materialism or
physicalism is lucidly discussed by J.W. Cornman, in the context of the
traditional mind-body problem, in [56] *Metaphysics, Reference and Language* (New Haven, Conn.: Yale Univ. Press, 1966). See also Cornman,
[*18] above; P. Nochlin, [57] "Reducibility and Intentional Words,"
JP 50 (1953), 625–638; and H.R. Schuford, [58] "Logical Behaviorism
and Intentionality," *Theoria* 32 (1966), 246–251. The general problem
of the relation between linguistic questions and philosophical questions
is discussed in a number of essays in R. Rorty, ed., [59] *The Linguistic
Turn* (Chicago: Univ. of Chicago Press, 1967).

Distinct forms of physicalism with which a defender of some form of
the intentionalist thesis is likely to take issue are represented in each of
the essays (by C.G. Hempel, R. Carnap, T. Nagel, and G. Ryle) included in Part II of the present volume. Hempel's essay [*60] "Logical
Analysis of Psychology," which contains one of the earliest and most
concise formulations of logical (analytical, philosophical) behaviorism,
advances essentially the same thesis as that advanced by R. Carnap in
[61] "Psychologie in physikalischer Sprache," *Erkenntnis* 3 (1932–33),
107–142 [English trans.: "Psychology in Physical Language," in A.J.
Ayer, ed., [62] *Logical Positivism* (New York: Free Press, 1959)], and
still earlier in [63] "Die Physikalische Sprache als Universalsprache der
Wissenschaft," *Erkenntnis* 2 (1931–32), 432–465 [English trans.: *The
Unity of Science* (London: Kegan Paul, 1934)]. Similar views were also
defended by A.J. Ayer in [64] *Language, Truth and Logic* (London:
Gollancz, 1936), and in his inaugural lecture [65] *Thinking and Meaning* (London: H.K. Lewis, 1947). Detailed references to the writings of
logical behaviorists and their critics may be found in the annotated bibliography at the end of ch. 3 of P. Edwards and A. Pap, eds., [66] *A Modern Introduction to Philosophy*, rev. ed. (New York: Free Press, 1966),
as well as in the excellent bibliography in Ayer, [62] above.

Both Carnap and Hempel have abandoned their earlier views in favor
of weaker forms of physicalism. The so-called method of "reduction
sentences," first proposed by Carnap in [67] "Testability and Meaning,"
POS 3–4 (1936–37), 419–471, 1–40, and expounded informally in
[*68] "Logical Foundations of the Unity of Science," provided a new
method for the "introduction" of psychological terms, dispositionally interpreted, into the physicalistic language. For critical discussion on this
phase of Carnap's philosophy see especially I. Scheffler, [69] *The Anatomy of Inquiry* (New York: Alfred A. Knopf, 1963), ch. 2; A. Pap, [70]
"Reduction Sentences and Disposition Concepts," in P.A. Schilpp, ed.,

[71] *The Philosophy of Rudolf Carnap* (La Salle, Ill.: Open Court, 1963). Further developments of Carnap's views are represented in Carnap, [72] "Meaning Postulates," *PS* 3 (1952), 65–73 (reprinted in Carnap, [200] below), and in Carnap, [73] "The Methodological Character of Theoretical Concepts," *MSPS I*. Hempel's most recent views have found expression in [74] "Reduction: Ontological and Linguistic Facets," in S. Morgenbesser *et al.*, eds., [75] *Philosophy, Science, and Method* (New York: St. Martin's Press, 1969), and in [76] "Logical Positivism and the Social Sciences," in P. Achinstein and S.F. Barker, eds., [77] *The Legacy of Logical Positivism* (Baltimore: The Johns Hopkins Press, 1969). The latter essay also contains a useful historical survey and appraisal of various physicalistic views. See also H. Feigl, [78] "The Mind-Body Problem in the Development of Logical Empiricism," in H. Feigl and M. Brodbeck, eds., [79] *Readings in the Philosophy of Science* (New York: Appleton-Century-Crofts, 1953); H. Feigl, [80] "The 'Mental' and the 'Physical'," *MSPS II*; and, also in *MSPS II*, other essays by P. Oppenheim and H. Putnam, C.G. Hempel, M. Scriven, A. Pap, and others. Feigl's essay [80] also contains a very comprehensive bibliography.

The type of physicalism defended by T. Nagel in [*81] "Physicalism" is a form of the so-called "identity theory" of mind developed especially by U.T. Place in [82] "Is Consciousness a Brain Process?," *British Journal of Psychology* 47 (1956), 44–50 (reprinted in Chappell [101] and in O'Connor [88]); by J.J.C. Smart in [83] "Sensations and Brain Processes," *PR* 68 (1959), 141–156 (reprinted in Chappell [101] and in O'Connor [88]); and by H. Feigl in [80] above. A highly original interpretation of the identity theory, involving the idea (shared by Carnap in [73]) that psychological terms be construed as theoretical terms, is given by Sellars in [84] "The Identity Approach to the Mind-Body Problem," *RM* 18 (1965), 430–451 (reprinted in O'Connor [88]); see also Sellars' [85] "Mind, Meaning and Behavior," *PS* 3 (1952), 83–95, and [86] "A Semantical Solution of the Mind-Body Problem," *Methodos* 5 (1953), 45–84. Three anthologies devoted entirely to a discussion of the identity theory are: C.V. Borst, ed., [87] *The Mind-Brain Identity Theory* (New York: St. Martin's Press, 1970); J. O'Connor, ed., [88] *Modern Materialism: Readings on Mind-Body Identity* (New York: Harcourt, Brace and World, 1969); and C.F. Presley, ed., [89] *The Identity Theory of Mind* (St. Lucia: Queensland: Univ. of Queensland Press, 1967). All three anthologies contain valuable bibliographical references. See also the annotated bibliography in ch. 3 of Edwards and Pap, [66] above.

Although Ryle would probably not wish to be classified as a physicalist, his views in [90] *The Concept of Mind* (London: Hutchinson, 1949) have often been regarded as expressing a form of behaviorism (see [*91] "Dispositions"). Selective bibliographical references to the vast critical literature generated by Ryle's views may be found in the annotated bibliography in ch. 3 of Edwards and Pap, [66] above. See, in addition, a detailed critique by B. Medlin, [92] "Ryle and the Mechanical Hypothesis," in Presley, [89] above, and chs. 3 and 4 of P. Geach, [104] below.

Significant philosophical alternatives to the physicalistic views represented in Part II of this anthology may be found in the writings of L. Wittgenstein, [93] *Philosophical Investigations* (New York: Macmillan, 1953); J.L. Austin, [94] *Philosophical Papers* (Oxford: Clarendon, 1961) and [95] *How To Do Things with Words* (Oxford: Oxford Univ. Press, 1965); P. Strawson, [96] *Individuals* (London: Methuen, 1959); J. Wisdom, [97] *Other Minds* (Oxford: Blackwell, 1965); S. Hampshire, [98] *Thought and Action* (London: Chatto and Windus, 1960). Of older vintage, but no less philosophically significant, is D.C. Broad's book [99] *Mind and Its Place in Nature* (London: Routledge and Kegan Paul, 1925). See also C.W. Morris' critical examination of various theories of mind dominant in the earlier part of the century, in [100] *Six Theories of Mind* (Chicago: Univ. of Chicago Press, 1932), especially ch. 4, entitled "Mind as Intentional Act." Other older studies of mind by such philosophers as G.F. Stout, G.B. Pratt, D. Drake, B. Blanshard and others, are mentioned in the bibliography in chapter 3 of Edwards and Pap, [66] above. Numerous influential papers by contemporary philosophers on various issues in the philosophy of mind, as well as additional bibliographical references, may be found in the following anthologies: V.C. Chappell, ed., [101] *The Philosophy of Mind* (Englewood Cliffs, N.J.: Prentice Hall, 1962); G.N.A. Vesey, ed., [102] *Body and Mind* (London: Allen and Unwin, 1964); and D.F. Gustafson, ed., [103] *Essays in Philosophical Psychology* (Garden City, N.Y.: Doubleday, 1964).

IV. INTENTIONALITY, THOUGHT, AND LANGUAGE

The medieval idea that thought and language are isomorphically related and that, consequently, the concept of the one can be understood as an "analogical extension" of the concept of the other, has recently been revived and developed (along different lines) by P. Geach and W. Sellars. In [104] *Mental Acts* (London: Routledge and Kegan Paul, 1957), Geach develops a theory of judgment as "inner speech" involving "mental utterances" related to one another in a way analogous to that in which *overt* utterances are capable of being related. The details of Geach's relational analysis of judgment are derived in part from B. Russell's analysis of judgment in [105] *The Problems of Philosophy* (London: Oxford Univ. Press, 1962), [106] "Knowledge by Acquaintance and Knowledge by Description," *ASP* 11 (1910–11), 108–128 (reprinted in

B. Russell, [107] *Mysticism and Logic and Other Essays* (New York: Longmans, 1918)), and [108] *Philosophical Essays*, rev. ed. (London: Allen & Unwin, 1966). Geach's theory has been further developed by A. Kenny in [109] *Action, Emotion and Will* (London: Routledge and Kegan Paul, 1963), chs. 10 and 11, and extended to apply to volitional acts such as wanting, wishing, etc.

A concise formulation of W. Sellars' analogical account of thought, in the context of a wide range of epistemological and metaphysical issues, is contained in his well-known essay [*110] "Empiricism and the Philosophy of Mind," first published in *MSPS I* and reprinted in Sellars' [111] *Science, Perception and Reality* (London: Routledge and Kegan Paul, 1963). Sellars has been primarily concerned with establishing the claim, disputed by Chisholm in [13], [*16], [17], and especially in [*112]. The Chisholm-Sellars Correspondence on Intentionality, that the concept of meaningful speech is to be taken as basic, and that the intentional features of thought are to be understood in terms of the "semantical" features of language—the latter in turn being explained in terms of linguistic "roles" and "rule-governed" behavior. A fair assessment of Sellars' thesis requires an examination of his account of "meaning", "rules", "theoretical explanation", and other related concepts which together constitute the framework of Sellars' system of "scientific realism". The broad outlines of this system are set out in [113] "Philosophy and the Scientific Image of Man," in [111], and the details are worked out in a wide range of papers, the most relevant for our purposes being: [114] "Some Reflections on Language Games," in [111]; [115] "Being and Being Known," in [111]; [116] "The Language of Theories," in [111]; [117] "Abstract Entities," *RM* 16 (1963), 627–671; [118] "Empiricism and Abstract Entities," in Schilpp [71]; [*119] "Notes on Intentionality"; [120] "Some Reflection on Thoughts and Things," *Noûs* 1 (1967), 97–121; [121] "Language as Thought and Communication," *PPR* 29 (1969), 506–527; and [122] "The Structure of Knowledge" (especially Part II), (forthcoming). Sellars' most recent systematic account of his views is contained in [123] *Science and Metaphysics* (New York: Humanities Press, 1968). For further threshing of issues see [124] the unpublished correspondence between Sellars and H-N. Castañeda on thoughts and self-knowledge,[2] and [*125] The Rosenthal-Sellars Correspondence on Intentionality. The latter, as well as D.M. Rosenthal's unpublished Ph.D. dissertation [126] "Intentionality: A Study of the Views of Chisholm and Sellars" (Princeton University, 1968), provide an excellent commentary to the Chisholm-Sellars exchange on intentionality in [*112]. See also, on the same topic,

---

[2] Copies of this correspondence may be obtained, with permission of Professors Sellars and Castañeda, by writing to the present editor.

the abstracts of the following two papers read by Chisholm and Sellars at a symposium on "The Problem of Intentionality" held at West Virginia University in May, 1965: Chisholm, [127] "The Problem of Intentionality"; Sellars, [128] "Intentionality: The Dialogue Continued."

Although Sellars' views have been very influential in contemporary philosophy (see, e.g., B. Aune, [129] *Knowledge, Mind, and Nature* (New York: Random House, 1967), especially chapter 8, [*130] "Thinking"), critical writing on Sellars' views have not, to date, been as numerous as might be desirable. A fine exposition of Sellars' views has been given by R. Bernstein in [131] "Sellars' Vision of Man-in-the-Universe," *RM* 20 (1966), 113–143, 290–316. A detailed criticism of certain aspects of Sellars' philosophy of "scientific realism" has recently been given by J.W. Cornman in [132] "Sellars, Scientific Realism, and Sensa," *RM* 23 (1970), 417–451. For critical papers on various aspects of Sellars' semantics see J.W. Cornman, [133] "Speak Your Thoughts," *JP* 61 (1964), 665–668 (written in reply to Sellars' [*119]); S. Thomas, [134] "Professor Sellars on Meaning and Aboutness," *PS* 13 (1962), 68–74 (in reply to Sellars' [209]); and G. Harman, [135] "Sellars' Semantics," *PR* 79 (1970), 404–419. See also the following book reviews: K. Lehrer, [136] Review of *Science, Perception and Reality*, *JP* 63 (1966), 266–277; G. H. Harman, [137] Review of *Philosophical Perspectives*, *JP* 66 (1969), 133–144; and B. Aune, [138] Review of *Science and Metaphysics*, *JP* 67 (1970), 251–256.[3] To be contrasted with Sellars' is G. Bergmann's analysis of linguistic meaning in terms of the intentionality of thought, given in [*139] "Intentionality" (reprinted in Bergmann's [140] *Meaning and Existence* (Madison, Wis.: Univ. of Wisconsin Press, 1960)) and in other essays in [140] as well as in his [141] *Logic and Reality* (Madison, Wis.: Univ. of Wisconsin Press, 1964). See also Bergmann's recent book [142] *Realism: A Critique of Brentano and Meinong* (Madison, Wis.: Univ. of Wisconsin Press, 1967), which also illustrates Bergmann's deep interest in the phenomenological movement. An account of meaning and intentionality in the same spirit as Bergmann's is also given by R. Grossmann in [143] *The Structure of Mind* (Madison, Wis.: Univ. of Wisconsin Press, 1965).

[3] A *festschrift* consisting of articles on various aspects of Sellars' philosophy and replies by the same is currently in preparation by H-N. Castañeda, to be published by Bobbs-Merrill. Two papers by A. Marras discussing certain aspects of Sellars' analogical account of the mental are forthcoming in *Noûs* and in the *Canadian Journal of Philosophy*, respectively: "Sellars on Thought and Language" and "On Sellars' Linguistic Theory of Conceptual Activity." A reply by Sellars to the latter paper is also forthcoming in the same journal. Sellars has recently replied to Cornman [132] in "Science, Impressions, and Sensa: A Reply to Cornman," *RM* 24 (1971), 391–447.

A pragmatic-behavioral approach to linguistic meaning (different from Sellars', especially in philosophical intent) is recommended by W.V.O. Quine in chs. 1 and 2 of [144] *Word and Object* (Cambridge, Mass.: The M.I.T. Press, 1960), and by A.J. Ayer in [145] "Meaning and Intentionality," *Proceedings of the XIIth International Congress of Philosophy*, vol. 1 (Florence: Sansoni Editore, 1960). (See also previous references to Carnap, [54] and [55].) The philosophical background of behavioral-pragmatic approaches to linguistic meaning is provided primarily by C.W. Morris' pioneering work, especially [146] *Foundations of the Theory of Signs* (Chicago: Univ. of Chicago Press, 1938), and [147] *Sign, Language and Behavior* (New York: Prentice Hall, 1946). From the point of view of theoretical psychology, perhaps the most influential account of linguistic meaning and language acquisition in terms of stimulus-response (S-R) behaviorism is given by B.F. Skinner in [148] *Verbal Behavior* (New York: Appleton-Century-Crofts, 1957). Skinner's views have not always been well received by philosophers and linguists, as witnessed by N. Chomsky's criticism of these views in [149] "A Review of B.F. Skinner's *Verbal Behavior*," *Language* 35 (1959), 26–58 (reprinted in Fodor and Katz, [157] below). Chomsky's original contributions to descriptive linguistics—see especially [150] *Syntactic Structures* (The Hague: Mouton and Co., 1957) and [151] *Aspects of the Theory of Syntax* (Cambridge, Mass.: The M.I.T. Press, 1965)—have emphasized the "generative" or developmental aspect of grammar and have tended to suggest that humans have some innate predisposition for language learning which is not capable of explanation in terms of traditional S-R theories. On this latter point, see esp. Chomsky's [152] *Cartesian Linguistics* (New York: Harper and Row, 1966), and [153] *Language and Mind* (New York: Harcourt, Brace and World, 1968). For other studies in the same spirit as Chomsky's see J.A. Fodor, J.J. Jenkins, and S. Saporta, [154] *An Introduction to Psycholinguistic Theory* (Englewood Cliffs, N.J.: Prentice Hall, 1969); J.J. Katz, [155] *The Philosophy of Language* (Englewood Cliffs: N.J.: Prentice Hall, 1965); P. Postal, [156] *Constituent Structure* (Bloomington, Ind.: Indiana University Press; The Hague: Mouton and Co., 1964); and other papers in J.A. Fodor and J.J. Katz, eds., [157] *The Structure of Language* (Englewood Cliffs, N.J.: Prentice Hall, 1965). Chomsky's influence on the psychological study of language acquisition and linguistic performance is especially evident in the work of current "psycholinguists" such as D. McNeill, G.A. Miller, T.G. Bever, D.I. Slobin, and others. For a critical discussion of current trends in psycholinguistics and for full bibliographical references, see S.M. Ervin-Tripp and D.I. Slobin, [158] "Psycholinguistics," *Annual Review of Psychology* 17 (1966), 435–474; see also G.A. Miller and D. McNeill, [159] "Psycholinguistics," in *Handbook of Social Psychology*, rev. ed. (Reading, Mass.: Addison-

Wesley, 1969). Other psychologists who have been questioning the Skinnerian approach to meaning and verbal behavior as conditioned response include C.D. Spielberger, D.E. Dulany, and R.F. Terwilliger; see, respectively, [160] "Theoretical and Epistemological Issues in Verbal Conditioning," in Rosenberg, [167] below; [161] "Awareness, Rules, and Propositional Control: A Confrontation with S-R Behavior Theory," in Horton and Dixon, [169] below; and [162] *Meaning and Mind: A Study in the Psychology of Language* (London: Oxford Univ. Press, 1968). For further sources in the psychology of language see: L.S. Vigotsky, [163] *Thought and Language* (Russian ed., 1934; English trans., Cambridge, Mass.: The M.I.T. Press, 1962)—a pioneering but still very interesting study; P. Henle, ed., [164] *Language, Thought and Culture* (Ann Arbor, Mich.: Univ. of Michigan Press, 1958); S. Saporta, ed., [165] *Psycholinguistics: A Book of Readings* (New York: Holt, 1961); J.B. Carroll, [166] *Language and Thought* (Englewood Cliffs, N.J.: Prentice Hall, 1964)—an elementary introduction; S. Rosenberg, ed., [167] *Directions in Psycholinguistics* (New York: Macmillan, 1965); F. Smith and G.A. Miller, eds., [168] *The Genesis of Language: A Psycholinguistic Approach* (Cambridge, Mass.: The M.I.T. Press, 1966); D. Horton and T. Dixon, eds., [169] *Verbal Behavior and S-R Behavior Theory* (Englewood Cliffs, N.J.: Prentice Hall, 1967); L.A. Jakobovits and M.S. Miron, eds., [170] *Readings in the Psychology of Language* (Englewood Cliffs, N.J.: Prentice Hall, 1967); D.C. Hildum, [171] *Language and Thought: An Enduring Problem in Psychology* (Princeton, N.J.: Princeton Univ. Press, 1967); and R.C. Oldfield and J.C. Marshall, eds., [172] *Language* (Harmondsworth, England: Penguin Books, 1968).

Other philosophical theories of meaning and language not represented in this anthology may be found in L. Wittgenstein, [93] and [173] *Tractatus Logico-Philosophicus*, 2nd English ed. (London: Routledge and Kegan Paul, 1961); B. Russell, [174] "The Philosophy of Logical Atomism," reprinted in Russell's [175] *Logic and Knowledge* (London: Allen and Unwin, 1956); G. Ryle, [176] "Ordinary Language" and [177] "The Theory of Meaning," both reprinted in Caton [182]; J.L. Austin, [94] and [95]; P. Strawson, [96] and [178] "Intention and Convention in Speech Acts," *PR* 73 (1964), 439–460; P. Grice, [179] "Meaning," *PR* 66 (1957), 377–388; and J.R. Searle, [180] *Speech Acts* (Cambridge: Cambridge Univ. Press, 1969). An excellent comprehensive collection of essays on various topics in the philosophy of language is T.M. Olshewsky, [181] *Problems in the Philosophy of Language* (New York: Holt, Rinehart, and Winston, 1969), which also contains an annotated bibliography after each chapter. The ordinary language approach is emphasized in the essays in C. Caton, ed., [182] *Philosophy and Ordinary Language* (Urbana, Ill.: Univ. of Illinois Press, 1963). See also L. Linsky's classic anthology [183] *Semantics and the Philosophy of Language* (Urbana, Ill.: Univ. of Illinois Press, 1952).

V. INTENTIONALITY, LOGIC, AND SEMANTICS

The adequacy of a semantical theory applicable to ordinary language has often been measured by its ability to provide an acceptable interpretation of contexts containing psychological terms. Gottlob Frege was probably the first modern philosopher to notice the paradoxical, or supposedly nonextensional, characteristics of psychological contexts (and, in general, of contexts containing indirect-speech constructions), and to attempt an "interpretation" of such contexts by undertaking what we might now call a semantical analysis of ordinary language. See [*184] "On Sense and Nominatum" and other essays in P. Geach and M. Black, eds., [185] *Translations from the Philosophical Writings of Gottlob Frege* (Oxford: Blackwell, 1960). See also the excellent collection of critical essays on various aspects of Frege's work edited by E.D. Klemke, [186] *Essays on Frege* (Urbana, Ill.: Univ. of Illinois Press, 1968); three essays by Frege not included in [185] are also included in the appendix of Klemke's volume.

Whereas Frege's strategy in dealing with psychological contexts was in terms of his distinction between the sense and the nominatum (reference) of names, together with the claim that sense and nominatum vary systematically according to the context, Russell's strategy was instead in terms of his now famous theory of descriptions, which had the effect of restricting the class of names to a class of "purely designating devices" ("logically proper names") whose occurrence in any context was supposed to be entirely unproblematic (extensional). The first formulation of Russell's theory of descriptions is found in [*186] "On Denoting"; an expanded formulation may be found in [187] *Introduction to Mathematical Philosophy* (London: Allen and Unwin, 1919). The logical and epistemological role of names as contrasted with descriptions is extensively discussed in Russell's [174] above. The critical literature on Russell's theory of descriptions is vast. Of special interest are: G.E. Moore, [188] "Russell's Theory of Descriptions," in P.A. Schilpp, ed., [189] *The Philosophy of Bertrand Russell* (Evanston and Chicago: Northwestern Univ. Press, 1944); P.T. Geach, [190] "Russell's Theory of Descriptions," A 10 (1949), 84–88; P. Strawson, [191] "On Referring," in Caton [182]; R.L. Clark, [192] "Presuppositions, Names and Descriptions," *PQ* 6 (1956), 145–154; K.S. Donnellan, [193] "Reference and Definite Descriptions," *PR* 75 (1966), 281–304; and L. Linsky, [194] *Referring* (New York: Humanities Press, 1967), esp. ch. 4. A detailed critique of Russell's treatment of belief-contexts is given by Linsky in [*195] "Substitutivity and Descriptions," reprinted as ch. 5 of [194]. On the subject of proper names see especially A.W. Burks, [196] "A Theory of Proper Names," *PS* 2 (1951), 36–45, and J.R. Searle, [197] "Proper Names," *Mind* 67 (1958), 166–173 (reprinted in Caton [182]).[4]

Frege's contribution, though acknowledged by Russell in [*186], was ignored for about fifty years until A. Church pointed out its importance in various reviews in *JSL* in the early 1940's. Church himself adopted and developed Frege's semantical method in [198] "A Formulation of the Logic of Sense and Denotation," in P. Henle *et al.*, eds., [199] *Structure, Method, and Meaning: Essays in Honor of H.M. Sheffer* (New York: Liberal Arts, 1951). A quasi-Fregean method is also developed by R. Carnap in [200] *Meaning and Necessity*, 2nd. ed. (Chicago: Univ. of Chicago Press, 1956). Carnap's notion of intensional isomorphism (see [*201] "The Analysis of Belief Sentences") was intended as an explication of synonymity as well as a criterion on identity for beliefs. Carnap's notion of intensional isomorphism, as well as his analysis of belief-sentences, generated a great deal of discussion, beginning with certain criticisms by B. Mates in [202] "Synonymity," *University of California Publications in Philosophy* 25 (1950), 201–226 (reprinted in Linsky [183]), and by Church in [203] "On Carnap's Analysis of Statements of Assertion and Belief," *A* 10 (1950), 97–99, reprinted in M. Macdonald, ed., [204] *Philosophy and Analysis* (Oxford: Blackwell, 1954). A reply to Mates' and Church's criticisms is given by H. Putnam in [*205] "Synonymity and the Analysis of Belief Sentences," where the author also provides a modified definition of intensional isomorphism, endorsed in principle by Carnap himself in [206] "On Belief Sentences: A Reply to Alonzo Church" (in Macdonald [204]; also reprinted in Carnap [200]), where Carnap, however, pursues an interpretation of psychological concepts along the new lines of Carnap [73]. In a reply to Putnam entitled [207] "Belief, Synonymity and Analysis," *PS* 4 (1955), 12–15, A. Pap dismisses Mates' criticism of Carnap as irrelevant and urges (against Carnap and in agreement with Church) an analysis of belief-sentences in terms of propositions rather than in terms of responses to sentences. (For details of Pap's own analysis of belief-sentences see [208] "Belief and Propositions," *POS* 24 (1957), 123–136.)[5] Further replies to Putnam and Mates and Church, respectively, are found in W. Sellars, [209]

---

[4] An excellent discussion of Searle's views is contained in M. McKinsey's recent paper "Searle on Proper Names," *PR* 80 (1971), 220–229.

[5] On the nature of propositions two papers by R. Cartwright are especially worth mentioning: "Propositions," in R.J. Butler, ed., *Analytical Philosophy*, First Series (Oxford: Blackwell, 1962), and "Propositions Again," *Noûs* 2 (1968), 229–246. An extension of the idea of propositions as objects of contents of belief for the intriguing case of demonstrative reference is contained in Castañeda [279] below.

"Putnam on Synonymity and Belief," A 15 (1954–55), 117–121; in A. Church, [210] "Intensional Isomorphism and Identity of Belief," PS 5 (1954), 65–73; and I. Scheffler, [211] "An Inscriptional Approach to Indirect Quotation," A 14 (1953–54), 83–90. For additional discussion see also I. Scheffler, [212] "On Synonymy and Indirect Discourse," PS 22 (1955), 39–44; K. Lambert, [213] "Synonymity Again," A 16 (1955–56), 68–71; L. Meckler, [214] "An Analysis of Belief Sentences," PPR 16 (1955–56), 317–330; and R. W. Beard, [215] "Synonymy and Oblique Contexts," A 26 (1965–66), 1–5.

Carnap's recommended analysis of belief-contexts in [200] was in terms of a metalanguage capable of an extensional interpretation. Such an interpretation, which relates believers to sentences rather than to propositions (or other intensional entities), is favored by those philosophers who, like Quine, find nonextensional languages semantically obscure. Quine's misgivings about admitting nonextensional contexts center around the claim that these contexts are "referentially opaque," in the sense that they are unable to accommodate our ordinary concept of singular reference as expressed by such fundamental principles as the principle of substitutivity ("Leibniz's law") and the principle of existential generalization. By Quine on referential opacity see especially [216] "Reference and Modality," in Quine's [217] *From a Logical Point of View*, 2nd ed. (Cambridge, Mass.: Harvard Univ. Press, 1961); [218] "Three Grades of Modal Involvement," reprinted in Quine's [219] *The Ways of Paradox and Other Essays* (New York: Random House, 1966); [*220] "Quantifiers and Propositional Attitudes"; and chs. 4 and 5 of [144]. Critics of Quine's concept of referential opacity include R. Carnap, in [221] "Modalities and Quantification," JSL 11 (1946), 33–64; A.F. Smullyan, in [222] "Modality and Description," JSL 13 (1948), 31–37 (see a comment by N.L. Wilson in [223] *The Concept of Language* (Toronto: Univ. of Toronto Press, 1959), ch. 2); F.B. Fitch, in [224] "The Problem of the Morning Star and the Evening Star," POS 16 (1949), 137–140, reprinted in I.M. Copi and J.A. Gould, eds., [225] *Contemporary Readings in Logical Theory* (New York: Macmillan, 1967); R.B. Marcus, in [226] "Modalities and Intensional Languages," *Synthese* 27 (1962), 303–322, reprinted in Copi and Gould [225] (see Quine's reply, [227] "Reply to Professor Marcus," *Synthese* 27 (1962), 323–330, reprinted in Copi and Gould [225]); P. Geach, in [228] "Quantification Theory and the Problem of Identifying Objects of Reference," APF 16 (1963), 41–52; A.N. Prior, in [229] "Is the Concept of Referential Opacity Really Necessary?," APF 16 (1963), 189–198; B. Rundle, in [230] "Modality and Quantification," in Butler [42]; and J. Hintikka, in [231] "Modality as Referential Opacity," *Ajatus* 20 (1957), 49–63. For further references, see the bibliography in G.E.

Hughes and M.J. Cresswell, [232] *An Introduction to Modal Logic* (London: Methuen, 1968).

In [*220] and [144] Quine also made a contribution of his own toward an extensional interpretation of psychological contexts by construing psychological (or "propositional") attitudes as relations between persons and ("eternal") sentences or sentential matrices and their arguments. An account in the same spirit as Quine's, but appealing to inscriptions or "utterance-events" instead of to eternal sentences or matrices, is recommended by I. Scheffler in [211], in [233] "Thoughts on Teleology," *British Journal for the Philosophy of Science* 9 (1959), 265–284, and in [69], 88–110. A difficulty in Scheffler's proposal is pointed out by Quine in [144], §44. On Quine's own account in [*220], see R. Severens, [234] "Psychological Contexts," *JP* 59 (1962), 95–100. On both Quine and Scheffler, see D. Davidson, [235] "On Saying That," in D. Davidson and J. Hintikka, eds., [236] *Words and Objections: Essays on the Work of W.V. Quine* (Dordrecht, Holland: Reidel, 1969). A very promising extension and reinterpretation of Quine's account has recently been given by D. Kaplan in [237] "Quantifying In," in Davidson and Hintikka [236]; see Quine's approving reply to Kaplan in the same volume. Probably the most systematic and detailed attempt to carry out an extensional interpretation of belief-contexts and indirect discourse generally has been made by R. Martin in [238] *Towards a Systematic Pragmatics* (Amsterdam: North-Holland, 1959), in [239] "Toward an Extensional Logic of Belief," *JP* 59 (1962), 169–172, and in [240] "On Knowing, Believing, Thinking," *JP* 59 (1962), 586–600. See A.R. Anderson's reply to the latter paper in [241] "On Professor Martin's Beliefs," *JP* 59 (1962), 600–607. A long and instructive exchange of views between A.N. Prior and L.J. Cohen on the possibility of an extensional interpretation of belief and indirect discourse originated with a paper by Cohen on a semantical paradox. See: Cohen, [242] "Can the Logic of Indirect Discourse Be Formalized?," *JSL* 22 (1957), 225–232; Prior, [243] "Epimenides the Cretan," *JSL* 23 (1958), 261–266; Cohen, [244] "Why Do Cretans Have to Say So Much?," *PS* 12 (1961), 72–78; Prior, [245] "Indirect Speech Again," *PS* 14 (1963), 12–15; Cohen, [246] "Indirect Speech: A Rejoinder to Prof. A.N. Prior," *PS* 14 (1963), 15–18; Prior, [247] "Indirect Speech and Extensionality," *PS* 15 (1964), 35–38; Cohen, [248] "Indirect Speech: A Further Rejoinder to Professor Prior," *PS* 15 (1964), 38–40.

An interesting interpretation of psychological contexts exploiting Russell's theory of descriptions and a thesis of Quine on the eliminability of singular terms is given by R. Chisholm in [249] "Leibniz's Law in Belief Contexts," in A. Tymienieka, ed., [250] *Contributions to Logic and Methodology in Honor of I.M. Bochenski* (Amsterdam: North-Holland, 1965). See, however, [251] D. Føllesdal's comments in his review of the same book in *PR* 76 (1967), 536–542, and a paper by R.J. Swartz, [252] "Leibniz's Law and Belief," *JP* 67 (1970), 122–137. An account close to Chisholm's is given by E. Sosa in [253] "Quantifiers, Beliefs, and Sellars," in J.W. Davis, D.J. Hockney, and W.K. Wilson, eds., [254] *Philosophical Logic* (Dordrecht, Holland: Reidel, 1969); Sosa's paper was written as a reply to an original contribution on the same subject by W. Sellars, [255] "Some Problems about Belief," in Davis, Hockney, and Wilson [254]. By Sosa see also [256] "Propositional Attitudes *de Dictu* and *de Re*," *JP* 67 (1970), 883–895. An important contribution has also been made by P. Geach who, in his book [257] *Reference and Generality* (Ithaca, N.Y.: Cornell Univ. Press, 1962), develops a theory of reference immediately relevant to the interpretation of indirect discourse. See also Geach's paper [258] "Intentional Identity," *JP* 64 (1967), 627–632, and replies by J.C. Cohen, [259] "Geach's Problem about Intentional Identity," *JP* 65 (1968), 329–335, D.C. Dennet, [260] "Geach on Intentional Identity," *JP* 65 (1968), 335–341, and J.G. Barense, [261] "Identity in Indirect Discourse," *JP* 66 (1969), 381–382.

In contrast with the Frege-Russell-Quine tradition, L. Linsky in [*195] "Substitutivity and Descriptions" questions the universal validity of the principle of substitutivity, and claims that this principle should be restricted to contexts determined by open sentences (predicates) expressing properties. Linsky, however, does not attempt to provide an (independent) criterion for an open sentence expressing a property. Two distinct attempts in this direction may be found in A. Marras, [262] "Identity and Existence in Intentional Contexts," *Methodology and Science* 2 (1968), 190–209, and [38]. Of help in this regard may also be N.B. Cocchiarella's notion of an "existence entailing attribute," in [263] "Existence Entailing Attributes, Modes of Copulation and Modes of Being in Second Order Logic," *Noûs* 3 (1969), 33–48. Replies to Linsky are found in R. Cartwright, [264] "Substitutivity," *JP* 63 (1966), 684–685, and in K. Donnellan, [265] "Substitutivity and Reference," *JP* 63 (1966), 685–688.

Recent work on the semantics of psychological contexts has been significantly enhanced as a result of some highly original contributions by J. Hintikka. Hintikka's semantics exploits the idea of a model-set ("possible world") employed in model-theories of modality, and is developed in the light of the claim that failure of (unrestricted) substitutivity and quantification in psychological contexts is due not to "referential opacity" (Quine) but to "referential multiplicity" (that is, to failure of uniqueness of reference, not to failure of reference altogether). Hintikka has developed and defended his views in [266] *Knowledge and Belief* (Ithaca, N.Y.: Cornell Univ. Press, 1962), where he also constructs an axiomatic system for the concepts of knowledge and belief, as well as in a number of papers, some of which Hintikka himself has recently collected in a volume entitled [267] *Models for Modalities: Selected Essays* (Dordrecht, Holland: Reidel, 1969). In the same volume, see especially [*268] "Semantics for Propositional Attitudes," and [269] "Existential and Uniqueness Presuppositions." See also [270] "Modality as Referential Multiplicity," *Ajatus* 20 (1957), 49–64. Hintikka's views have been discussed by H-N. Castañeda who, in his [271] Review of *Knowledge and Belief, JSL* 29 (1964), 132–134, and then again in [272] "On the Logic of Self-Knowledge," *Noûs* 1 (1967), 9–21, raises some doubts about Hintikka's ability to express in his system statements about self-knowledge. Hintikka's reply, [273] "Individuals, Possible Worlds, and Epistemic Logic," *Noûs* 1 (1967), 33–62 (see also [274] " 'Knowing Oneself' and Other Problems in Epistemic Logic," *Theoria* 32 (1966), 1–13), prompted a rejoinder by Castañeda, [275] "On the Logic of Attributions of Self-Knowledge to Others," *JP* 65 (1968), 439–456, to which Hintikka has again replied; see [276] "On Attributions of Self-Knowledge," *JP* 67 (1970), 73–87. See also Castañeda's related papers, [278] " 'He': The Logic of Self-Consciousness," *Ratio* 8 (1966), 130–157, [279] "Indicators and Quasi-Indicators," *APQ* 4 (1967), 85–100, [280] "On the Phenomeno-Logic of the I," in *Proceedings of the XIV International Congress of Philosophy* (Vienna: Herder, 1970).[6] Further objections to Hintikka's views are raised by R. Chisholm in [281] "The Logic of Knowing," *JP* 60 (1963), 773–795; in particular, Chisholm questions the meaningfulness of the identification of individuals "across" possible worlds. See also, by Chisholm, [282] "Identity through Possible Worlds," *Noûs* 1 (1967), 1–8, and Hintikka's reply in [273]. R.L. Sleigh, in the spirit of Quine, has also raised some questions about quantification into psychological contexts: see [283] "On Quantifying into Epistemic Contexts," *Noûs* 1 (1967), 23–31, and Hintikka's reply, [273]. By Sleigh see also [284] "A Note on an Argument of Hintikka's," *PS* 18 (1967), to which Hintikka has replied in [285] "Partially Transparent Senses of Knowing," *PS* 20 (1969), 5–8. On the Sleigh-Hintikka exchange in *Noûs* see an interesting paper by G. Stine, [286] "Hintikka on Quantification and Belief," *Noûs* 3 (1969), 399–408. Discussion of certain problems connected with the substitutivity principle in psychological contexts is contained in an exchange between Hintikka and D. Føllesdal: Føllesdal, [287] "Knowledge, Identity, and Existence," *Theoria* 33 (1967), 1–27; Hintikka, [288] "Existence and Identity in Epistemic Contexts," *Theoria* 33 (1967), 138–147. Further discussion

---

[6] In another recent paper entitled "On Knowing (or Believing) that One Knows (or Believes)," *Synthese* 21 (1970), 187–203, Castañeda has also argued that the relation of alternativeness characteristic of psychological modalities cannot be strictly transitive, as Hintikka had supposed in [266]. Castañeda's paper was part of a symposium on 'Knowing that One Knows', which also included contributions by R. Hilpinen, K. Lehrer, J. Hintikka, and C. Ginet.

on other aspects of Hintikka's account of knowledge and belief may be found in C. Pailthorp, [289] "Hintikka and Knowing that One Knows," *JP* 64 (1967), 491–497; M. Deutscher, [290] "Hintikka's Conception of Epistemic Logic," *Australasian Journal of Philosophy* 47 (1969), 205–208; and R. de Sousa, [291] "Knowledge, Consistent Belief, and Self-Consciousness," *JP* 67 (1970), 63–73. Hintikka has recently applied his semantical method also to an analysis of the modalities of perception; see [292] "On the Logic of Perception," in N.S. Care and R.M. Grimm, eds., [293] *Perception and Personal Identity* (Cleveland: Case Western Reserve Univ. Press, 1969). See also, in the same volume, R.L. Clark's comments and Hintikka's reply. A distinctive feature of Hintikka's semantics is that it requires a logic which is free of the "existential presuppositions" of traditional (Russellian) quantification theory. On this, see especially Hintikka's [294] "Existential Presuppositions and Existential Commitments," *JP* 56 (1959), 125–137, and [269]. Systems of "free logic" have been proposed by other logicians such as H.S. Leonard, in [295] "The Logic of Existence," *PS* 7 (1956), 49–64; H. Leblanc and T. Hailperin, in [296] "Nondesignating Singular Terms," *PR* 68 (1959), 239–243; B.C. van Fraassen, in [297] "Singular Terms, Truth-Value Gaps, and Free Logic," *JP* 63 (1966), 481–495; R.K. Meyer and K. Lambert, in [298] "Universally Free Logic and Standard Quantification Theory," *JSL* 33 (1968), 8–26. Also, for background on the model theoretical interpretation of modal logics see esp. S. Kanger, [299] *Provability in Logic* (Stockholm: Almqvist and Wiksell, 1957), and S. Kripke, [300] "Semantical Considerations on Modal Logic," *APF* 16 (1963), 83–94.

An exploratory study of some basic methodological problems involved in the axiomatization of a logic for belief statements is undertaken by N. Rescher in [301] "The Problem of a Logical Theory of Belief Statements," *POS* 27 (1960), 88–95. Rescher's paper was written in part as a response to an earlier attempt by A. Pap in [208] to provide a set of postulates for a system of belief statements. A discussion of Rescher's paper is found in M. Fisher, [302] "Remarks on a Logical Theory of Belief Statements," *PQ* 14 (1964), 165–169. A pioneering attempt to axiomatize a logic of "practical reasoning," employing such concepts as judging, deciding, and doing (causing to be), is contained in R.W. Binkley's interesting essay [303] "A Theory of Practical Reason," *PR* 74 (1965), 423–448. Among numerous other formal and informal studies of various aspects of the logic of psychological statements, the following ought to be mentioned: A.N. Prior, [304] "Some Exercises in Epistemic Logic," in C.D. Rollins, ed., [305] *Knowledge and Experience* (Pittsburgh: Pittsburgh Univ. Press, 1964); J. Wallace, [306] "Propositional Attitudes and Identity," *JP* 66 (1969), 145–152; R.C. Stalnaker, [307] "Wallace on Propositional Attitudes," *JP* 66 (1969), 803–806; B. Freed,

[308] "Beliefs about Objects," *PS* 21 (1970), 41–47, and [309] "Saying of and Saying That," *JP* 67 (1970), 969–978; F.I. Dretske, [310] "Epistemic Operators," *JP* 67 (1970), 1007–23; J. Hintikka, [311] "Objects of Knowledge and Belief: Acquaintances and Public Figures," *JP* 67 (1970), 869–883; as well as Chisholm's papers [27] and [*32], previously mentioned.

Index of Names